# The Vortex State

# NATO ASI Series

## Advanced Science Institutes Series

*A Series presenting the results of activities sponsored by the NATO Science Committee, which aims at the dissemination of advanced scientific and technological knowledge, with a view to strengthening links between scientific communities.*

The Series is published by an international board of publishers in conjunction with the NATO Scientific Affairs Division

| | |
|---|---|
| **A Life Sciences** | Plenum Publishing Corporation |
| **B Physics** | London and New York |
| | |
| **C Mathematical** | Kluwer Academic Publishers |
| **and Physical Sciences** | Dordrecht, Boston and London |
| **D Behavioural and Social Sciences** | |
| **E Applied Sciences** | |
| | |
| **F Computer and Systems Sciences** | Springer-Verlag |
| **G Ecological Sciences** | Berlin, Heidelberg, New York, London, |
| **H Cell Biology** | Paris and Tokyo |
| **I Global Environmental Change** | |

### NATO-PCO-DATA BASE

The electronic index to the NATO ASI Series provides full bibliographical references (with keywords and/or abstracts) to more than 30000 contributions from international scientists published in all sections of the NATO ASI Series.
Access to the NATO-PCO-DATA BASE is possible in two ways:

– via online FILE 128 (NATO-PCO-DATA BASE) hosted by ESRIN,
Via Galileo Galilei, I-00044 Frascati, Italy.

– via CD-ROM "NATO-PCO-DATA BASE" with user-friendly retrieval software in English, French and German (© WTV GmbH and DATAWARE Technologies Inc. 1989).

The CD-ROM can be ordered through any member of the Board of Publishers or through NATO-PCO, Overijse, Belgium.

Series C: Mathematical and Physical Sciences - Vol. 438

# The Vortex State

edited by

## Nicole Bontemps
Laboratoire de Physique de la Matière Condensée,
Ecole Normale Supérieure,
Paris, France

## Yvan Bruynseraede
Laboratorium voor vaste Stof-Fysika en Magnetisme,
Katholieke Universiteit Leuven,
Leuven, Belgium

## Guy Deutscher
School of Physics and Astronomy,
Tel Aviv University,
Tel Aviv, Israel

and

## Aharon Kapitulnik
Department of Applied Physics,
Stanford University,
Stanford, California, U.S.A.

SPRINGER-SCIENCE+BUSINESS MEDIA, B.V.

Proceedings of the NATO Advanced Study Institute on
Vortices in Superfluids
Cargèse, Corsica, France
July 19–31, 1993

A C.I.P. Catalogue record for this book is available from the Library of Congress.

ISBN 978-0-7923-2971-8     ISBN 978-94-011-0974-1 (eBook)

DOI 10.1007/978-94-011-0974-1

*Printed on acid-free paper*

# TABLE OF CONTENTS

# A tale of whirls

The story of quantized vortices is a splendid fairy tale -with powerful magicians (Onsager, Feynman), artful jewellers (many of which show up in this book) and a few dark witches (eg the Oxford school of the late forties, refusing to recognize the importance of Shubnikov's work).

Our wandering knights of science have now migrated from the peaceful forest of low $T_c$'s to the rugged land of high $T_c$'s. Their basic creed has not changed : it is still based on Cooper pairs. But the landscape around them is strange. The new materials have very short coherence lengths, and this leads to consequences which (I think) none of us would have suspected ten years ago. An outstanding example is the melting of flux lattices -lucidly described in the present book by David Nelson.

Of course, the book does not compile all the adventures of our great knights : for instance, the behavior under magnetic fields of loosely connected, granular, superconductors, near their percolation threshold -a story written ten years ago by Shlomo Alexander. But the papers in the book are adjusted to the main current questions : the role of a strong anisotropy, the statistics of depinning, ...

In this saga, there is an underlying quest : a Holy Grail ; namely, the microscopic origin of the pairing. Some inspired magicians tell us that it is hidden in unknown territories. Others search it in the neighboring province of magnetism. And others, yet, firmly believe that we do not need any quest. The Grail, they say, is standing in front of us, and can be seen with simple eyes... This part of the story lies beyond the present book, but it is present in the mind of all actors.

It may be that high $T_c$ superconductors are conceptually not different from classical BCS materials -a somewhat disappointing Grail. But, even if it is, the beautiful saga of vortices will remain untouched.

P.G. de Gennes
Paris, March 1994

# EDITORS' NOTE

This book collects the essence of the lectures which were given at the NATO Advanced Study Institute held in Cargèse, Corsica from July 19th to July 31st, 1993.

One of the direct and most striking consequences of the description of the superfluid condensate in superfluid He or in superconductors as a single macroscopic quantum state is the quantization of circulation. This results in quantized vortex lines that play a role in limiting the superfluid flow velocity. With the discovery of high-temperature superconductors and its peculiar vortex state, a spectacular progress has been made in the past six years in our understanding of this state in both, superconductors and superfluids. We therefore felt that it was a timely task to give a "status report" on this field. Indeed, we think that this Cargese Summer Institute has been a great success in this respect. Leading experts in the field both experimentalists and theoreticians gathered and presented a series of lectures from the very introductory level to the most advanced and provocative ideas.

Pr. A.A. Abrikosov gave the introductory talks on the necessary background about type II superconductors: the reader is referred to his well known text book : "Fundamentals of the theory of metals", and also to "Superconductivity of metals and alloys", by Pr P.G. de Gennes, and "Introduction to superconductivity", by Pr M. Tinkham.

The first chapters recall the basic properties of superfluid helium, layered superconductors, the remarkable properties of the vortex pinning and melting, and two dimensional superfluids. More advanced lectures describe the experimental and theoretical aspects of the vortex state.

Our hope is that these proceedings will offer an up-to-date overview of the fields of vortices in superfluids and make explicit some of their unusual consequences on the properties and applications especially in high $T_c$ superconductors.

We wish to acknowledge the generous support of The North Atlantic Treaty Organization (NATO), the help of the Centre National de la recherche Scientifique, France (CNRS) and of the Direction de la Recherche et des Etudes Techniques, France (DRET).

Nicole Bontemps
Yvan Bruynseraede
Guy Deutscher
Aharon Kapitulnik

# AN INTRODUCTION TO SUPERFLUIDITY AND VORTICES IN SUPERFLUID $^4$He

W F VINEN
*School of Physics and Space Research*
*University of Birmingham*
*Birmingham B15 2TT*
*England*

## 1.  Introduction

These lectures have two aims:  to provide an elementary general introduction to superfluidity in liquid $^4$He (experimental facts and theoretical understanding);  and more specifically to discuss quantized vortex lines - what they are, the role they play in superfluid helium, what we know about them, and what we do not know about them.

Much of the content of these lectures is contained in the following textbooks, to which the reader should refer for further detail.

D R Tilley & J Tilley: *Superfluidity and Superconductivity*
R J Donnelly: *Quantized Vortices in Helium II*
I M Khalatnikov: *Introduction to the Theory of Superfluidity*

References to original papers will be given only if they are not given in one of these general references.

## 2.  The Phase Diagram of Condensed $^4$He

The solid phase exists only at pressures exceeding 25 atmospheres;  at lower pressures zero point energy prevents the atoms getting sufficiently close together to form a solid.  There is therefore no conventional triple point at which the solid, liquid and vapour phases are in equilibrium.  The liquid exists between the solid phase at high pressures and a conventional vapour pressure line, the critical point being at about 5.2 K.  There are two liquid phases:  helium I, which exists at temperatures above about 2.17 K at the vapour pressure;  and helium II which exists at lower temperatures.  Helium I is a conventional liquid, albeit with a density much less than one would expect from close packing;  helium II is a superfluid.

## 3.  The Observed Properties of the Superfluid Phase

Liquid in the superfluid phase can flow through a very narrow channel without friction up to a certain critical velocity that depends on the channel width but is typically of order 1 cm s$^{-1}$.  A striking example is provided by the phenomenon of film flow, which allows an open vessel containing helium to empty by a kind of frictionless syphoning action through a very thin film that forms a result of van der Waals attraction on the whole of the surface of the vessel that is in contact with the liquid.

1

*N. Bontemps et al. (eds.), The Vortex State*, 1–23.
© 1994 *Kluwer Academic Publishers.*

The superfluid phase exhibits a very large and anomalous heat conductivity. If the helium is confined in a very narrow channel, width d, with a small temperature gradient $\nabla T$, the heat current density along it is proportional to the temperature gradient, $\nabla T$, and to $d^2$, while in the opposite case it is independent of d and proportional to $(\nabla T)^{1/3}$; in a conventional thermal conductor it would be independent of d and proportional to $\nabla T$.

The liquid exhibits a large thermomechanical effect (fountain effect). If two vessels containing superfluid helium are connected by a very narrow channel, and a temperature gradient is established along the channel, then a large pressure gradient also appears, given by

$$\nabla p = \rho s \nabla T, \tag{3.1}$$

where $\rho$ is the density of the liquid and s is its entropy per unit mass.

The Andronikashvili experiment shows that only a part of the liquid takes part in frictionless flow. A pile of closely spaced discs attached in a horizontal plane to a spindle is suspended in the liquid by a torsion fibre, and its effective móment of inertia is measured from its observed period of oscillation. The spacing is such that in helium I viscosity ensures that all the fluid between the discs is dragged round by the discs and so contributes to the moment of inertia. As the helium is cooled below the superfluid transition temperature the moment of inertia falls, indicating that a fraction of the fluid - the superfluid fraction - is no longer dragged by the discs. This fraction rises from zero at the transition temperature to unity at $T = 0$.

Closely associated with the anomalous heat conduction is the fact that a time dependent temperature in the bulk liquid obeys the wave equation

$$\nabla^2 T = (1/c^2)\partial^2 T/\partial t^2 \tag{3.2}$$

rather than the thermal diffusion equation

$$\nabla^2 T = (\rho c/K)\partial T/\partial t. \tag{3.3}$$

It is therefore possible to propagate lightly damped temperature waves in the liquid, which are called second sound.

We have seen that in the Andronikashvili experiment only a fraction of the superfluid phase rotates in an oscillatory way with the disc system. However, if the helium is contained in a *steadily* rotating vessel, it behaves differently and the whole liquid rotates. This can be established by using a partially filled vessel and observing the shape of the liquid meniscus. But if one propagates second sound in the rotating liquid, it is found to suffer an extra attenuation, proportional to the angular velocity of rotation. This suggests that the structure of the steadily rotating superfluid phase differs in some way from the non-rotating phase.

## 4.    Heat Capacity near the Superfluid Transition Temperature

The heat capacity is observed to have a sharp peak at the transition temperature, the shape of which has led to the name $\lambda$-point for this transition. We know that such $\lambda$-points are characteristic of "order-disorder" transitions, such as that occurring at the Curie point of a ferromagnetic, where the elementary magnetic dipoles (electron spins) start to become spontaneously ordered and so lead to a spontaneous magnetic moment. We conclude that superfluidity must be associated with some kind of ordering in the liquid helium. Remember that the liquid can exist over a range of pressures at $T = 0$; the Third Law of Thermodynamics requires that ordering must then be complete.

## 5.    Bose Condensation in an Ideal Bose Gas

Helium atoms are bosons. This suggests that in a search for the ordering mechanism responsible for superfluidity we examine the predicted behaviour of a hypothetical ideal Bose gas. An analysis based on the Bose distribution function

$$n_i = [\exp\{(\varepsilon_i - \mu)/k_B T\} - 1]^{-1}; \ \Sigma n_i = N \qquad (5.1)$$

shows that above a certain critical temperature, $T_c$, the particles of the gas are distributed evenly over a wide range of energy levels, but that below $T_c$ a finite fraction of them, $n_0/N$, occupy the lowest energy or momentum state. This fraction (the 'condensate') stays constant as the size of the system increases at constant density, and it tends to unity as T tends to zero. There is an ordering in momentum space. That this prediction may be relevant to the liquid helium problem is confirmed by the fact that the predicted heat capacity of the gas has a sharp cusp (peak) at $T_c$, qualitatively similar to the $\lambda$-point in helium.

## 6.    Bose Condensation in a Liquid

It is now believed that a form of Bose condensation can occur also in a liquid. Below a critical temperature there is again an accumulation of particles in the lowest momentum state, in the sense that $n_0/N$ tends to a constant non-zero value as $N \to \infty$ at constant density, where $n_0 = \langle \psi | a^+(0)a(0) | \psi \rangle$ and the operator $a(0)$ destroys a particle in the lowest momentum state. (Equivalently, the single-particle density matrix, $\langle \psi^+(r)\psi(r') \rangle$ tends to a non-zero value as $| r' - r | \to \infty$.) The liquid differs from the ideal gas in that there is 'depletion' of the condensate as a result of interactions between the particles: $n_0/N$ tends to a value that is less than unity even at $T = 0$. At $T = 0$ this depletion does not destroy the single quantum state of the system as a whole, and so does not lead to any 'disordering'.

This form of Bose condensation is believed to occur in liquid helium and to be responsible for superfluidity. Both theory and a special form of neutron scattering (deep inelastic neutron scattering) indicate that in liquid helium at low temperatures and low pressures $n_0/N$ is probably of order $0.1$.

## 7.    The Condensate Wave Function

Informally (phenomenologically) we can introduce a "condensate wave function", analogous to the Ginsburg-Landau order parameter for a superconductor,

$$\Psi = a(r) \exp[iS(r)], \qquad (7.1)$$

where $a(r)$ is a measure of the local condensate fraction and $S(r)$ is the phase of the quantum state into which condensation has occurred. We have allowed $a$ and $S$ to depend on position. More formally the condensate wave function can be defined in terms of the single particle density matrix

$$\langle \psi^+(r)\psi(r') \rangle \to \Psi^*(r)\Psi(r') \text{ as } | r - r' | \to \infty. \qquad (7.2)$$

Owing to the fact that the condensate contains a macroscopic number of particles, the condensate wave function behaves like a classical coherent field, somewhat like a coherent electromagnetic field.

The ordered state of any system that undergoes an order-disorder transition always requires for its thermodynamic description an extra thermodynamic parameter that measures the degree of ordering. This parameter is called the order parameter. The condensate wave function $\Psi(r)$ is the appropriate order parameter for superfluid helium; we see that it is a *two-component* order parameter, involving both the condensate fraction and the phase of the state into which condensation has occurred.

## 8. Bose Condensation and Superflow

Suppose that condensation takes place into a uniformly moving condensate: velocity $v = \hbar K/m_4$ (or we simply view the helium from a moving frame of reference). $m_4$ is the mass of a helium atom. Then

$$\Psi = a_0 \exp\{iK.r\}, \qquad (8.1)$$

We can presumably generalise to non-uniform condensate motion

$$\Psi = a(r) \exp[iS(r)] \qquad (8.2)$$

where the local condensate velocity is

$$v_S = (\hbar/m_4) \nabla S(r) \qquad (8.3)$$

We call $v_S$ the *superfluid velocity*. We guess that $a(r) \simeq a_0$, unless $v_S$ varies rapidly with position.

Suppose that the helium with its moving condensate is situated in a stationary tube. Can the moving condensate "slow down", thereby reducing its energy? Or, more precisely, can the value of $v_S$ change? Atoms of the helium can of course be scattered at the walls of the tube. As a result some atoms in the condensate may be scattered out of the condensate. However, removing a few atoms from the macroscopically occupied condensate will not destroy the condensate; $\Psi$ will remain in existence, albeit with a slightly reduced amplitude, and the phase $S(r)$, together with the velocity $v_S$, will be unchanged. A change in $v_S$ would require a simultaneous change in quantum state of the macroscopic number of atoms contained in the condensate, and this will have a very low probability (just how low we shall discuss later). The moving condensate, and with it the velocity $v_S$, is *metastable*.

The metastable state that we have described has a minimum in the free energy subject to the condition that the phase $S(r)$ of the condensate is maintained, and with it the velocity $v_S$. This local minimum is separated from the absolute minimum with $v_S = 0$ by free energy barrier (we discuss later the height and width of this barrier). It is reasonable to assume that this metastable state has associated with it a mass current density, $J_S$, which for small $v_S$, is proportional to $v_S$:

$$J_S = \rho_S v_S. \qquad (8.4)$$

We call $\rho_S$ ($\leq \rho$) the *superfluid density*. We identify the state we have described with the observed state of frictionless flow through a narrow tube.

Note that maintaining a particular (metastable) value of $v_S$ breaks the symmetry of the system, just as maintaining a ferromagnetic moment in a particular direction breaks the symmetry of a ferromagnetic system.

Since $v_S$ is given by (8.3), the superfluid velocity field must be irrotational:

$$\text{curl } v_S = 0. \qquad (8.5)$$

## 9.    The Two-Fluid Model

Suppose that a volume of liquid helium as a whole is suddenly set moving along a tube with velocity **V**, simply by pushing it, and then allowed to reach (metastable) equilibrium with the walls of the (stationary) tube. The value of $v_s$ is initially equal to **V** and must remain so. Since the metastable state involves the mass current density (8.4), where $\rho_s$ is generally less that $\rho$, part of the fluid must slow down. We refer to this part as the *normal fluid*. The normal fluid has density $\rho_n = \rho - \rho_s$, and in general it will have a velocity $v_n$. The part of the helium that moves persistently with velocity $v_s$ (density $\rho_s$) is called the *superfluid component*.

In the Andronikashvili experiment only the normal fluid rotates with the discs. Measurement of the moment of inertia of the discs therefore measures $\rho_n$. In pure $^4$He, the normal fluid fraction, $\rho_n/\rho$, is found to fall steadily from unity at the $\lambda$-point to zero at T = 0. In superfluid $^3$He-$^4$He mixtures $\rho_n$ remains finite at T = 0.

Remember, however, that in a *steadily* rotating vessel all the helium is observed to rotate. How does the superfluid component manage to rotate, in apparent violation of (8.5)? We return to this question later.

## 10.    Thermally Excited States of Liquid Helium

The nature of these states was first considered by Landau, who showed how they are connected with the normal fluid. He guessed that at low temperatures they can be described in terms of a gas of weakly interacting excitations. He also guessed that the spectrum of these excitations (energy $\varepsilon$ *versus* momentum p) has the form shown in figure 1, and this has been confirmed directly by

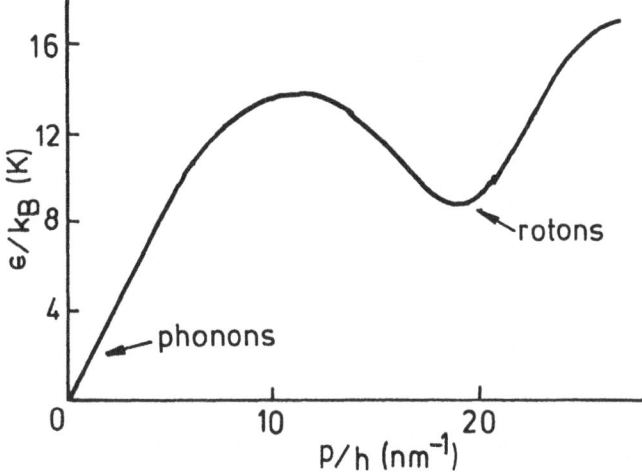

*Figure 1.    Excitation spectrum for liquid $^4$He.*

inelastic neutron scattering (the excitations involve density fluctuations and therefore interact with neutrons). At the lowest temperatures only the quantized sound waves, or phonons, are excited. At somewhat higher temperatures the excitations with p $\sim$ $\hbar/a$, where a is the interatomic spacing, are also excited; these are called rotons and are probably more or less free particles with an accompanying backflow.

Landau showed that we can use this picture of the excited states of the helium to understand the nature of the normal fluid and to calculate $\rho_n$. To follow Landau's argument we need first to understand how the excitation distribution function is modified in the presence of a superfluid velocity.

## 11.   Excitations in the Presence of a Superfluid Velocity

Suppose that the superfluid is flowing with uniform velocity $v_s$ through a tube. In the frame of reference in which $v_s = 0$ the energy required to create an excitation of momentum p is ε. A straightforward argument, based on a Galilean transformation, shows that the energy required to create the same excitation measured in the frame of reference at rest relative to the walls of the tube is given by

$$\varepsilon' = \varepsilon + \mathbf{p}.\mathbf{v}_s. \tag{11.1}$$

## 12.   The Normal Fluid Density

It follows that the excitations (which are bosons) have the following distribution function when they are allowed (by scattering) to come into equilibrium with the walls of the tube.

$$f(\mathbf{p},\mathbf{v}_s) = [\exp\{\varepsilon(\mathbf{p}) + \mathbf{p}.\mathbf{v}_s\}/k_sT - 1]^{-1}. \tag{12.1}$$

For small $v_s$

$$f(\mathbf{p},\mathbf{v}_s) = f_0(\mathbf{p},0) + (\partial f_0/\partial\varepsilon)(\mathbf{p}.\mathbf{v}_s) + 0(v_s^2). \tag{12.2}$$

It follows that the total momentum density in the frame of reference in which $v_s = 0$ is given by

$$\mathbf{P}_n = \int \mathbf{p} \, f(\mathbf{p}) \, d^3p = \int \mathbf{p} \, (\partial f_0/\partial\varepsilon)(\mathbf{p}.\mathbf{v}_s) \, d^3p + 0(v_s^2) \tag{12.3}$$

where we have used the fact that there is no net momentum associated with $f_0(\mathbf{p})$. Hence we find that

$$\mathbf{P}_n = - \rho_n \, \mathbf{v}_s + 0(v_s^2), \tag{12.4}$$

where

$$\rho_n = - \int 2\pi p^4 \, (\partial f_0/\partial\varepsilon) \cos^2\theta \, \sin\theta \, d\theta \, dp \tag{12.5}$$

Therefore the momentum density in the frame of reference at rest relative to the tube is

$$\mathbf{P} = \rho \, \mathbf{v}_s - \rho_n \, \mathbf{v}_s = \rho_s \, \mathbf{v}_s + 0(v_s^2), \tag{12.6}$$

where we have used the fact that in the absence of excitations $\mathbf{P}$ is simply $\rho \, \mathbf{v}_s$.

For the excitation spectrum shown in figure 1, $\rho_n$ calculated from (12.5) is less than $\rho$ at sufficiently low temperatures, so that $\rho_s$, introduced in equation (8.4), is non-zero. At low temperatures the values of $\rho_n$ calculated from (12.5) agree well with those measured in the Andronikashvili experiment. At higher temperatures the density of excitations becomes high, interactions between them cannot be neglected, and (12.5) no longer holds.

It must be emphasized that the superfluid fraction $\rho_s/\rho$ is not related in any simple way to the condensate fraction $n_0/N$. It is true that when $n_0/N$ vanishes at $T_\lambda$, so also does $\rho_s/\rho$, but the proof of this requires a more sophisticated analysis. In superconductors with a large coherence length, the order parameter analogous to $\Psi$ obeys the Ginsburg-Landau equation, and then $\rho_s$ is proportional to $\Psi\Psi^*$. For helium the situation is much more complicated because a very short coherence length leads to large fluctuations in $\Psi$, especially near $T_\lambda$.

## 13. The Landau Critical Velocity: Flow through a Tube

If $v_s$ becomes so large that $(\varepsilon + \mathbf{p}.\mathbf{v}_s)$ is negative for any excitations, then these excitations will be created without limit, and superfluidity must be destroyed $(\rho_s = 0)$. It follows that there is a critical superfluid velocity, equal to the minimum value of $|\varepsilon(p)/p|$, at which superfluid must be destroyed. For the excitation spectrum of figure 1 this Landau critical velocity involves roton creation and is equal to approximately 60 m s$^{-1}$ at low pressure.

In equations (12.2) to (12.6) we ignored terms $O(v_s^2)$. If we do not do this, we find that $\rho_s$ effectively decreases with increasing $v_s$ and vanishes at the Landau critical velocity. (Strictly speaking we are ignoring the interaction between excitations, but interaction probably reduces their energy, so our calculated Landau critical velocity is probably an overestimate.) In practice, as we shall see, frictionless superfluid flow through a tube breaks down at velocities much less than the Landau critical velocity, which is not therefore observable.

## 14. The Landau Critical Velocity: Motion of a Particle in Helium

A different manifestation of the Landau critical velocity does turn out to be observable. Suppose that a particle of mass M moves through superfluid helium with velocity $\mathbf{v}$ at $T = 0$. It will suffer drag only if it can lose energy by creating an excitation $(\varepsilon, \mathbf{p})$. Conservation of energy and momentum in such a process requires that

$$\tfrac{1}{2}M(v'^2 - v^2) = \varepsilon(p) \tag{14.1}$$

and $\quad M(\mathbf{v}' - \mathbf{v}) = \mathbf{p}$ (14.2)

where $\mathbf{v}'$ is the velocity of the particle after emission of the excitation. Therefore

$$\tfrac{1}{2}(\mathbf{v}' + \mathbf{v}).\mathbf{p} = \varepsilon(p) \tag{14.3}$$

If M is large, in comparison with the mass of a helium atom, $\mathbf{v}' \approx \mathbf{v}$, and then

$$\mathbf{v}.\mathbf{p} = \varepsilon(p). \tag{14.4}$$

Hence the particle can slow down by creation of excitations $\varepsilon(p)$ if, and only if,

$$v > |\,\varepsilon(p)/p\,|_{min} \tag{14.5}$$

which is again the Landau critical velocity.

### 15. Ions in Liquid Helium

Two types of charged particle, or ion, can be created in liquid helium by either field ionization or field emission at a sharp tungsten tip carrying a high voltage. The *negative ion* is an electron in a bubble of radius about 1.9 nm (effective mass about 237 $m_4$). The *positive ion* is a $He^+$ ion embedded in a small sphere of solid helium (radius about 0.55 nm; effective mass about 30 $m_4$). The ions form small, more or less spherical, particles in the helium, and are useful probes of its properties.

At low drift velocities the mobilities of the ions are limited by the scattering of phonons and rotons: theory and experiment agree well.

At low temperatures, and high pressures, the negative ion is observed to suffer excess drag if its velocity exceeds the Landau critical value. At low pressures this process is masked by another, involving the production of vortex rings (discussed later).

### 16. The Fluid Mechanics of the Two-Fluid Model

We derive the equation of motion for the superfluid component, analogous to the Euler equation for an ideal fluid.

Consider the time dependence of the phase of the condensate wave function.

$$\Psi = \Psi_0 \exp(iS) = \Psi_0 \exp(iS_0) \exp(-i\varepsilon_c t/\hbar) \tag{16.1}$$

We make the reasonable assumption that $\varepsilon_c$ is the energy required to add one atom to the condensate. To obtain this energy we need to do some thermodynamics.

For an ordinary fluid the extensive thermodynamic variables, such as the internal energy E, are functions of three variables such as the entropy S, the volume V, and the mass M. In the case of superfluid helium we must add another variable which reflects the fact that specification of the macroscopic state of the system requires a knowledge of the (metastable) superfluid velocity. This other variable is conveniently taken as $P_n$, the momentum of the normal fluid relative to the superfluid. We then obtain the following generalization of a standard thermodynamic equation

$$dE = TdS - pdV + \mu dM + (v_n - v_s).dP_n \tag{16.2}$$

where $\mu$ is the chemical potential per unit mass of liquid. Addition of an atom to the condensate does not change the state of the excitations, which determines the entropy and the normal fluid momentum; it must not change the volume if we are to be sure that no extra energy enters or leaves the system through external work. It follows that

$$\varepsilon_c = m_4 \, (\partial E/\partial M)_{s,v,pn} = m_4 \, \mu. \tag{16.3}$$

Using equations (8.3), (16.1) and (16.3) we find that

$$\partial v_s/\partial t = - \nabla\mu \tag{16.4}$$

If the superfluid is moving in the laboratory frame of reference, we must replace $\mu$ by $\mu + 1/2 \, v_s^2$.

The appropriate Gibbs free energy is

$$G = E - TS + pV - (v_n - v_s).P_n, \tag{16.5}$$

so that

$$\mu = e - Ts + p/\rho - (v_n - v_s).p_n \tag{16.6}$$

where lower case letters refer to unit mass. It follows that

$$\partial v_s/\partial t + 1/2 \, \nabla(v_s^2) = -(1/\rho)\nabla p + s\nabla T + 1/2(\rho_n/\rho)\nabla(v_n - v_s)^2 \tag{16.7}$$

which is the required equation of motion.

We mention two applications. In the thermomechanical effect (section 3), $v_s = v_n = 0$, and equation (3.1) follows immediately. Later we shall need the equation describing the Bernouilli effect for steady superflow, without normal flow ($v_n = \nabla T = \partial/\partial t = 0$):

$$p + 1/2 \, \rho_s \, v_s^2 = \text{constant} \tag{16.8}$$

An equation of motion for the normal fluid can also be derived. We refer to Khalatnikov's book for the details. In the absence of dissipation the result is

$$\rho_n \partial v_n/\partial t + \rho_n(v_n.\nabla)v_n = -(\rho_n/\rho)\nabla p - \rho_s s\nabla T - (\rho_n\rho_s/2\rho)\nabla(v_n - v_s)^2 \tag{16.9}$$

The effect of dissipation is to add terms involving one coefficient of first viscosity, three independent coefficients of second viscosity, and a coefficient of thermal conductivity for the normal fluid. For example, the effect of first viscosity is to add the term $\eta_n \nabla^2 v_n$ to the right hand side of (16.9).

## 17.   Heat Flow;  First and Second Sound

We can apply the equations of motion of the two fluids to the problem of heat flow in liquid helium. Note first that entropy is carried only by the excitations and therefore only by the normal fluid. It follows that flow of the normal fluid leads to a heat flux given by

$$Q = \rho s T v_n \tag{17.1}$$

This "convective" heat flux is generally much larger than any associated with the process of normal thermal conduction in the normal fluid.

Heat flow down a tube, with no net mass flow, occurs by counterflow of the two fluids. At the level of presentation so far, the only dissapative process limiting the heat flow is that associated with viscous (Poiseuille) flow of the normal fluid along the tube. The heat current is predicted to be proportional to $\nabla T$, but with an unconventional dependence on the cross-section of the tube. This agrees with experiment only for flow through very narrow tubes at small values of $\nabla T$.

The fact that helium is described by a two-fluid model means that two types of longitudinal wave propagation are possible. In *first sound* the two fluids move together, temperature fluctuations are very small, and the speed of propagation is equal to $(\partial p/\partial \rho)^{1/2}$, as with ordinary sound in an ordinary liquid. In *second sound* there is no net mass flow (the two fluids moving in antiphase), pressure fluctuations are very small, and the speed of propagation is equal to $(Ts^2\rho_s/c\rho_n)^{1/2}$, where c is the heat capacity of the liquid per unit mass. This predicted speed of propagation is in good agreement with experiment.

## 18. Critical Velocities

The two-fluid model works in the simple form that we have described only at low velocities: typically velocities less than of order 1 cm s$^{-1}$, but depending on the geometry. Note that these velocities are much less than the Landau critical velocity for roton creation. Above these low critical velocities there is extra dissipation.

The form of this extra dissipation is most clearly seen in heat flow experiments. The experimental results (section 3) can be described phenomenologically as implying the existence of a "mutual friction" between the two fluids of the form

$$F_{sn} = A\rho_s\rho_n |v_s - v_n|^2(v_s - v_n) \tag{18.1}$$

per unit volume (the "Gorter-Mellink" force).

If second sound propagates in the presence of the heat current, it suffers an extra attenuation; this attenuation is described by the mutual friction (18.1), provided that the factor $|v_s - v_n|^2$ is taken to be due to the heat current alone. This means that second sound alone is not attenuated by the mutual friction, even when $(v_s - v_n)$ due to the second sound wave itself is quite large.

This behaviour of second sound suggests that the factor multiplying $(v_s - v_n)$ in (18.1) takes time to be established or to change from one magnitude to another. This has been confirmed by observing directly how the factor builds up with time after the heat current is switched on. It is found to build up quite slowly; the time involved depends on the magnitude of the heat current and is typically of order 1 s. Furthermore, the time is not reproducible and is dependent on the history of the helium; it is reduced if the helium has carried a heat current in the recent past. This suggests very strongly that the mutual friction is associated with turbulence in the helium, the faster growth rate being due to vorticity remaining from the earlier turbulence. There is no reason why turbulence in the normal fluid should give rise to mutual friction, so we come to the conclusion that in spite of (8.5) turbulence is possible in the superfluid component. Historically, this was the first experimental indication that (8.5) can be violated with interesting consequences.

## 19. Irrotational Superflow in a Multiply-Connected Geometry

We turn to now to the possibility that the superfluid can indeed rotate. Suppose first that the helium is contained in a vessel in the form of a toroid, so that there is a hole through the centre of the helium (figure 2). We define the hydrodynamic *circulation* in the superfluid as

$$\kappa_s = \oint v_s.dr \tag{19.1}$$

where the integral is taken round any circuit, the perimeter of which lies in the fluid. It is easily seen that we can have a circulatory flow of superfluid round the hole through the middle of the helium, without violating (8.5) if $\kappa$ is a non-zero constant for any circuit that encloses the hole and is zero for any circuit that does not enclose the hole. We have described an irrotational circulation of the superfluid round the hole.

## 20. The Quantization of Circulation

However, the magnitude of the superfluid circulation cannot take any value. Using equation (8.3) for $v_s$, we find that

$$\kappa_s = (\hbar/m_4) \oint \nabla S.dr \qquad (20.1)$$

The condensate wave function must presumably be single-valued, so that $\oint \nabla S.dr$ must be an integral multiple, n, of $2\pi$. Hence we find that

$$\kappa_s = n\, h/m_4 \qquad (20.2)$$

The superfluid circulation must be quantized in units of $h/m_4$. Note that this is a consequence of the existence of a condensate wave function with long-range phase coherence.

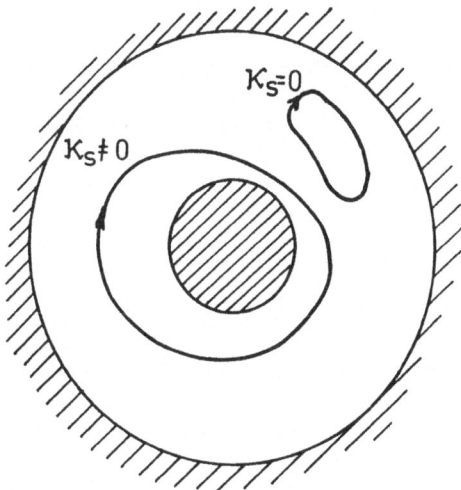

*Figure 2.    Illustrating irrotational flow in a multiply-connected geometry.*

The first experimental verification of (20.2) used an apparatus in which the hole in figure 2 was replaced by a fine wire. In the absence of a circulation, the modes of transverse vibration of the wire are plane polarized; ideally, at least, there are then two degenerate modes. A circulation round the wire gives rise to a transverse *Magnus force* on the moving wire, as described later in these notes. The Magnus force lifts the degeneracy, producing two circularly polarized modes with a splitting that is proportional to the circulation. Since the quantum of circulation is macroscopically large, the splitting is quite large and easily measurable. The validity of (20.2) was confirmed.

## 21.    Free Quantized Vortex Lines

Imagine that we draw a line through a volume of superfluid helium and establish a quantum of superfluid circulation round the line. If the line is straight, the magnitude of the resulting superfluid velocity at a radial distance r from the line is equal to $\hbar/m_4 r$. The divergence in $v_s$ as r tends to zero is unphysical and something must happen. Note that the pressure in the liquid is given by (equation (16.8))

$$p = p_0 - 1/2\, \rho_s v_s^2 = p_0 - \rho_s \hbar^2/2m_4^2 r^2 \qquad (21.1)$$

If the helium were a classical ideal incompressible fluid with surface tension $\sigma$ a hole would appear when $p < -\sigma/r$; i.e. for $p_0 \approx 0$, when

$$r < a_0 = \rho_s \kappa_s^2/8\pi^2 \sigma \qquad (21.2)$$

Substituting the known value of $\sigma$ for helium and taking $\rho_s = \rho$ (low temperature), we find that $a_0 = 0.05$ nm. We have described a free quantized vortex line in the superfluid helium. The "core", radius $a_0$, is microscopic in size. The appearance of such a line allows the superfluid component to rotate, even in a simply-connected volume.

Note that at the core surface $v_s = \hbar/m_4 a_0 = 300$ m s$^{-1}$, which is greater than the Landau critical velocity for roton creation. It is possible therefore that the core is really a region of normal liquid rather than a hole. The structure of the core is really not known, although experiments mentioned later show that its radius is indeed of order 0.1 nm. Very near $T_\lambda$ the core must become larger (of order a coherence length, $\xi$, which goes to infinity at $T_\lambda$), and then the core must be more like normal liquid.


## 22.    Rotation of the Superfluid in a Simply-Connected Vessel

Quantized vortex lines allow the superfluid component to rotate in a simply connected vessel. Suppose, for example, that a cylindrical vessel with circular cross section is filled with superfluid containing a uniform array of rectilinear lines, N per unit area, each having one quantum of circulation with the same sense. Then the superfluid circulation round a circular path of radius R, concentric with the axis of the cylinder, will be equal to $\kappa_s$ times the number of lines threading the path; i.e. equal to $\pi R^2 N \kappa_s$. If the superfluid were in a state of solid body rotation with angular velocity $\Omega$, the circulation would be $2\pi R^2 \Omega$. It follows that, if density of vortex lines is high, the gross features of the velocity field due to the array are similar to those for solid body rotation at angular velocity $\Omega$, provided that $N = 2\Omega/\kappa_s$. If $\Omega = 1$ s$^{-1}$, $N = 10^7$ m$^{-2}$, and the vortex spacing is about 0.2 mm.

An array of vortex lines provides a good imitation of solid body rotation at large $\Omega$ and in a large vessel; in this case the array of lines represents the equilibrium state of the superfluid in the rotating vessel. We now see how the superfluid can appear to rotate with a steadily rotating containing vessel. As we shall see later, vortices take time to be generated, and it is for this reason that the superfluid appears not to rotate with the oscillating disc system in the Andronikashvili experiment.


## 23.    Critical Angular Velocities

At low angular velocities, when the vortex spacing would be comparable with the size of the vessel, the situation needs more careful consideration. We now show that if the containing vessel rotates with angular velocity less than a critical value, $\Omega_{c1}$, the equilibrium state of the superfluid contains no vortices and the superfluid does not rotate at all. (This is the analogue of the Meissner effect in a superconductor, and $\Omega_{c1}$ is the analogue of the lower critical field.)

We calculate $\Omega_{c1}$ for a cylindrical vessel of radius b. The equilibrium state of a fluid in a vessel rotating at angular velocity $\Omega$ is given by minimizing the free energy $F' = F - L.\Omega$, where L is the angular momentum of the liquid. Therefore the equilibrium state of a superfluid will contain no vortex unless

$$\varepsilon - \ell\Omega < 0 \tag{23.1}$$

where $\varepsilon$ is the energy per unit length of a vortex along the axis of the cylinder, and $\ell$ is its angular momentum.

## 24.  Energy and Angular Momentum of a Vortex:  Calculation of $\Omega_{c1}$

It is easily seen that the kinetic energy associated with a vortex along the axis of a vessel of radius b is

$$1/2\ \rho_s \int v_s^2\ d\tau = 1/2\ \rho_s \int_{a_0}^b (\kappa_s/2\pi r)^2\ 2\pi r\ dr = (\rho_s\kappa_s^2/4\pi)\ \ln(b/a_0).$$

The energy associated with the core can be shown to be $\eta\ \rho_s\kappa_s^2/4\pi$, where $\eta$ is of order unity. Hence

$$\varepsilon = (\rho_s\kappa_s^2/4\pi)[\ln(b/a_0) + \eta] \tag{24.1}$$

Similarly

$$\ell = \rho_s\kappa_s b^2/2 \quad (b >> a_0) \tag{24.2}$$

Therefore

$$\Omega_{c1} = (\kappa_s/2\pi b^2)[\ln(b/a_0) + \eta] \tag{24.3}$$

Note that there is also an analogue of $H_{c2}$. $\Omega_{c2}$ is the angular velocity at which the cores of the flux lines overlap and superfluidity is completely lost. It is given by $\Omega_{c2} \sim \kappa_s/2a_0^2 \sim 10^{14}$!!
For b = 1 cm, $\Omega_{c1} = 3 \times 10^{-3}$ s$^{-1}$. $\Omega_{c1}$ is difficult to observe in practice because it is difficult to ensure that the helium is in equilibrium.

## 25.  Observation of Vortex lines in Uniformly Rotating Helium

The most direct observation has been made with ions (Packard *et al*). As we have seen , ions are small spheres, and they are attracted to the core of a vortex line because they displace fluid and reduce the kinetic energy of flow. Below a certain temperature (0.9 K for positive ions;  1.7 K for negatives) they can be trapped on the core of the line.   The experiment consists of trapping negative ions on the cores of an array of lines in a slowly rotating vessel.  By applying a suitable voltage to the helium the electrons in the negative-ion bubble states are extracted from the helium in a direction parallel to the lines.  The electrons are pulled onto a phosphor, where they cause the emission of flashes of light, from which an image of the vortex array can be formed.  For details see Donnelly's book.
A small dimple should form at the point where a vortex emerges at a free surface of superfluid helium.  In practice this dimple is too small to see.  It is larger, and therefore might be more easily

visible, at the interface formed by phase separation of a $^3$He-$^4$He mixture, the lower, $^4$He rich, phase being superfluid (Sonin & Manninen: *Phys Rev Lett* **70** 2585 (1993)).

## 26. Quantized Vortices and Mutual Friction

As we explain in more detail later, the normal fluid interacts with a vortex core, giving rise to a force that effectively acts between the two fluids. Such a mutual friction force should therefore be present in the uniformly rotating liquid, the force per unit volume being proportional to length of vortex line per unit volume, which is proportional to the angular velocity of rotation. This mutual friction will attenuate second sound in the uniformly rotating liquid. Historically, the observation of this attenuation provided the first experimental evidence for the existence of vortex lines.

## 27. The Dynamics of Quantized Vortices at T = 0

### 27.1 THE CLASSICAL DYNAMICS OF A CYLINDER WITH CIRCULATION IN A CLASSICAL IDEAL FLUID

Consider a solid cylinder, round which there is a circulation $\kappa$, moving (perpendicularly to its length) with velocity V relative to a fixed ideal incompressible fluid. The Bernoulli effect gives rise to a *Magnus force* per unit length of cylinder

$$F_M = \rho \, \kappa \times V \qquad (27.1.1)$$

independent of the cross section of the cylinder. Hence the equation of motion of the cylinder under the influence of a steady external transverse force (**f** per unit length) is

$$M^* d^2 r/dt^2 = f + \rho \, \kappa \times dr/dt \qquad (27.1.2)$$

where $M^*$ is the effective mass of the cylinder per unit length (bare mass plus mass of fluid displaced). This is very similar to the equation of motion of a charged particle in a static crossed E, B field

$$m \, d^2 r/dt^2 = qE - q \, \mathbf{B} \times dr/dt \qquad (27.1.3)$$

Hence the motion of the cylinder is the superposition of a "cyclotron orbit" at frequency $\rho\kappa/M^*$, and a steady translation at speed $f/\rho\kappa$ in a direction perpendicular to **f**.

### 27.2 THE DYNAMICAL BEHAVIOUR OF A FREE RECTILINEAR VORTEX IN HELIUM AT T=0

We guess that the motion is similar to that of the cylinder in a classical fluid, with $M^* = \pi\rho a_0^2$. We note, however, that the "cyclotron angular frequency" is about $10^{13}$ s$^{-1}$, so that the cyclotron motion will be heavily damped by the emission of excitations and can be neglected. Strictly speaking helium is not incompressible, but we guess that this has little effect. Therefore we conclude that the motion of a rectilinear free vortex in helium under the influence of an external force **f** per unit length acting on its core is a steady motion relative to the superfluid with velocity $v_L$ given by

$$f = - \rho \, \kappa_s \times v_L \qquad (27.2.1)$$

(In practice **f** could be an electrostatic force if the core were charged with ions). If, as is the usual case at T = 0, **f** = 0, then $v_L$ = 0, and we find that *the vortex must move with the local superfluid velocity*. For example, a pair of parallel vortices (spacing d), with equal but oppositely directed circulations, will move in a direction perpendicular to the line joining them at a speed $\kappa/2\pi d$.

## 27.3 RECTILINEAR VORTEX ARRAYS IN UNIFORMLY ROTATING HELIUM AT T = 0

In Packard's experiments on the direct visualization of vortex lines in a rotating vessel (section 25), the angular velocities of rotation were very small and the vortex densities correspondingly small. Under these conditions the vortex array tends not to be in full equilibrium, so the experiments do not tell us what the equilibrium configuration ought to be.

A theoretical study of the equilibrium configuration of vortices at high angular velocities was carried out by Tkachenko. His work was based essentially on the principles of section 27.2. He showed that the stable configuration of lines has the form of a triangular lattice. Furthermore his calculations revealed that there ought to be oscillatory propagating modes of the vortex lattice, in which the lines remain straight and the wavevector **k** is perpendicular to the angular velocity $\Omega$; the dispersion relation for these modes is

$$\omega = c_r k ,$$  (27.3.1)

where $c_T^2 \sim \Omega \kappa_s$. The modes are partly transverse and partly longitudinal. The probable experimental observation of Tkachenko waves is described in Donnelly's book.

## 28. The Dynamics of Bent Superfluid Vortices

### 28.1 THE BEHAVIOUR OF A VORTEX RING

So far we have considered only the dynamical behaviour of rectilinear vortices. The essential features of the behaviour of a bent vortex can be seen by considering the motion of a vortex ring (well-known in the context of smoke rings).

A circular vortex ring (radius R) in an otherwise undisturbed superfluid moves with unchanged form in a direction at right angles to the plane of the ring with a velocity $v_T$ that increases with decreasing R. There are two ways of viewing this motion. The more rigorous way is to recognize that each part of the ring moves with the local superfluid velocity (section 27.2), as generated by the rest of the ring. The motion then arises in a way that is similar to that operating with the pair of rectilinear vortices (section 27.2). However, the local superfluid velocity is hard to compute, because in the case of a ring part of it is due to closely neighbouring parts of the line. It is therefore convenient to use an alternative view, in which we think of there being a vortex line tension, given by the energy per unit length (equation 24.2), which gives rise to a force on the line in the direction of the centre of the ring (force per unit length $\varepsilon/R$), which in turn gives rise to the motion of the ring by the Magnus effect. We see that the velocity of translation is then

$$v_T = \varepsilon/\rho_s \kappa_s R = (\kappa_s/4\pi R)[\ln(b/a_0) + \eta]$$  (28.1.1)

Presumably b in the log term ought to be replaced by something like R (a rigorous treatment indicates that it ought to be 8R).

Quantized vortex rings were seen in superfluid helium in the classic experiments of Rayfield and Reif. They injected ions (section 15) into the helium at a low temperature, when they suffer little drag from the thermal excitations, and accelerated them through a known electrostatic potential so that they acquired a known energy E. For sufficiently large E they found from the subsequent time of flight of the ions that the ions were moving with a speed that was *inversely* proportional to E. Such behaviour is characteristic of a vortex ring, so that an accelerated ion must have created a vortex ring, into the core of which it was then trapped (section 25). Detailed measurements of the velocity of translation of the rings as a function of the energy confirm this interpretation and lead to a value of $a_0$ of about 0.1 nm, as we expect (section 21).

## 28.2 KELVIN WAVES

That it is possible to propagate waves on a vortex was recognized long ago by Lord Kelvin. Kelvin waves can propagate on an isolated vortex, and they can also exist in a modified form in an array of vortices. The waves have wave vectors that are parallel to the lines, in contrast to the Tkachenko waves, for which the wave vector is perpendicular to the lines.

We give the theory of Kelvin waves on an isolated vortex that lies, when undisturbed, along that z-axis. We denote displacements of the line by $\xi(z,t)$ and $\eta(z,t)$ along the x and y axes respectively. We make the reasonable assumption that the displaced line moves at any point in a direction perpendicular to the local radius of curvature (R) at a speed $\varepsilon/\rho_s\kappa_s R$, as in the case of a vortex ring. Therefore for small displacements

$$\varepsilon\, \partial^2\xi/\partial z^2 = \rho_s\kappa_s\, \partial\eta/\partial t; \text{ and } \varepsilon\, \partial^2\eta/\partial z^2 = -\rho_s\kappa_s\, \partial\xi/\partial t \qquad (28.2.1)$$

A solution of these equations is

$$\xi = \xi_0 \exp[i(kz - \omega t)]; \quad \eta = \mp \xi; \quad k^2 = \pm (\rho_s\kappa_s/\varepsilon)\omega \qquad (28.2.2)$$

The resulting waves are circularly polarized. If the upper signs apply the waves are undamped and are Kelvin waves with

$$\omega = (\kappa_s k^2/4\pi)[\ln(b/a_0) + \eta] \qquad (28.2.3)$$

where b ought presumably to be replaced by $1/k$. If the lower signs apply the waves are evanescent.

Suppose that a horizontal disc is placed in helium that is rotating about a vertical axis. If the disc oscillates in its own plane, its effective moment of inertia can be modified by excitation of vortex waves, especially if the disc is roughened so the ends of the vortices are fixed to points on the disc. Observation of this change in moment of inertia was the basis of the first experimental verification of vortex wave propagation in rotating helium by Hall.

## 29.  The Dynamics of Free Quantized Vortices at a Finite Temperature

At a finite temperature normal fluid excitations will be present and can be scattered by a vortex core. The resulting frictional force modifies the dynamical behaviour of the vortex.

We write the force per unit length acting on a vortex due to the scattering of thermal excitations as

$$f = D(v_R - v_L) + D'\, \hat{\kappa} \times (v_R - v_L) \qquad (29.1)$$

where $v_L$ is the velocity of the line, $v_R$ is the velocity of the excitations close to the core, and $\hat{\kappa}$ is a unit vector along the length of the line. Scattering cannot transfer momentum parallel to the line, but owing to the sense of rotation of the circulation, symmetry does not demand that $D' = 0$. At low temperatures, when the density of excitations is low, D and D' must be obtained from detailed scattering calculations. Such calculations are hampered by our lack of knowledge of the core structure, although interesting progress has been made recently in theoretical study of that part of the scattering that is due to the $p.v_s$ interaction (section 11): see Samuels & Donnelly: *Phys Rev Lett* **65** 187 (1990); Ferrell & Bumsoo Kyung: *Phys Rev Lett* **67** 1003 (1991); and Donnelly's book.

We assume that the vortex line responds to the force **f** by moving in accord with the Magnus formula (27.2.1); i.e.

$$f = \rho_s \, (v_L - vs) \times \kappa_s \tag{29.2}$$

where $v_s$ is the local superfluid velocity. Equations (29.1) and (29.2) serve to determine the motion of the vortex.

## 30.   Mutual Friction in the Uniformly Rotating Liquid

The frictional force (29.1) is responsible for the force of mutual friction in uniformly rotating helium (section 26). The quantity that is measured in experiments is the mutual friction per unit volume

$$F_{sn} = (2\Omega/\kappa_s) \, f \tag{30.1}$$

The use of equations (29.1) and (29.2) allows one to express **f** in terms of $v_R$ and the local value of $v_s$. To compare with experiment we need to know $F_{sn}$ in terms of $\bar{v}_s$ and $\bar{v}_n$, where the bars denote spatial averages. The superfluid velocities $v_s$ and $\bar{v}_s$ are equal, but $v_R$ and $\bar{v}_n$ differ because there is local dragging of the normal fluid near the line. Detailed calculations are given in Donnelly's book, and the result is that

$$F_{sn} = - \, B \, (\rho_s \rho_n/2p) \, \Omega \times [\hat{\Omega} \times (\bar{v}_s - \bar{v}_n)] - B' \, (\rho_s \rho_n/2\rho)[\Omega \times (\bar{v}_s - \bar{v}_n) \tag{30.2}$$

where B and B' are complicated functions of D, D', $\rho_s$, $\rho_n$, $\kappa_s$, the frequency of the measurement, and the viscosity and mean free path in the normal fluid. B can be obtained from experimental measurements of the attenuation of second sound in rotating helium. To obtain experimental values of B', which is non-dissipative, it is necessary to use a square second sound resonator which has two degenerate second sound modes that propagate at right angles to one another; the term in B' couples the two modes together, and the resulting frequency splitting can be used to deduce B'.

## 31.   The Lifetime of a Vortex Ring

The drag on a moving vortex ring causes it to shrink: a nice example of the operation of the Magnus effect. The effect can be seen in the Rayfield-Reif experiment (section 28.1). Observations of the rate of shrinkage, or rate of loss of energy, yield values of the friction coefficient D.

## 32.   Superfluid Turbulence

Mutual friction is present not only in uniformly rotating superfluid helium but also in a sufficiently large steady heat current (section 18). There is evidence that this friction is associated with turbulence in the superfluid. Any turbulence must involve a non-vanishing curl **v**, so that

18

superfluid turbulence must involve vortex lines. These vortex lines will produce the mutual friction. Presumably the turbulence takes the form of a random tangle of vortex lines, and the form of equation (18.1) shows that the length of line per unit volume must be proportional to $|\bar{v}_s - \bar{v}_n|^2$. We discuss the reason later.

## 33. Pinned Vortices

So far we have assumed that the vortices in the helium are free to move and are not pinned in any way. In practice the vortices can be pinned at the walls of a vessel, especially if the walls are rough. The walls can then exert a force on the vortex, which can then remain stationary even in the presence of a Magnus force or a friction force. Consider two examples.

A vortex is pinned at two points on opposite sides of a channel down which there is a superflow ($\bar{v}_s \neq 0$; $v_n = 0$). There is a Magnus force per unit length of line equal to $-\rho_s\kappa_s v_s$. The vortex bends into a loop with radius of curvature R given by equating this Magnus force to the tension force $\varepsilon/R$ (section 28.1). The Magnus force, integrated along the length of the vortex, is transferred to the pins.

If the same vortex is subject to the flow of the normal fluid only ($v_s = 0$; $v_n \neq 0$) the force per unit length of line due to friction from the normal fluid must again be balanced by a force $\varepsilon/R$. Again the force is transferred to the pins.

Note that in neither case is there any force that is tending to accelerate or decelerate the superfluid. If there is to be a force on the superfluid the vortices must move.

## 34. Vortices and Phase Slip

The connection between vortex movement and a force on the superfluid can be seen clearly by thinking in terms of vortex movement producing "phase slip" in the condensate wave function (Anderson).

Consider a row of free vortices (spacing b) across a channel down which the superfluid flows with average velocity $v_s$ (figure 3).

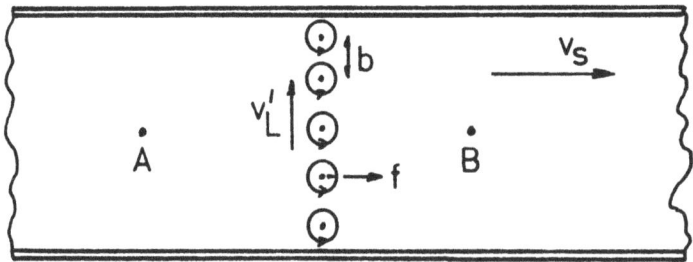

*Figure 3. Vortices moving across a channel.*

An external force f per unit length of vortex is applied to each vortex in a direction parallel to the length of the channel. As a result each vortex moves with velocity $v_L$ such that the Magnus force $- \rho_s(v_L - v_s) \times \kappa_s$ balances the force f. Since f is in the direction of $v_s$, the component of $v_L$ at right angles to $v_s$ ($v_L'$) must therefore be given by

$$f = \rho_s v_L' \kappa_s. \tag{34.1}$$

Now consider two points, A and B, on either side of the row of vortices. Going completely round any path that encloses a vortex leads to a phase change in the condensate wave function equal to $2\pi$. It follows that as each vortex crosses the channel the phase difference between the points A and B must also change by $2\pi$. Therefore as the row of vortices moves with the velocity $v_L'$ the phase at A must slip continuously with respect to the phase at B at the rate $2\pi v_L/b$. But the time-dependence of the phase of the order parameter is (section 16).

$$dS/dt = - m_4 \mu/\hbar. \tag{34.2}$$

Therefore there must be a difference in chemical potential given by

$$\Delta\mu = 2\pi\hbar v_L'/bm_4 \tag{34.3}$$

between A and B.

Consider two cases:
1. The channel forms part of a toroidal tube, so that the superfluid is flowing round the torus. Motion of the vortices changes the circulation round the torus, and leads therefore to acceleration or deceleration of the superfluid. In this case the jump in chemical potential $\Delta\mu$ must lead to a steady gradient in $\mu$ round the rest of the torus. This gradient in $\mu$ changes $v_s$ at just the rate required to produce the required rate of change of circulation round the torus.
2. The channel connects two large reservoirs, C and D, between which a steady superflow is maintained by means of a difference in chemical potential ($\mu_C - \mu_D$). In this case a steady flow of superfluid in the region remote from the vortices requires that there be no $\nabla\mu$ in this region, and therefore the jump in chemical potential $\Delta\mu$ is equal to ($\mu_C - \mu_D$). The difference ($\mu_C - \mu_D$) in chemical potential provides the force on the superfluid that exactly balances the force f integrated over the total length of vortex line.

## 35.   The AC Josephson Effect

The movement of vortices across the channel under the action of the difference in chemical potential ($\mu_C - \mu_D$) has associated with it a characteristic angular frequency $\omega$ given by

$$\hbar\omega = m_4(\mu_C - \mu_D). \tag{35.1}$$

This is the analogue of the AC Josephson effect. The observation of discrete events involving phase slips of $2\pi$ in superflow through a narrow constriction has been made in the elegant and delicate experiments of Varoquaux and Avenel: *Phys Rev Lett* **55** 2704 (1985).

## 36.  Critical Velocities Associated with Vortex Creation

Frictionless superflow through a channel can take place only below a certain, relatively small, critical velocity. We can tentatively suppose that this critical velocity corresponds to the velocity at which vortices are nucleated and move across the flow. Let us examine how such a process can actually occur.

Suppose that the helium in the channel contains initially no vortices. We set the superfluid into motion at a velocity ($\sim 1$ cm s$^{-1}$) that ought according to experiment to lead to the generation of vortices. How are these vortices generated? The probability that a macroscopic length of vortex appears suddenly in the middle of the fluid must be very small, because such a process would involve a transition in which a macroscopic number of condensate atoms change their quantum state simultaneously (see section 8). But we must now examine just how small it actually is.

In practice the vortex line must appear as a loop attached to the wall of the channel, or as a ring in the middle of the liquid.

## 37.  Vortex Nucleation in Superflow through a Straight Channel

Let us consider the nucleation of a vortex loop (figure 4). The loop is semicircular, radius R, and has its normal in the direction along the channel. We can satisfy the boundary condition that the

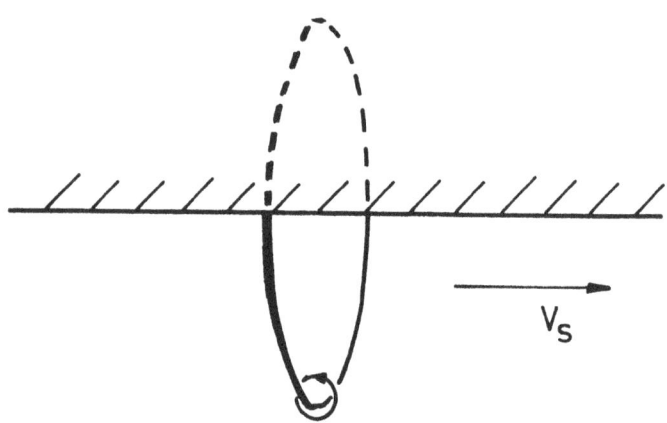

*Figure 4.    Nucleation of a vortex loop.*

normal component of $v_s$ be zero at the wall by adding an image in the form of the rest of the circular loop. In the absence of any net superflow down the channel, the energy of the loop is

$$E \approx (\rho_s \kappa_s^2 R/4)[\ln(8R/a_0) + \eta] \tag{37.1}$$

(see equation (24.1) and the comment following equation (28.1.1)). The momentum of the loop (strictly speaking its hydrodynamic impulse) can be shown to be given by

$$P = \pi R^2 \rho_s \kappa_s/2, \tag{37.2}$$

pointing, as shown, in a direction normal to the plane of the loop. Now add a superflow down the tube, in the direction shown, with superfluid velocity $V_s$. Using equation (11.1) we find that the energy of the loop becomes

$$E' \approx (\rho_s \kappa_s^2 R/4)[\ln(8R/a_0) + \eta] - \pi R^2 \rho_s \kappa_s V_s/2. \tag{37.3}$$

A plot of E' against R for given $V_s$ shows that the creation of the loop is opposed by an energy barrier of height $\rho_s \kappa_s^3 K^2/32\pi V_s$ and width $\kappa_s K/2\pi V_s$, where K is a complicated function that varies only slowly with $V_s$ and is of order 10. The probability that the superfluid can slow down is therefore related to the probability that this barrier can be surmounted.

We note in passing that loops that are smaller than that corresponding to the maximum in E' move relative to the normal fluid (assumed to be at rest) in such a way that mutual friction causes the loop to contract; loops that are larger move in such a way that mutual friction causes them to expand.

In principle, the barrier can be overcome either by thermal excitation or by quantum tunnelling. If $V_s = 1$ cm s$^{-1}$ and $T \approx 0$, the barrier height is about $1.4 \times 10^{-17}$ J ($10^6$ K) and its width is about 16 $\mu$m. At this relatively low $V_s$ the barrier is very high and very wide, and the probability of its being overcome must be exceedingly small. A proper calculation of the probability is a typical nucleation problem; vortex nucleation by thermal activation was first worked out properly by Iordanskii (*Zh Eksp Theor Fiz* **19** 560 (1967), and his calculations confirm that under most conditions the probability is indeed very small. It is this fact that ensures that superflow can be metastable.

## 38. The Role of Remanent Vorticity

However, in practice, values of $V_s$ as low as 1 cm s$^{-1}$ do often seem to be sufficient for the creation of vortex line. It seems very unlikely that this line could be produced by the type of *intrinsic nucleation* process that we have been considering, and we are therefore forced to the conclusion that the line is produced by the stretching of small lengths of line present initially in the helium (remanent vorticity). That such remanent vorticity is usually present has been shown in a number of experiments, including the vibrating wire experiment (section 20) and experiments with ions (Awschalom & Schwarz, *Phys Rev Lett* **52** 49 (1984)). The process by which a remanent vortex can expand (extrinsic nucleation) under the influence of a supercurrent exceeding a critical value has been considered by Schwarz (*Phys Rev Lett* **64** 1130 (1990)). It seems likely that most observed critical velocities are of this extrinsic type and are therefore of no fundamental significance.

## 39. Intrinsic Nucleation of Vortices

We ask whether experimental conditions can be found where vortices do appear as a result of intrinsic nucleation.

There are two possibilities. First, near the λ-point, where $\rho_s$ is small, so that barrier heights are small. And secondly, at lower temperatures, and where for some special reason high superfluid velocities can be generated in very limited volumes where remanent vorticity is unlikely to be found.

Extensive studies of vortex nucleation near the λ-point were carried out by Reppy *et al* (*Progress in Low Temp Physics* **6** 1 (1970)). Nucleation rates involving a barrier proportional to $\rho_s$ were observed, as expected from equation (37.3). There was, however, always some doubt whether the processes involved were really intrinsic, or whether they involved the freeing of a remanent vortex by thermal excitation, which would also involve a barrier proportional to $\rho_s$. The experiments often involved superflow through some sort of fine powder, so that the geometry was not well defined, and a complete analysis of the experiments was therefore difficult.

Two types of experiment have been performed that involve vortex nucleation in regions of high localized superfluid velocity. In the first, nucleation by rapidly moving ions was observed (Hendry *et al*; *Phil Trans Roy Soc* **A322** 387 (1990)). The high superfluid velocity exists only in the immediate neighbourhood of the ion, and the geometry is very well defined. Both thermal nucleation and quantum nucleation were observed, and a detailed theory (Muirhead *et al*; *Phil Trans Roy Soc* **A311** 433 (1984)) gives a good account of the experimental results, although we lack a proper theory of the quantum process. In the second type of experiment, already mentioned in section 35, the superflow is through a very small orifice and vortex nucleation gives rise to individually-observed discrete $2\pi$ phase slips. Again both the thermally activated and the quantum processes have been observed. Although the geometry in these experiments is less well defined than in the ion experiments ingenious arguments have been employed to deduce absolute superfluid velocities at the nucleation site (Ihas *et al*: *Phys Rev Lett* **69** 327 (1992); Davis *et al*: *Phys Rev Lett* **69** 323 (1992); Varoquaux *et al*: *Phys Rev Lett* **70** 2114 (1993)).

### 40. Vortices in Thermal Equilibrium

We have discussed the thermal excitation of vortex line in the presence of a superflow. We can ask whether vortex lines can exist in significant lengths in thermal equilibrium in the absence of superflow.

Thermally generated line is certainly important in thin films, where it is responsible for the Kosterlitz-Thouless transition (bound vortex pairs form at low temperatures, where they contribute to the normal fluid density; above the KT transition temperature they dissociate and superfluidity is effectively destroyed). This important topic is covered in the lectures of Professor Young.

There is no doubt that in some sense vortex line must be generated at high temperatures in three dimensional helium (simply as a result of fluctuations in the order parameter), and it has been suggested that this line plays an important part in the loss of superfluidity at the λ-point. Thermally generated vortex line is not explicitly mentioned in the very successful conventional theories of the λ-transition, so that it is clear that explicit inclusion of vortex lines is not necessary.

### 41. Superfluid Turbulence

The generation of vortex line in superflow through a channel often leads to a random array of lines, which constitutes what we call *superfluid turbulence*. Superfluid turbulence occurs in what is probably its simplest form in a steady heat current in a wide channel. There is strong evidence that the turbulence is then homogeneous, so that it must be maintained by the relative motion of the two fluids.

Some progress can be made in understanding this homogeneous turbulence by scaling arguments. It is reasonable to assume that the arrangement of the lines in the turbulence is always geometrically similar except for a scaling factor that ensures that the fluctuations in the turbulent velocity field are proportional to the average value of $(v_s - v_n)$. It is then easy to show that the length of line per unit volume must be proportional to the square of this average value, and hence that the mutual friction must have the form of equation (18.1), as is observed (more refined versions of this argument are given in Donnelly's book).

Further understanding has emerged from detailed computer simulations carried out by Schwarz. These simulations are based on the basic laws governing vortex motion, as explained in these lectures, with the additional feature that when two vortex lines come close they can reconnect in such a way that lines AB and CD become lines AC and BD. The computations start with some arbitrarily defined vortex tangle and then work out how it will evolve with time. For fixed average $(v_s - v_n)$ it is found to evolve into a dynamic steady state, involving a constant average line density, which is dynamically self-sustaining, and which leads again to equation (18.1). In a channel of finite width the interaction of the vortex lines with the walls of the channel must be taken into account. The vortex tangle then ceases to be self-sustaining below a certain critical velocity, as is observed.

There is much about superfluid turbulence that is not understood. It is probable that the basic laws governing the vortex motion are known, but, as in classical turbulence, these basic laws (the Navier-Stokes equation in the classical case) can lead to an incredibly rich variety of behaviour, which offers scope for almost endless investigation. Unfortunately, the experimental tools at our disposal are still quite crude. Local densities and anisotropies of the vortex line tangle can be probed with good sensitivity by measuring second sound attenuation, but we lack a means for measuring local velocity fields (either $v_s$ or $v_n$).

## 42. Conclusions

Quantized vortices play an important role in superfluid $^4$He for two reasons: they allow the superfluid to rotate; and their formation is the mechanism by which frictionless superflow normally breaks down at a high enough velocity.

The quantized vortices still present us with unsolved problems. The structure of their core is not fully understood; aspects of the nucleation of vortex line are not yet fully understood, particularly nucleation by quantum tunnelling; and we still lack a full understanding of superfluid turbulence in many regimes.

# ANISOTROPIC SUPERCONDUCTORS: FUNDAMENTALS OF VORTICES IN LAYERED SUPERCONDUCTORS

JOHN R. CLEM
*Ames Laboratory and Department of Physics and Astronomy*
*Iowa State University*
*Ames, Iowa 50011*
*U.S.A.*

ABSTRACT. Some fundamentals of the magnetic structure in anisotropic superconductors are presented. The approach focuses on the vortex structure in layered superconductors, the most important example being the high-$T_C$ cuprates. The Ginzburg-Landau theory is discussed, as well as its extensions to include anisotropy via an anisotropic effective mass tensor. The Lawrence-Doniach theory is introduced, and it is used to show that a vortex parallel to the layers has a Josephson core confined to the interlayer junction. Two-dimensional (2D) pancake vortices are discussed, and the fields and currents generated by them are presented. The Josephson coupling energy between 2D pancake vortices in adjacent layers is estimated. Some basics of flux pinning in the layered superconductors are examined. Finally, an explanation is given of why the activation energy is much larger than the single-pancake pinning energy in strongly coupled superconductors but is approximately equal to the single-pancake pinning energy in weakly coupled superconductors.

## 1. Introduction

The purpose of this paper is to briefly discuss a number of fundamental ideas on how to imagine the structure of vortices in layered superconductors, especially the high-temperature copper-oxide superconductors. Such considerations provide the key to a good understanding of the field and temperature dependencies of the critical-current density in these superconductors.

I shall begin in Sec. 2 by discussing the Ginzburg-Landau theory for isotropic superconductors. It is then relatively simple to see how anisotropy can be accounted for with the help of an anisotropic effective mass tensor. I then point how that difficulties arise when trying to use this anisotropic continuum theory in cases when the coherence length becomes smaller than the layer spacing. These difficulties can be handled with the help of the Lawrence-Doniach theory (Sec. 3), which I use to discuss the properties of a vortex parallel to the layers. In Sec. 4, I examine the properties of two-dimensional (2D) pancake vortices, and in Sec. 5, I give a procedure for estimating the Josephson coupling between 2D pancakes in different layers. Some elementary considerations about flux pinning in the high-temperature superconductors are presented in Sec. 6. In particular, I estimate the single-pancake-vortex pinning energy for optimal pinning, and I use this to estimate the highest possible critical-current density that could be achieved in the limit of zero field and zero temperature. I also consider Josephson interactions between pancake vortices and use

25

a simple model to show why the activation energy for strong interlayer coupling is generally much higher than that for weak interlayer coupling. Finally, I present a brief summary in Sec. 7.

## 2. Ginzburg-Landau Theory

A good starting point for a theoretical understanding of the structure of vortices in type-II superconductors is the isotropic Ginzburg-Landau (GL) theory [Ginzburg and Landau (1950), Werthamer (1969), Fetter and Hohenberg (1969)]. The essential components of the theory are two coupled equations, which complement Maxwell's equations. One of the GL equations governs the spatial variation of the superconducting order parameter, which is a quantity similar to the wave function in ordinary quantum mechanics. The other equation governs the spatial variation of the supercurrent density $\mathbf{j}$, the vector potential $\mathbf{a}$, and the local magnetic flux density $\mathbf{b}$.

Appearing in the Ginzburg-Landau equations for an isotropic superconductor are two characteristic lengths of superconductivity, the weak-field penetration depth $\lambda$ and the coherence length $\xi$. When a weak magnetic field is applied to a superconductor, currents that screen the field from the superconductor's interior are automatically generated, and the exponential decay length of these screening currents is $\lambda$. The coherence length $\xi$ is the characteristic length scale over which the order parameter can vary. The Ginzburg-Landau parameter is defined as the ratio of these two lengths: $\kappa = \lambda/\xi$. Type-II superconductors, the subject of this paper, are superconductors for which $\kappa > 1/\sqrt{2}$. The two characteristic lengths of superconductivity in the Ginzburg-Landau theory also are related to the bulk thermodynamic critical field $H_c$ of the superconductor via $H_c = \phi_0/2^{1/2}2\pi\xi\lambda$, where $\phi_0 = hc/2e = 2.07 \times 10^{-7}$ Gcm$^2$. The condensation energy density, the free energy difference between the normal and superconducting states, is $H_c^2/8\pi$.

As shown from the Ginzburg-Landau theory by Abrikosov [Abrikosov (1957)], when a magnetic field H is applied parallel to a type-II superconducting sample in the shape of a long cylinder or slab, a weak applied field penetrates only to a depth $\lambda$, as mentioned above, but when the applied field increases to a characteristic value $H_{c1}$, the lower critical field, it becomes energetically favorable for magnetic flux to be present deep in the superconductor in the form of singly quantized vortices, each carrying magnetic flux $\phi_0$. An accurate expression for $H_{c1}$ when $\kappa \gg 1$ is [Hu (1972)]

$$H_{c1} = (H_c/\kappa\sqrt{2})(\ln\kappa + 0.50).\tag{1}$$

The order parameter is zero on the vortex axis. The magnitude of the order parameter rises linearly with radial coordinate r on the scale of $\xi$ and, for an isolated vortex, becomes nearly constant outside $\xi$. The central region of the vortex, where the order parameter is suppressed, is called the Abrikosov core. For a number of purposes this core can be thought of as a normal region of cross-sectional area $\pi\xi^2$. The local magnetic field generated by the vortex is roughly constant inside the core but drops off outside the core, initially as $\ln(1/r)$ and then approximately as $\exp(-r/\lambda)$ at distances $r > \lambda$. The associated vortex supercurrent density initially rises linearly with r, reaches a maximum close to the depairing current density at $r \approx \xi$, then drops off as $1/r$ and finally as $\exp(-r/\lambda)$ at distances $r > \lambda$.

As the applied magnetic field increases, the equilibrium density n of vortices increases and their field contributions overlap, but the average flux density $B = n\phi_0$ remains less than H, so that the magnetization M remains negative ($-4\pi M = H - B$). With increasing H, the vortices pack closer together, and when the vortices are roughly a distance $\xi$ apart, bulk superconductivity is quenched at a field $H_{c2}$, the upper critical field. Abrikosov first showed that the field at which this occurs is $H_{c2} = \phi_0/2\pi\xi^2$. An equivalent expression, which will be more convenient for later use, is $H_{c2} = \kappa\sqrt{2}\, H_c$.

Included among the results of the Ginzburg-Landau theory, in addition to the above equations for the critical fields $H_c$, $H_{c1}$, and $H_{c2}$, are various useful expressions for the free energy density and the magnetization M as a function of H or B. As was shown by Gor'kov [Gor'kov (1959)], the Ginzburg-Landau theory can be derived from the microscopic BCS [Bardeen, Cooper, and Schrieffer (1957)] theory, but such a derivation is justified only very close to the superconducting transition temperature $T_c$. Nevertheless, because of their simplicity, the results of the Ginzburg-Landau theory are commonly used well outside their domain of validity. However, the errors in doing this are not large, usually less than 25% [Fetter and Hohenberg (1969)].

Anisotropic superconductors, such as the layered superconductors, are not well described by the above isotropic Ginzburg-Landau theory. A convenient way to modify the theory for layered superconductors is to incorporate anisotropy via an effective mass tensor [von Laue (1952), Caroli et al. (1963), Gor'kov and Melik-Barkhudarov (1964), Tilley et al. (1964), Tilley (1965), Katz (1969 and 1970), Takanaka (1975), Klemm and Clem (1980), Kogan (1981), Kogan and Clem (1981 and 1987), Balatskii et al. (1986), Bulaevskii et al. (1988)]. In a reference frame aligned with the principal axes $\hat{x}_i$ (i = 1, 2, 3 = a, b, c = x, y, z), this mass tensor is diagonal, and it is most convenient to normalize the diagonal elements $m_i$ such that $m_1 m_2 m_3 = 1$. Whereas in the isotropic case there are two characteristic lengths of superconductivity, $\lambda$ and $\xi$, in the anisotropic case are six, $\lambda_i$ and $\xi_i$, where i = 1, 2, 3. These six lengths are not independent of each other, however. The penetration depth associated with $j_i$, the component of the screening current that flows along $\hat{x}_i$, is $\lambda_i = \lambda\sqrt{m_i}$. As a function of the distance x from the surface, a weak magnetic field penetrating into an anisotropic superconductor then decays as $\exp(-x/\lambda_i)$ if the induced screening currents flow parallel to the principal axis $\hat{x}_i$. The length scale that characterizes the spatial variation of the superconducting order parameter along the principal axis $\hat{x}_i$ is the coherence length $\xi_i = \xi/\sqrt{m_i}$. Because $m_1 m_2 m_3 = 1$, we see that $(\lambda_1 \lambda_2 \lambda_3)^{1/3} = \lambda$ and $(\xi_1 \xi_2 \xi_3)^{1/3} = \xi$. The bulk thermodynamic critical field $H_c$, the scalar quantity entering the expression $H_c^2/8\pi$ for the superconducting condensation energy density, can be written in terms of $\lambda$ and $\xi$ as $H_c = \phi_0/2^{1/2}2\pi\lambda\xi = \phi_0/2^{1/2}2\pi\lambda_i\xi_i$, as in the isotropic theory. The isotropic Ginzburg-Landau parameter is still $\kappa = \lambda/\xi$.

When the magnetic field is aligned along principal axis $\hat{x}_i$, it is possible to introduce a coordinate transformation that transforms the original anisotropic Ginzburg-Landau equations back into isotropic form [Klemm and Clem (1980)]. One then obtains expressions for the energy density, vortex energy, magnetization, upper and lower critical fields, etc., by starting from the corresponding Ginzburg-Landau quantity (expressed in terms of $H_c$ and $\kappa$) and simply replacing $\kappa$ by $\kappa_i = \kappa/\sqrt{m_i}$. For example, the upper critical field along $\hat{x}_i$ is $H_{c2}(\| \hat{x}_i) = \kappa_i \sqrt{2}\, H_c$, and the lower critical field is, to good approximation when $\kappa_i \gg 1$,

$$H_{c1}(\| \hat{x}_i) = (H_c/\kappa_i\sqrt{2})(\ln\kappa_i + 0.50). \qquad (2)$$

Note that the largest $H_{c1}$ and the smallest $H_{c2}$ are found along the axis with the largest $m_i$ (smallest $\kappa_i$), while the smallest $H_{c1}$ and the largest $H_{c2}$ are found along the axis with the smallest $m_i$ (largest $\kappa_i$).

When the vortices are at an arbitrary angle $(\theta,\phi)$ relative to the principal axes, the analogous scaling procedure, that of replacing $\kappa$ by $\tilde{\kappa}(\theta,\phi)$, where

$$\tilde{\kappa}(\theta,\phi) = \kappa/(m_a \sin^2\theta \cos^2\phi + m_b \sin^2\theta \sin^2\phi + m_c \cos^2\theta)^{1/2}, \qquad (3)$$

is not in general exact [Kogan (1981), Kogan and Clem (1981)], but for high-$\kappa$ superconductors it is an excellent approximation in fields well above the lower critical field [Hao et al. (1991a), Hao and Clem (1991 and 1992)] and becomes exact at the upper critical field. For example, while the isotropic Ginzburg-Landau theory gives $H_{c2} = \kappa\sqrt{2}\, H_c$, in the anisotropic case the upper critical field at angle $(\theta,\phi)$ is $H_{c2}(\theta,\phi) = \tilde{\kappa}(\theta,\phi)\sqrt{2}\, H_c$. This can be written as

$$H_{c2}(\theta,\phi) = [(H_{c2a})^{-2} \sin^2\theta \cos^2\phi + (H_{c2b})^{-2} \sin^2\theta \sin^2\phi + (H_{c2c})^{-2} \cos^2\theta]^{-1/2}, \qquad (4)$$

where $H_{c2a}$, $H_{c2b}$, and $H_{c2c}$ are the upper critical fields along the a, b, and c axes.

Along the axis of each vortex is an Abrikosov core, a region of suppressed order parameter, since in this theory the order parameter still must be precisely zero on the vortex axis. For example, for a vortex centered on the a axis in the high-temperature superconductor $YBa_2Cu_3O_{7-\delta}$, the core is elliptical in cross section, with the semimajor axis approximately $\xi_b$ along the b axis and the semiminor axis $\xi_c$ along the c axis. Measurements yield values for the ratio $\xi_b/\xi_c = \lambda_c/\lambda_b = (m_c/m_b)^{1/2}$ in the range 5-8. In $YBa_2Cu_3O_{7-\delta}$, the characteristic lengths of superconductivity order as $\xi_c < \xi_a < \xi_b \ll \lambda_b < \lambda_a < \lambda_c$. As can be seen from the above equations, the largest $H_{c1}$ and the smallest $H_{c2}$ are found along the c axis, while the smallest $H_{c1}$ and the largest $H_{c2}$ are found along the b axis, since $m_b < m_a < m_c$ and $\kappa_c < \kappa_a < \kappa_b$.

An implicit assumption of the Ginzburg-Landau theory is that all the characteristic lengths of superconductivity are much larger than the interatomic spacings. The smallest characteristic length of superconductivity in the anisotropic Ginzburg-Landau theory for the layered high-temperature superconductors is $\xi_c$, and it is important to compare this length with the relevant layer periodicity length s. In $YBa_2Cu_3O_{7-\delta}$, for example, the layers responsible for superconductivity evidently are the double $CuO_2$ layers surrounding the Y layers. The characteristic superconducting layer periodicity s then corresponds to the $YBa_2Cu_3O_{7-\delta}$ lattice parameter c. (Similarly, for the case of $Bi_2Sr_2CaCu_2O_8$, which has two formula units per unit cell, the appropriate layer periodicity is s = c/2, the center-to-center distance between adjacent $CuO_2$ bilayers.) When all the characteristic lengths of superconductivity are much larger than s (i.e., when $\xi_c >> s$), it is sensible to use the anisotropic Ginzburg-Landau theory. On the other hand, it makes little sense to use such a continuum theory to describe the vortex structure when the coherence length $\xi_c < s$, a condition that holds at temperatures below around 75 K in $YBa_2Cu_3O_{7-\delta}$ and at practically all temperatures in the Bi- and Tl-based high-temperature superconductors. For purposes of describing phenomena that involve the vortex-core dimensions, the Ginzburg-Landau theory therefore must be replaced by one that explicitly incorporates the discreteness of the atomic layers.

## 3. Lawrence-Doniach Model

The model introduced by Lawrence and Doniach (1971), who approximated a layered superconductor as a stack of weakly coupled superconducting layers, remains applicable when $\xi_c < s$. The Lawrence-Doniach theory treats superconductivity within the layers via the Ginzburg-Landau theory, but treats the flow of current between adjacent layers via the Josephson effect. This theory has been applied [Bulaevskii (1973), Clem and Coffey (1990), Clem et al. (1991), Koshelev (1993)] to calculate the structure of a vortex whose axis is centered on the Josephson junction between two superconducting layers. The resulting field and supercurrent distributions resemble those of the anisotropic Ginzburg-Landau theory. However, there is no Abrikosov core close to the axis, because the order parameter remains high on all superconducting layers. Instead, for a vortex parallel to the a axis, there is a Josephson core.

At distances well outside the vortex core, the discrete equations that describe transport in the z direction perpendicular to the layers can be linearized, and the equations governing the supercurrents and local magnetic flux density reduce to those of the anisotropic Ginzburg-Landau theory. In the vicinity of the core, however, details of the Josephson coupling become important. The current density across each Josephson junction between adjacent superconducting layers has only a z component, given by $J_z = J_0 \sin \Delta\gamma$, where $\Delta\gamma$ is the local gauge-invariant phase difference across the junction. For weak interlayer coupling, the penetration depth $\lambda_c$ for currents in the c direction is related to the maximum Josephson current density $J_0$ via $\lambda_c = (c\phi_0/8\pi^2 sJ_0)^{1/2}$ [Clem (1989)]. The current density across the central junction (z = 0) is maximized at $y = y_{max} \equiv (s/2)(\lambda_c/\lambda_b)$, where $J_z(y_{max},0) = J_0$, the maximum Josephson current density. Since the current density reaches its maximum magnitude along the y axis at $y = \pm y_{max} = \pm(s/2)(\lambda_c/\lambda_b)$ and along the z axis at $z = \pm s/2$, it is helpful to think of the Josephson core as having an elliptical cross section with semimajor axis $(s/2)(\lambda_c/\lambda_b)$ parallel to the layers and semiminor axis s/2 perpendicular to the layers. The ratio of the semimajor axis to the semiminor axis of the Josephson core is therefore $\lambda_c/\lambda_b = (m_c/m_b)^{1/2}$, exactly the same ratio as for an Abrikosov core for a vortex centered on the a axis.

The lower critical field $H_{c1}$ of a Josephson vortex parallel to the layers has been calculated approximately by Bulaevskii (1973), Clem et al. (1991), and Koshelev (1993). The result is similar to that in the anisotropic Ginzburg-Landau theory, but the main correction is that $\xi_c$ is replaced by the layer periodicity s. Note that $\xi_c$ physically cannot enter into any theoretical expressions when $\xi_c < s$, because s becomes the smallest physical length scale in the problem.

When vortices are inclined at an angle $\theta$ relative to the z axis (or the c axis) of strongly anisotropic layered superconductors with $m_c >> m_a$ or $m_b$, as in the high-temperature superconductors, calculations using the anisotropic Ginzburg-Landau theory show that the supercurrents flowing around the vortex axis tend to flow not in circular paths in planes perpendicular to the vortex axis, as in isotropic superconductors, but in roughly elliptical paths that are nearly parallel to the layers [Kogan (1981), Kogan and Clem (1981)]. If one now takes discreteness of the atomic layer structure into account using the Lawrence-Doniach theory, the corresponding picture of a tilted vortex now becomes that of a tilted stack of two-dimensional (2D) pancake vortices, connected by Josephson strings. Pancake vortices, to be described in the next section, are characterized by circular (or slightly elliptical) supercurrent patterns confined to the superconducting layers, while Josephson strings are short segments of Josephson vortices, whose axes are confined to the insulating regions between the superconducting layers.

## 4. 2D Pancake Vortices

In the limit of extreme anisotropy, a vortex line threading through a stack of superconducting layers can be regarded as a stack of magnetically coupled two-dimensional (2D) pancake vortices, with one pancake vortex in each layer. In this limit, for which the Josephson coupling between layers is zero and the penetration depth $\lambda_c$ is infinite, it is useful to calculate the magnetic field generated throughout space by a single pancake vortex. An analytic solution can be obtained [Artemenko and Kruglov (1990), Clem (1991), Fischer (1991)]. Using this solution and the principle of linear superposition, it is then possible to calculate the magnetic field throughout all space produced by an arbitrary distribution of pancake vortices in various layers. The case of a straight stack, with one pancake in each layer, is the simplest example of such a procedure. It is also simple to calculate the magnetic interaction forces between pancake vortices in the same or different layers.

Before considering the field generated by a pancake vortex in the central layer of an infinite stack of superconducting films, let us consider the field generated by a vortex in an isolated thin film with no other superconducting layers above and below it. When the film is thin ($d << \lambda$, where d is the thickness and $\lambda$ is the bulk penetration depth), a vortex in such a film is characterized by a supercurrent density j that varies as $1/r$ from the vortex core radius $\xi$ to approximately the two-dimensional screening length $2\lambda^2/d$ [Pearl (1964 and 1965), de Gennes (1966)]. Beyond this distance, j in the film varies as $1/r^2$. At distances $r > 2\lambda^2/d$, the magnetic field b in the space above the film appears as if produced by a magnetic monopole; flux $\phi_0$ spreads out into $2\pi$ steradians, so that $b_r \approx \phi_0/2\pi r^2$. In the space below the film, the field appears as if is produced by a negative magnetic monopole, so that $b_r \approx -\phi_0/2\pi r^2$.

Let us now imagine starting with a vortex in an isolated thin film of thickness d and then bringing in, from above and below, an infinite set of identical superconducting films containing no pancake vortices, such that the layer periodicity length in the z direction is s. The screening effect of these other superconducting films is such that the field distribution generated by the pancake vortex in the central layer is now strongly distorted. The sheet current density in the central layer centered at $z = 0$ is [Clem (1991)] $K(\rho,0) = \hat{\phi}\, K_\phi(\rho,0)$, where $K_\phi(\rho,0) \approx c\phi_0/4\pi^2\Lambda\rho$ for all $\rho > \xi$. [Here we use cylindrical coordinates $\rho = (x^2 + y^2)^{1/2}$, $\phi = \tan^{-1}(y/x)$, and $z$ with unit vectors $\hat{\rho} = \hat{x}\cos\phi + \hat{y}\sin\phi$, $\hat{\phi} = \hat{y}\cos\phi - \hat{x}\sin\phi$, and $\hat{z}$.] We assume for simplicity that the films are isotropic in the xy (or ab) planes, so that $\lambda_{ab} = \lambda_a = \lambda_b = (s/d)^{1/2}\lambda$ [Clem (1989)]. The relevant 2D screening length is $\Lambda = 2\lambda_{ab}^2/s \gg \lambda_{ab}$, since $s \ll \lambda_{ab}$. (We consider values of $s \sim 10\text{-}20$ Å and $\lambda_{ab} \sim 1000\text{-}2000$ Å that are characteristic of high-temperature superconductors and artificial multilayers of these materials.)

In all other superconducting layers the sheet current direction is reversed, since these layers are attempting to screen out the magnetic field generated by the central layer, and the magnitude of the sheet current is much smaller. We find $K(\rho,z) = \hat{\phi}K_\phi(\rho,z_n)$, where $z_n = ns$, $n = \pm 1, \pm 2, \pm 3, ...$,

$$K_\phi(\rho,z_n) = -(c\phi_0\lambda_{ab}/4\pi^2\Lambda^2\rho)(e^{-|z_n|/\lambda_{ab}} - e^{-r_n/\lambda_{ab}}), \qquad (5)$$

and $r_n = (\rho^2 + z_n^2)^{1/2}$.

The magnetic field generated by the pancake vortex is $b = \hat{\rho}b_\rho(\rho,z) + \hat{z}b_z(\rho,z)$, where

$$b_\rho(\rho,z) = (\phi_0/2\pi\Lambda\rho)[\,(z/|z|)\, e^{-|z|/\lambda_{ab}} - (z/r)\, e^{-r/\lambda_{ab}}], \qquad (6)$$

$$b_z(\rho,z) = (\phi_0/2\pi\Lambda r)\, e^{-r/\lambda_{ab}}. \qquad (7)$$

The magnetic flux $\Phi_z(\rho,z) = 2\pi\rho a_\phi(\rho,z)$ up through a layer at height $z$ within a circle of radius $\rho$ is

$$\Phi_z(\rho,z) = (s/2\lambda_{ab})\phi_0\, (e^{-|z|/\lambda_{ab}} - e^{-r/\lambda_{ab}}), \qquad (8)$$

such that the total flux up through a layer at height $z$ is $\Phi_z(\infty,z) = (s/2\lambda_{ab})\phi_0 \exp(-|z|/\lambda_{ab})$. In particular, the flux up through the central layer, $\Phi_z(\infty,0) = (s/2\lambda_{ab})\phi_0$, is much less than $\phi_0$ (recall that $s \ll \lambda_{ab}$). It is important to note that fluxoid quantization (not flux quantization) holds in superconductors, and in this case the main contribution to the fluxoid is the line integral of the current density around the vortex. Since $j$ here varies as $1/\rho$ at large distances, it makes a constant contribution, nearly equal to $\phi_0$.

Forces between pancake vortices in the same and different layers easily can be calculated from the above equations using the Lorentz force expression, $F_\rho = K_\phi\phi_0/c$. Given a pancake vortex at the origin ($\rho = 0$, $z = 0$), the force on a second pancake vortex in the same layer at a distance $\rho$ is repulsive, with value $F_\rho \approx \phi_0^2/4\pi^2\Lambda\rho$. The corresponding interaction energy, $U(\rho) \approx -(\phi_0^2/4\pi^2\Lambda)\ln\rho$, is logarithmic to all distances. By contrast, the force [see Eq. (5)]

$$F_\rho(\rho,z) = -(\phi_0{}^2\lambda_{ab}/4\pi^2\Lambda^2\rho)(e^{-|z|/\lambda_{ab}} - e^{-r/\lambda_{ab}}), \tag{9}$$

on a second pancake vortex in a <u>different</u> layer at $(\rho,z)$ [$r = (\rho^2 + z^2)^{1/2}$] is attractive, with a magnitude that is much smaller (by at least a factor of $s/2\lambda_{ab}$) than that between vortices in the same layer. Since pancake vortices cannot leave the layers they belong to, the interaction force, which is directed along the layer, tends to make the pancake vortices coaxial.

The above results tell us that if one 2D pancake vortex is placed in each layer, the zero-temperature configuration (neglecting zero-point fluctuations) is a straight stack aligned parallel to the z axis. If such a stack is aligned along the z axis, the magnetic field at any arbitrary point $(\rho,z)$ can be calculated by superposition of the fields given in Eqs. (6) and (7). The summation can be converted to an integral, since the characteristic length scale for spatial variation of the fields ($\lambda_{ab}$) is much smaller than the separation (s) of the pancakes, and the resulting integral yields

$$\mathbf{b} = \hat{z}(\phi_0/2\pi\lambda_{ab})^2 K_0(\rho/\lambda_{ab}), \tag{10}$$

the well-known London-model result [Fetter and Hohenberg(1969)] for a straight vortex in a superconductor with penetration depth $\lambda_{ab}$. (Here $K_0$ is the modified Bessel function). The simple form of the interaction forces between pancake vortices makes possible a number of calculations involving thermal fluctuations and the energies of pancakes disturbed from equilibrium [Clem (1991)].

From Eq. (10) it can be shown that the zero-temperature equilibrium lattice of a pancake-vortex array corresponding to flux density B perpendicular to the layers is a triangular lattice of straight pancake stacks, perpendicular to the layers, with nearest-neighbor distance $a = (2\phi_0/B\sqrt{3})^{1/2}$ [Fetter and Hohenberg(1969)]. In the limit of zero interlayer Josephson coupling, this arrangement of pancakes, with the stacks perpendicular to the layers, is still the minimum-energy configuration when the flux density B has not only a component $B_\perp$ perpendicular to the layers but also a component $B_{||}$ parallel to the layers. The perpendicular component gives rise to a triangular lattice with nearest-neighbor spacing $a = (2\phi_0/B_\perp\sqrt{3})^{1/2}$, while the parallel component, which is unscreened in the absence of Josephson coupling, penetrates uniformly between the layers [Kes et al. (1990)].

## 5. Josephson-Coupled Pancake Vortices

The previous section treated the case of vortex structure in a layered superconductor in the limit of infinite anisotropy, for which the Josephson coupling between layers is zero and the pancake vortices couple only magnetically. When the Josephson coupling is nonzero, however, Josephson strings, whose axes thread through the Josephson junctions between superconducting layers, stretch from the center of a 2D pancake vortex in one superconducting layer to the center of the 2D pancake vortex in the next layer [Bulaevskii (1973), Clem and Coffey (1990), Feinberg and Villard (1990a and 1990b), Feigel'man et al. (1990), Clem et al. (1991), Glazman and Koshelev (1991), Kapitulnik (1992)]. The energy associated with such Josephson vortices gives rise to additional attractive forces between the pancake vortices.

A fully developed Josephson core (a long Josephson string) forms parallel to the layers only if the distance $\Delta x$ between the 2D pancake vortices it connects exceeds $\lambda_J = \gamma s$, where $\gamma = \lambda_c/\lambda_{ab} = (m_c/m_{ab})^{1/2}$. Its energy (aside from logarithmic corrections) is approximately [Clem (1993)]

$$E_{long}(\Delta x) \approx (\phi_0/4\pi)^2 \Delta x/\lambda_{ab}\lambda_c, \quad \Delta x > \lambda_J. \tag{11}$$

When $\Delta x < \lambda_J$, we have only a short Josephson string, whose energy can be estimated by calculating the integral of the Josephson coupling energy per unit area $(\hbar J_0/2e)(1 - \cos \Delta\gamma)$, where $\Delta\gamma$ is the gauge-invariant phase difference across the interlayer Josephson junction. Along the line connecting the two pancake vortices, we have $\Delta\gamma = \pi$, and around the circle of radius $\Delta x/2$ centered on the midpoint of this line, we have $\Delta\gamma = \pm\pi/2$. Multiplying the maximum Josephson coupling energy per unit area $\hbar J_0/2e = \phi_0^2/8\pi^3 s\lambda_c^2$ by the circle's area $\pi(\Delta x/2)^2$ gives the estimated cost in Josephson coupling energy from the area immediately between the two pancake vortices. There is, however, a roughly equal contribution from regions distant from the pancakes. Adding the two contributions gives (aside from logarithmic corrections) the energy [Clem (1993)]

$$E_{short}(\Delta x) \approx (\phi_0/4\pi)^2 \Delta x^2/s\lambda_c^2, \quad \Delta x < \lambda_J, \tag{12}$$

for the energy of a short Josephson string of length $\Delta x$. Note that the approximate expressions for the long and short Josephson string ($E_{long}$ and $E_{short}$) become equal when $\Delta x = \lambda_J = \gamma s$. Note also that the attractive Josephson-string interaction between pancake vortices on adjacent layers becomes very small when the Josephson coupling strength $J_0$ between superconducting layers becomes weak and $\lambda_c$ becomes large.

## 6. Flux Pinning

The above ideas help to provide an understanding of some of the numerous experiments that have been done on the high-temperature superconductors. Many experiments have been done to examine the dependence of the transport properties, such as the effective resistivity $\rho$ and the critical-current density $J_c$, on the angle $\theta$ of the vortices relative to the c axis. At the most elementary level, one expects scaling of $\rho$ and $J_c$ as a function of the magnitude B and angle $(\theta,\phi)$ of the flux density **B** to be similar to that of the thermodynamic properties [Hao et al. (1991b), Hao and Clem (1992), Blatter et al. (1992), Hao and Clem (1993), Blatter et al. (1993)], and such scaling has been found experimentally [Raffy et al. (1991), Schmitt et al. (1991)]. In other words, whereas $J_c$ is a function of the reduced field $B/B_{c2}$ in isotropic superconductors [Campbell and Evetts (1972)], one expects $J_c$ to be a function of the reduced field $B/B_{c2}(\theta,\phi)$ in anisotropic superconductors, where (in Gaussian units) $B_{c2}(\theta,\phi) = H_{c2}(\theta,\phi)$ is given by Eq. (4). For uniaxial superconductors, in which $m_a = m_b = m_{ab}$, Eq. (4) simplifies to

$$B_{c2}(\theta,\phi) = [(B_{c2ab})^{-2} \sin^2\theta + (B_{c2c})^{-2} \cos^2\theta]^{-1/2}, \tag{13}$$

where $B_{c2ab}$ and $B_{c2c}$ are the upper critical fields parallel to the ab plane (parallel to the layers) and the c axis (perpendicular to the layers), respectively. For highly anisotropic superconductors, for which $m_c \gg m_{ab}$ and $(B_{c2c}/B_{c2ab})^2 = m_{ab}/m_c \ll 1$, the reduced field thus becomes $B_\perp/B_{c2c}$, where $B_\perp = B \cos \theta$, to good approximation except for angles $\theta$ very close to $\pi/2$ (fields very nearly parallel to the layers). Thus, even in the context of the anisotropic Ginzburg-Landau theory one expects $J_c$ to be a function of the reduced field $B_\perp/B_{c2c}$, where $B_\perp = B \cos \theta$ is the component of the flux density perpendicular to the layers.

In the case of extremely strong anisotropy, a description of the vortices in terms of two-dimensional pancake vortices with negligible interlayer coupling should be valid. As discussed above, the flux density component $B_\parallel$ then easily penetrates between the layers. For transport of currents parallel to the layers, however, the key factor is the pinning of 2D pancake vortices, whose density is simply $n = B_\perp/\phi_0$. The critical value of $B_\perp$ that quenches superconductivity in the layers is $B_{c2c}$, and thus one again expects $J_c$ to depend only on the reduced field $B_\perp/B_{c2c}$, where $B_\perp = B \cos \theta$. There have been numerous experiments showing this kind of scaling in the critical-current anisotropy [see, for example, Schmitt et al. (1991)].

For currents flowing parallel to the layers, the highest critical-current density and the strongest possible pinning are expected to be those arising from vortex-core pinning. The most effective pinning center for a single 2D pancake vortex evidently would be a void or nonsuperconducting region of radius somewhat larger than the vortex-core radius ($\approx \xi_{ab}$). Since superconductivity is strongly suppressed in such a region, a pancake vortex centered on the pinning center can save both condensation energy and kinetic energy. The energy saved per pancake vortex can be estimated to be approximately

$$U_0 = (\phi_0/4\pi)^2 s/\lambda_{ab}^2. \tag{14}$$

One may think of the vortex as residing in a pinning potential $U(x)$, whose value is $U(0) = -U_0$ when $x = 0$, and which approaches zero for $x \gg \xi_{ab}$. The maximum of the restoring force, $F(x) = -dU/dx$, occurs when $x \approx \xi_{ab}$, and its magnitude is approximately $F_p \approx U_0/\xi_{ab} = (\phi_0/4\pi)^2 s/\xi_{ab}\lambda_{ab}^2$. The corresponding depinning current density $J_c^{ab}(0,0)$ (at $B = 0$ and $T = 0$) for currents in the ab direction is given by $F_p = J_c^{ab} s\phi_0/c$, or

$$J_c^{ab}(0,0) \approx c\phi_0/16\pi^2\xi_{ab}\lambda_{ab}^2 = cH_c/2^{1/2}4\pi\lambda_{ab}, \tag{15}$$

which approaches the theoretical maximum supercurrent density that can be carried by the superconductor, the depairing current density $J_{cd}$. Equation (15) thus suggests that the zero-temperature critical-current density could, in principle, achieve very high values in materials for which the pinning is optimized. For example, using the estimates $B_c \approx 1$ T and $\lambda_{ab} \approx 1,400$ Å for YBCO, Eq. (15) yields the value $J_c^{ab}(0,0) \approx 4 \times 10^8$ A/cm$^2$ for well-separated, independently pinned vortices. As B and the vortex density increase, the repulsive interactions among vortices make a triangular vortex array energetically most favorable. A random array of pinning centers is not consistent with such a triangular lattice, and the competition between the pinning forces and the repulsive intervortex forces generally leads to a reduction of $J_c^{ab}(B,0)$ as B increases. If the pinning centers could be arranged on a triangular lattice, however, there are a number of flux densities $B_m$ at which the triangular vortex lattice would have lattice points coincident with those of the pinning-center lattice and the repulsive interactions between vortices would exactly cancel. At such flux densities, one would expect matching peaks in $J_c^{ab}(B,0)$ versus B, at which $J_c^{ab}(B_m,0)$ could achieve values approaching $J_c^{ab}(0,0)$.

For currents flowing perpendicular to the layers, the supercurrent density can never exceed the Josephson critical current $J_0$, which is related to $\lambda_c$ via [Clem (1989)] $J_0 = c\phi_0/8\pi^2 s\lambda_c^2$. For this current direction, any voltage along the c direction must arise from phase slips corresponding to the motion of Josephson strings parallel to the layers. Thus the critical-current density $J_c^c$ for currents in the c direction is that corresponding to the depinning of the Josephson vortices. The current-density scale for such depinning is set by the Josephson current density $J_0$. Taking $J_c^c(0,0) = J_0$, we obtain from this and Eq. (15) the result that $J_c^c(0,0)/J_c^{ab}(0,0) = (2/\gamma^2)(\xi_{ab}/s)$, where $\gamma = \lambda_c/\lambda_{ab}$. Since $\xi_{ab}(T=0) < s$ for all the known high-temperature superconductors, we expect $J_c^c(0,0)/J_c^{ab}(0,0) << 1$ for optimal pinning in the most anisotropic superconductors for which $\gamma >> 1$.

At nonzero temperatures, experiments on the high-temperature superconductors, even those in thin films, generally have yielded critical-current values much less than those estimated above. One reason for this is that the energies of superconductivity, such as the condensation energy $H_c^2/8\pi$, are decreasing functions of the temperature T, vanishing linearly [as $(T_c - T)$] at the critical temperature $T_c$; equivalently, the characteristic lengths of superconductivity (the penetration depths and coherence lengths) are increasing functions of T, diverging as $(T_c - T)^{-1/2}$ at $T_c$. The depth of a deep pinning potential well for a pancake vortex, $U_0 = (\phi_0/4\pi)^2 s/\lambda_{ab}^2$, thus decreases with temperature and vanishes as $(T_c - T)$ near $T_c$. Pinning forces also must vanish at $T_c$ for similar reasons. For example, generalization of Eq. (15) yields an estimate of $J_c(0,T)$ that vanishes as $(T_c - T)^{3/2}$ as $T \rightarrow T_c$. A more important effect in the high-temperature superconductors, however, is that thermal energies ($\sim k_B T$) can approach the depths $U_0$ of the pinning potential wells, often giving rise to a considerable amount of flux creep when $U_0/k_B T < 20$. The single-pancake-vortex pinning energy $U_0 = (\phi_0/4\pi)^2 s/\lambda_{ab}^2$ becomes, for $s \approx 12$ Å and $\lambda_{ab} \approx$ 1,400 Å (values appropriate to YBCO), $U_0(0) \approx 1.7 \times 10^{-13}$ erg $\approx 100$ meV, or $U_0(0)/k_B$ $\approx 1,200$ K. In YBCO at 77 K, the combination of the reduction of $U_0$ via the temperature dependence of $\lambda_{ab}$ and the flux-creep effect leads to an appreciable suppression of the effective critical-current density estimates such as that given in Eq. (15). The suppression of the critical-current density at 77 K is even greater in more anisotropic superconductors, such as $Bi_2Sr_2CaCu_2O_8$.

As suggested above, the highest values of $J_c$ are expected if the pinning centers are nonsuperconducting regions comparable in size with that of the vortex core. L. Civale et al. (1991) have shown that flux pinning in single crystals of $YBa_2Cu_3O_{7-\delta}$ can be significantly enhanced by columnar defects produced by ion irradiation, in this case by 580 MeV $^{116}Sn^{30+}$ ions. In these experiments, the ion tracks produced nearly cylindrical amorphous regions of radius somewhat larger than the vortex-core radius. The single-pancake pinning energy $U_0$ thus can be estimated from Eq. (14), and the critical-current density $J_c(0,0)$ from Eq. (15).

As the vortex density increases, the repulsive interaction between pancake vortices makes it energetically costly for all vortices (which prefer a triangular lattice) to take advantage of the pinning centers, which are randomly distributed in space. From the second Ginzburg-Landau equation or the London equation it can be shown that the energy required to move a pancake vortex a distance r away from an equilibrium position in a lattice of magnetic flux density B is $U_B(r) = \phi_0 B s r^2 / 16\pi \lambda_{ab}^2$. When the intervortex spacing $a_0$ is large by comparison with the average spacing $L_p$ between pinning centers, the energy cost to move a pancake from an equilibrium position onto a pinning site is therefore approximately $U_B(L_p) = \phi_0 B s L_p^2 / 16\pi \lambda_{ab}^2$. Since this energy cost exceeds $U_0$ at flux densities $B > B_p = \phi_0 / \pi L_p^2$, this suggests that at low temperatures $J_c(B)$ should begin to drop off significantly from $J_c(0)$ for values of $B > B_p = \phi_0 / \pi L_p^2$, an effect which has been seen experimentally [Civale et al. (1991)].

At higher temperatures it is found in both $YBa_2Cu_3O_{7-\delta}$ [Civale et al. (1991)] and $Bi_2Sr_2CaCu_2O_8$ [Thompson et al. (1992)] that the effective critical-current density drops rapidly with temperature from the high values of $J_c$ observed at low temperature. The decrease with temperature is much more pronounced in $Bi_2Sr_2CaCu_2O_8$ [Thompson et al. (1992)] than in $YBa_2Cu_3O_{7-\delta}$ [Civale et al. (1991)]. This evidently is an effect of thermal activation: There is some characteristic activation energy U required to push a pancake vortex a certain distance a out of its pinning site. As the temperature T increases, the rate of its escape, which is proportional to $\exp(-U/k_B T)$, increases to the point that a typical vortex is no longer pinned on the time scale of the experiment.

The activation energy U is not in general equal to the single-pancake vortex energy $U_0$, but instead it depends strongly on the strength of the interlayer coupling. We can estimate U using the following procedure [Clem (1993)]. To be specific, let us think of a as being some fraction of the vortex lattice parameter $a_0$, so that $a \propto (\phi_0/B)^{1/2}$. Let us assume that $a < \lambda_J$, so that we can use the short-Josephson-string expression [Eq. (12)] to estimate the interaction of pancake vortices in adjacent layers. The energy expression is analogous to the energy $(1/2)k\Delta x^2$ stored in a spring of spring constant k. Here, however, the effective Josephson spring constant is approximately $k_J = \phi_0^2 / 8\pi^2 s \lambda_c^2$. For weak interlayer coupling [i.e., when $(1/2)k_J a^2 \ll U_0$], which occurs in strongly anisotropic high-temperature superconductors such as $Bi_2Sr_2CaCu_2O_8$ and $Bi_2Sr_2Ca_2Cu_3O_{10}$, the total energy cost to move a pancake vortex out of the well to a distance a is

$$U_{weak}(a) = U_0 + k_J a^2 \approx U_0, \quad (1/2)k_J a^2 \ll U_0. \qquad (16)$$

For strong interlayer coupling, however, [i.e., when $(1/2)k_J a^2 \gg U_0$], which occurs in the least anisotropic high-temperature superconductors such as $YBa_2Cu_3O_{7-\delta}$, the Josephson-string-energy cost of moving just one pancake out of the well is approximately $k_J a^2$, which is too expensive. On the other hand, it also is far too expensive to reduce the Josephson-string-energy cost to zero by moving an infinite number of the pancakes out of the well. The energy cost can be minimized, however, by moving approximately $(2k_J a^2/U_0)^{1/2} \gg 1$ pancakes out of their wells. This equalizes the total well-excitation-energy $(U_0)$ and Josephson-string-energy $[(1/2)k_J \Delta x^2]$ costs and yields a minimum excitation energy of

$$U_{strong}(a) = 2(2k_Ja^2/U_0)^{1/2}U_0 \gg U_0, \ (1/2)k_Ja^2 \gg U_0. \tag{17}$$

Note that $U_0 \ll U_{strong}(a) \ll (1/2)k_Ja^2$.

The above considerations and the results of Eqs. (16) and (17) provide useful insight into why the activation energy U in strongly coupled high-temperature superconductors, such as $YBa_2Cu_3O_{7-\delta}$, is much larger than the single-pancake pinning energy $U_0$, while U $\approx U_0$ in weakly coupled high-temperature superconductors, such as $Bi_2Sr_2CaCu_2O_8$ and $Bi_2Sr_2Ca_2Cu_3O_{10}$. In turn, because of the importance of thermal activation and the appearance of U in the Boltzmann factor $exp(-U/k_BT)$, this explains why the effective critical-current density $J_c^{ab}$ at 77 K is generally much higher in $YBa_2Cu_3O_{7-\delta}$ than in $Bi_2Sr_2CaCu_2O_8$ and $Bi_2Sr_2Ca_2Cu_3O_{10}$.

## 6. Summary

In this paper I have presented a brief overview of what I feel are the key ideas needed for an understanding of the anisotropic electrodynamic properties of the high-temperature superconductors. A fundamental starting point is the Ginzburg-Landau theory, modified to take into account anisotropy with the help of an anisotropic effective mass tensor. For describing the vortex core, however, this theory fails when $\xi_c$, the coherence length perpendicular to the layers, becomes smaller than the periodicity length s. The Lawrence-Doniach theory, which incorporates atomic-level discreteness, then becomes a useful model. In this theory, a vortex parallel to the layers is found to have a Josephson core when $\xi_c \ll s$ and an Abrikosov core when $\xi_c \gg s$. Another useful picture, especially useful for the most anisotropic superconductors, is a description of a vortex line as a succession of two-dimensional pancake vortices, which reside in the superconducting layers, and Josephson strings, which thread through the insulating layers and link the cores of the pancake vortices. This model enables estimates of pinning forces, critical currents, and activation energies to be made, and sheds light on the differences between strongly coupled high-temperature superconductors, such as $YBa_2Cu_3O_{7-\delta}$, and those with weak interlayer coupling, such as $Bi_2Sr_2CaCu_2O_8$ and $Bi_2Sr_2Ca_2Cu_3O_{10}$.

### Acknowledgments

I thank M. Benkraouda, E. H. Brandt, L. N. Bulaevskii, D. K. Christen, M. W. Coffey, Z. Hao, A. Kapitulnik, V. G. Kogan, D. R. Nelson, S. Ryu, and J. R. Thompson for stimulating discussions. Ames Laboratory is operated for the U.S. Department of Energy by Iowa State University under Contract No. W-7405-Eng-82. This research was supported in part by the Director for Energy Research, Office of Basic Energy Sciences, and in part by the Midwest Superconductivity Consortium through D.O.E. grant #DE-FG02-90ER45427.

38

References

Abrikosov, A. A. (1957) Zh. Eksp. Teor. Fiz. 32, 1442 [Sov. Phys. JETP 5, 1174 (1957)].

Artemenko, S. N., and Kruglov, A. N. (1990) Phys. Lett. A 143, 485.

Balatskii, A. V., Burlachkov, L. I., and Gor'kov, L. P. (1986) Sov. Phys. JETP 63, 866.

Bardeen, J., Cooper, L. N., and Schrieffer, J. R. (1957) Phys. Rev. 108, 1175.

Blatter, G., Geshkenbein, V. B., and Larkin, A. I. (1992) Phys. Rev. Lett. 68, 3875.

Blatter, G., Geshkenbein, V. B., and Larkin, A. I. (1993) Phys. Rev. Lett. 71, 302.

Bulaevskii, L. N., Ginzburg, V. L., and Sobyanin, A. A. (1988) Sov. Phys. JETP 68, 1499.

Bulaevskii, L. N. (1973) Zh. Eksp. Teor. Fiz. 64, 2241 [Sov. Phys. JETP 37, 1133 (1973)].

Campbell, A. M., and Evetts, J. E. (1972) Adv. Phys. 21, 199.

Caroli, C., de Gennes, P. G., and Matricon, J. (1963) Phys. Kondens. Mater. 1, 176.

Civale, L., Marwick, A. D., Worthington, T. K., Kirk, M. A., Thompson, J. R., Krusin-Elbaum, L., Sun, Y., Clem, J. R., and Holtzberg, F. (1991) Phys. Rev. Lett. 67, 648.

Clem, J. R. (1989) Physica C 162-164, 1137.

Clem, J. R. (1991) Phys. Rev. B 43, 7837.

Clem, J. R. (1993) Physica A (to be published).

Clem, J. R., and Coffey, M. W. (1990) Phys. Rev. B 42, 6209.

Clem, J. R., Coffey, M. W., and Hao, Z. (1991), Phys. Rev B 44, 2732.

De Gennes, P. G. (1966) Superconductivity of Metals and Alloys, Benjamin, New York, p. 60.

Feinberg, D., and Villard, C. (1990a) Mod. Phys. Lett. B4, 9.

Feinberg, D., and Villard, C. (1990b) Phys. Rev. Lett. 65, 919.

Feigel'man, V, Geshkenbein, V. B., and Larkin, A. (1990) Physica C 167, 177.

Fetter, A. L., and Hohenberg, P. C. (1969) in R. D. Parks (ed.), Superconductivity, Vol. 2, Dekker, New York, p. 817.

Fischer, K. H. (1991) Physica C 178, 161.

Ginzburg, V. L., and L. D. Landau, L. D. (1950) Zh. Eksp. Teor. Fiz. 20, 1064 [English translation in Men of Physics: L. D. Landau, edited by D. ter Haar (Pergamon, New York, 1965), Vol. 1, pp. 138-167].

Glazman, L. I., and Koshelev, A. E. (1991) Physica C 173, 180.

Gor'kov, L. P. (1959) Sov. Phys. JETP 9, 1364.

Gor'kov, L. P., and Melik-Barkhudarov, T. K. (1964) Sov. Phys. JETP 18, 1031.

Hao, Z., Clem, J. R., McElfresh, M. W., Civale, L., Malozemoff, A. P., and Holtzberg, F. (1991a) Phys. Rev. B 43, 2844.

Hao, Z., Clem, J. R., Cho, J. H., and D. C. Johnston, D. C. (1991b) Physica C 185-189, 1871.

Hao, Z., and Clem, J. R. (1991) Phys. Rev. B 43, 7266.

Hao, Z., and Clem, J. R. (1992) Phys. Rev. B 46, 5853.

Hao, Z., and Clem, J. R. (1993) Phys. Rev. Lett. 71, 301.

Hu, C.-R. (1972) Phys. Rev. B6, 1756.

Kapitulnik, A. (1992) in Phenomenology and Applications of High-Temperature Superconductors, edited by K. S. Bedell, M. Inui, D Meltzer, J. R. Schrieffer, and S. Doniach (Addison-Wesley, Reading), p. 34.

Katz, E. I. (1969) Sov. Phys. JETP 29, 897.

Katz, E. I. (1970) Sov. Phys. JETP 31, 787.

Kes, P. H., Aarts, J., Vinokur, V. M., and van der Beek, C. J. (1990) Phys. Rev. Lett. 64, 1063.

Klemm, R. A., and Clem, J. R. (1980) Phys. Rev. B 21, 1868.

Kogan, V. G. (1981) Phys. Rev. B 24, 1572.

Kogan, V. G., and Clem, J. R. (1981) Phys. Rev. B 24, 2497.

Kogan, V. G., and Clem, J. R. (1987) Jpn. J. Appl. Phys. 26 (Suppl. 26-3), 1159.

Koshelev, A. E. (1993) Phys. Rev. B 48, 1180.

Lawrence, W. E., and Doniach, S. (1971) in Proc. of the Twelfth International Conference on Low Temperature Physics, edited by E. Kanda (Academic Press of Japan, Kyoto), p. 361.

Pearl, J. (1964) App. Phys. Lett. 5, 65.

Pearl, J. (1965) in Proceedings of the Ninth International Conference on Low Temperature Physics, edited by J. G. Daunt, D. V. Edwards, F. J. Milford, and M. Yaqub, (Plenum Press, New York), Part A, p. 566.

Raffy, H., Labdi, S., Laborde, O., and Monceau, P. (1991) Phys. Rev. Lett. 66, 2515.

Schmitt, P., Kummeth, P., Schultz, L., and Saemann-Ischenko, G. (1991) Phys. Rev. Lett. 67, 267.

Takanaka, K. (1975) Phys. Status Solidi B 68, 623.

Thompson, J. R., Sun, Y. R., Kerchner, H. R., Christen, D. K., Sales, B. C., Chakoumakos, B. C., Marwick, A. D., Civale, L. and Thomson, J. O. (1992) Appl. Phys. Lett. 60, 2306.

Tilley, D. R. (1965) Proc. Phys. Soc. London 85, 1977; 86, 289.

Tilley, D. R., Van Gurp, G. J., and Berghout, C. W. (1964) Phys. Lett. 12, 305.

von Laue, M. (1952) Theory of Superconductivity, Academic Press, New York.

Werthamer, N. R. (1969), in R. D. Parks (ed.), Superconductivity, Vol. 1, Dekker, New York, p. 321.

# Pinning, Fluctuations and Melting of Superconducting Vortex Arrays*

David R. Nelson
*Lyman Laboratory of Physics*
*Harvard University*
*Cambridge, Massachusetts  02138*

ABSTRACT. We give a brief introduction to the analysis of vortex line fluctuations in high-temperature superconductors. We treat flux lines as classical objects, described by a string tension, their mutual repulsion, and interactions with pinning centers. The classical partition function, however, is isomorphic to the imaginary time path integral description of quantum mechanics. This observation is used to determine the thermal renormalization of critical currents in the presence of columnar pins and the flux lattice melting temperature at low and moderate field strengths. We review as well why flux lattice melting is expected to be a first order transition and argue that a continuous smectic-like freezing transition is likely for field orientations perpendicular to an array of columnar pins.

## 1. Introduction

The past five years have been a time of great ferment in the study of high-temperature superconductors. Although no general consensus about the microscopic mechanism for superconductivity in the cuprate materials has emerged, there is now considerable understanding of the remarkable behavior of these materials in a magnetic field. The theoretical analysis of vortex fluctuations in these extreme Type II materials requires only an underlying Ginzburg-Landau theory with a BCS-like order parameter. The basic conclusions are *independent* of the precise microscopic mechanism, and rely instead on the remarkably different Ginzburg-Landau parameters (coherence length, temperature range and anisotropy) which distinguish the cuprates from their low $T_c$ counterparts.

Figure 1 shows a schematic temperature-magnetic field phase diagram for cuprate superconductors subjected only to weak point disorder in the form of oxygen vacancies. Throughout this paper we assume for simplicity a field oriented along the $c$ axis, perpendicular to the copper-oxide planes. The magnetic field $B$ (proportional to the vortex density) is plotted because demagnetizing corrections in the usual slab-like experimental geometry insure that $B$ rather than the magnetic induction $H$ is held fixed in most experimental situations. The famous Abrikosov flux lattice, which exists for all temperatures below the upper critical field $B_{c2}(T)$ in mean field theory [1], appears here only below a much lower "melting line" $T_m(B)$. Above

* This article is similar to the one entitled "Vortex Line Fluctuations in Superconductors from Elementary Quantum Mechanics", appearing in the book: *Phase Transitions in Systems with Competing Energy Scales*, edited by T. Riste and D. Sherrington, Proceedings of a 1993 NATO spring school in Geilo, Norway, published by Kluwer Academic Publishers.

*N. Bontemps et al. (eds.), The Vortex State*, 41–61.
© 1994 *Kluwer Academic Publishers*.

this line, melted vortex arrays entangle in a novel "flux liquid." The line $B_{c2}^0(T)$ marks the onset of enhanced diamagnetism but is not expected to be a sharp phase boundary. The crossover fields $B_{c1}$ and $B_x$ are discussed in Sec. 3. There is now considerable evidence [2–5] that the Abrikosov lattice in clean (twin-free) single crystals of yttrium barium copper oxide (YBCO) melts at $T_m(B)$ via a first-order phase transition [6] into a flux liquid in which the quantized vortex filaments presumably wander and entangle in a complicated fashion. The melting curve $T_m(B)$ in Figure 1 should actually broaden into a region of two-phase coexistence (very narrow at high fields) with vertical tie lines when $B$ rather than $H$ is used as the vertical axis. Weak point disorder significantly alters the properties of the flux liquid only below the dotted line [7]. Note that a sliver of flux liquid may exist above this line and below the melting curve down to quite low temperatures. It is still unclear whether point disorder alone is sufficient to produce a distinct thermodynamic "vortex glass" phase [8] below the dotted line or within the crystalline region.

Figure 2 shows another striking development—a remarkable shift in the "irreversibility line" of a highly anisotropic thallium-based compound engineered via the deliberate introduction of columnar pins [9]. Similar pinning centers have been injected by many groups via heavy ion irradiation [10–13] with sufficient energy to produce long tracks of damaged material. The "irreversibility line" $T_{ir}(B)$ is the boundary below which the dynamics of field cooled materials slows down drastically [14]. In samples *without* correlated disorder the melting curve and irreversibility line may in fact be almost identical.

Since $T_{ir}(B)$ coincides approximately with the temperature at which the resistivity becomes unmeasurably small, the large upward shift in Fig. 2 is of considerable technological as well as intellectual interest. Note for example that the effective critical temperature for superconductivity at $B = 2.7$ tesla shifts from 53 K to 87 K upon irradiation. The critical current at liquid nitrogen temperature (77 K) increases by three orders of magnitude [9]. It is believed that after irradiation the irreversibility line becomes a locus of thermodynamically sharp "bose glass" transitions, below which the flux lines are localized on the columnar defects [15]. Other forms of correlated disorder may be important even in unirradiated samples. Twin boundaries, for example, appear to be responsible for the apparently continuous transition observed in twinned YBCO samples with the field aligned parallel to the $c$ axis [3].

In these lectures, we highlight these developments by describing how flux lines interact with correlated pinning centers and with each other. To account for thermal fluctuations, one must average over vortex configurations. Because the configuration sums bear a strong resemblance to the imaginary time path integral formulation of quantum mechanics [16], wave functions and binding energies which appear in simple quantum problems (such as the square well and the harmonic oscillator) have important implications for flux lines. With this identification we can quickly compute the free energy and localization length of vortices in the bose glass phase, the corresponding critical currents, and determine as well the line of melting temperatures in clean systems [15]. We review as well old arguments predicting that the melting transition is first order [6], and discuss a situation in which continuous freezing transition to a smectic-like vortex density wave is possible.

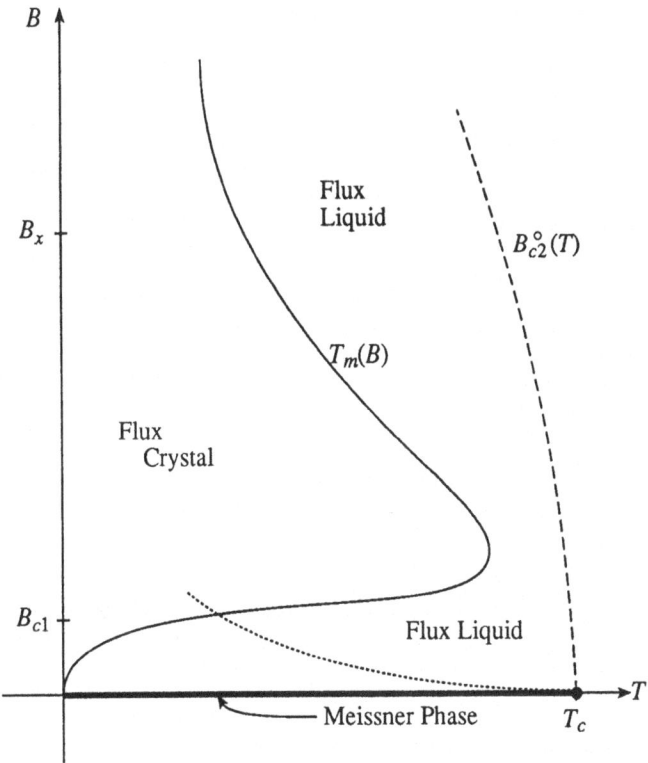

Figure 1. Schematic phase diagram of a high temperature superconductor in the clean limit. Although quantized vortex lines appear below $B_{c2}^0(T)$, the Abrikosov crystal phase appears only below $T_m(B)$. The Meissner phase collapses to the heavy line along the temperature axis in this representation. Point disorder only affects the flux liquid below the dotted line. Ref. [8] presents arguments that the maximum in the melting curve actually bends down closer to $T_c$ than indicated here.

The idea of using Schröedinger's equation to solve problems in classical statistical mechanics is not new. It was used by S.F. Edwards, P.G. de Gennes and others starting in the 1960s to treat the conformations of flexible polymer chains [18]. One can go further with this analogy for flux lines, however, because these are equivalent to a system of *directed* polymers, with a common average orientation. As a result, distant self-interactions along the filaments can usually be neglected, in contrast to isotropic self-avoiding polymer solutions. The essential physics is captured if one introduces a line tension to control vortex wandering and allows interactions between *different* polymers as well as interactions with the relevant pinning centers. Although we focus here primarily on single line problems, many flux line statistical mechanics (with multiple columnar pins) is in fact equivalent to the many body quantum mechanics of bosons in two dimensions [15, 17]. This system differs from the otherwise closely related problem of helium films on disordered substrates [19]

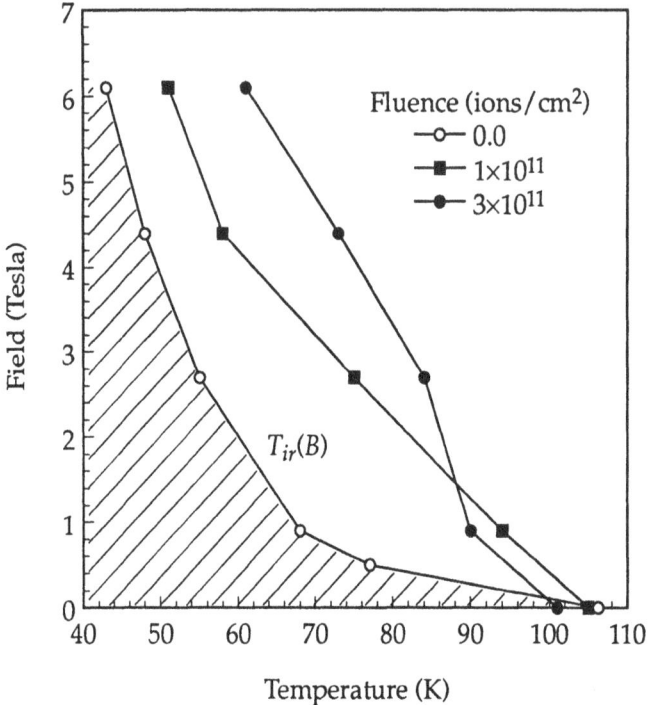

Figure 2. Effect of columnar pins on the irreversibility line of $Tl_2Ba_2Ca_2Cu_3O_{10}$ [9]. In the absence of irradiation, the resistivity only becomes unmeasurably small in the shaded region below $T_{ir}(B)$. The reentrant low field vortex liquid regime of Fig. 1 appears for $B \lesssim 10^{-2}$ T, and is hence not visible on this scale. Figure courtesy of R. Budhani, Brookhaven National Laboratory.

because vortices behave dynamically like *charged* bosons, and are easily manipulated in experiments by the injection of supercurrents. The richness and complexity of the *many* flux line problem rivals the physics of correlated electrons in semiconductors and metals. For details, readers are referred to Refs. 15 and 17, and a recent review [20].

The first three sections of this short introduction were adapted from lectures first presented at the 1993 NATO Advanced Study Institute on "Phase Transitions and Relaxation in Systems with Competing Energy Scales" in Geilo Norway [21]. Ref. 21 contains discussions of quantum double wells and Hubbard models for vortex lines, as well as discussions of the consequences of vortex entanglement which are not included here.

## 2. Correlated Pinning and Quantum Bound States

### 2.1. MODEL-FREE ENERGY

We start with a model-free energy $F_N$ for $N$ flux lines in a sample of thickness $L$, defined by their trajectories $\{\vec{r}_j(z)\}$ as they traverse a sample with both columnar pins and external magnetic field aligned with the $z$-axis, i.e., in the direction perpendicular to the $CuO_2$ planes,

$$F_N = \frac{1}{2}\tilde{\epsilon}_1 \sum_{j=1}^{N} \int_0^L \left|\frac{d\vec{r}_j(z)}{dz}\right|^2 dz + \frac{1}{2}\sum_{i \neq j} \int_0^L V(|\vec{r}_i(z) - \vec{r}_j(z)|)\, dz$$

$$+ \sum_{j=1}^{N} \int_0^L V_D[\vec{r}_j(z)]\, dz \tag{2.1a}$$

with

$$V_D(\vec{r}) = \sum_{k=1}^{M} V_1(\vec{r} - \vec{R}_k)\,. \tag{2.1b}$$

Here $V(|\vec{r}_i - \vec{r}_j|) = 2\epsilon_0 K_0(r/\lambda_{ab})$, is the interaction potential between lines with in-plane London penetration depth $\lambda_{ab}$ and the random potential $V_D(\vec{r})$ arises from a $z$-independent set of $M$ disorder-induced columnar pinning potentials $V_1(\vec{r})$ centered on sites $\{\vec{R}_k\}$. The tilt modulus $\tilde{\epsilon}_1 \approx (M_\perp/M_z)\epsilon_0 \ln(\lambda_{ab}/\xi_{ab})$, where the material anisotropy is embodied in the effective mass ratio $M_\perp/M_z \ll 1$, and $\epsilon_0 \approx (\phi_0/4\pi\lambda_{ab})^2$ is the energy scale for the interactions. The potential $V_D(\vec{r})$ arises from identical cylindrical traps with average spacing $d$ assumed for simplicity to pass completely through the sample with well depth per unit length $U_0$ and effective radius $b_0$. The parameter $b_0 \approx \max\{c_0, \xi_{ab}\}$ where $c_0 \approx 25 - 40$ Å is the radius of the columnar pins and $\xi_{ab}$ is the superconducting coherence length in the $ab$-plane [15].

A complete analysis of the many-line statistical mechanics associated with Eqs. (2.1) requires multiple path integrals over vortex trajectories weighted by $e^{-F_N/T}$ and subject to a complicated random pinning potential. See Fig. 3. Our goal here is to illuminate the essential physics by studying a few simple problems involving one flux line trapped by various configurations of columnar pins or interactions with its near neighbors.

### 2.2. ONE FLUX LINE AND ONE COLUMNAR PIN

Consider a vortex line confined by a single columnar pin with well depth $U_0$ and radius $b_0$ parallel to $z$ in an otherwise defect-free sample of thickness $L$, as shown in Fig. 4. A quantity of considerable physical interest is the binding *free* energy per unit length

$$U(T) = U_0 - TS\,, \tag{2.2}$$

46

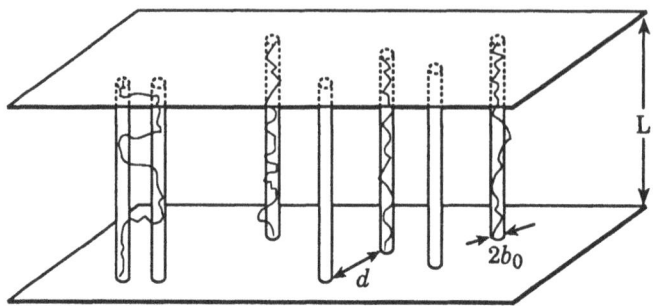

Figure 3. Schematic of columnar pins and pinned vortex lines.

where $S$ is the entropy reduction due to confinement. This free energy is given by a path integral,

$$e^{U(T)L/T} = \frac{\int \mathcal{D}\vec{r}(z)\exp\left[-\frac{\tilde{\epsilon}_1}{2T}\int_0^L \left(\frac{d\vec{r}}{dz}\right)^2 dz - \frac{1}{T}\int_0^L V_1[\vec{r}(z)]\, dz\right]}{\int \mathcal{D}r(z)\exp\left[-\frac{\tilde{\epsilon}_1}{2T}\int_0^L \left(\frac{d\vec{r}}{dz}\right)^2 dz\right]}, \tag{2.3}$$

where the denominator is required to subtract off the entropy of an unconfined line far from the pin. The cylindrically symmetric confining potential $V_1(r)$, indicated in Fig. 4b, tends to zero as $|\vec{r}| \to \infty$. The path integrals in (2.3) follow from standard statistical methods which express them in terms of the eigenvalues of a transfer matrix. In the limit $L \to \infty$, the smallest eigenvalue dominates, and $U(T) = -E_0(T)$, where $E_0(T)$ is the ground state energy of a two-dimensional "Schröedinger equation" (see Appendix A),

$$\left[-\frac{T^2}{2\tilde{\epsilon}_1}\nabla_\perp^2 + V_1(\vec{r})\right]\psi_0(\vec{r}) = E_0\psi_0(\vec{r}). \tag{2.4}$$

Here and henceforth, all vectors $\vec{r}$ will refer to positions in the plane perpendicular to $\hat{z}$ Note that $T$ plays the role of the Planck parameter $\hbar$ and $\tilde{\epsilon}_1$ plays the role of mass $m$ in this quantum mechanical analogy.

The ground state wave function $\psi_0(\vec{r})$ determines the localization length $\ell_\perp(T)$ displayed in Fig. 4a. As shown in Appendix B, the probability $P(\vec{r})$ of finding a point on the vortex at transverse displacement $\vec{r}$ relative to the center of the pin is independent of $z$ and given by the square of $\psi_0(\vec{r})$, just as in elementary quantum mechanics,

$$\mathcal{P}(\vec{r}) = \psi_0^2(\vec{r}) \left/ \int d^2r\,\psi_0^2(\vec{r}) \right. . \tag{2.5}$$

Because (2.4) is unchanged under complex conjugation, $\psi_0(\vec{r})$ can always be chosen to be real. We then define the localization length as

$$\ell_\perp^2 = \int d^2r\; r^2\psi_0^2(r) \left/ \int d^2r\; \psi_0^2(r) \right. . \tag{2.6}$$

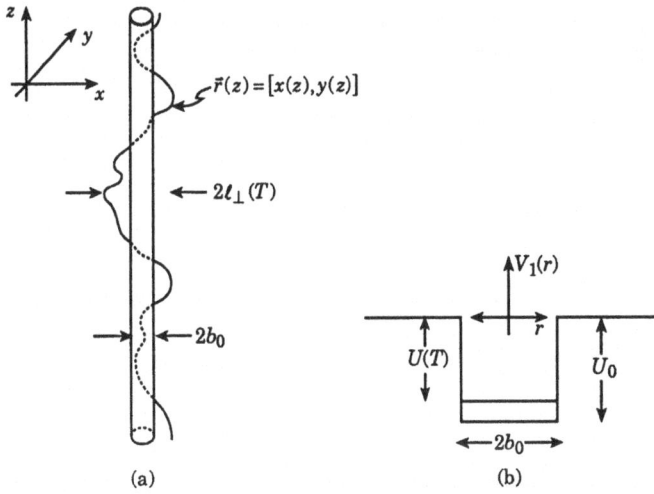

Figure 4. One columnar pin and one vortex line, indicating (a) the localization length $\ell_\perp(T)$ and (b) the thermally renormalized pinning potential $U(T)$.

The properties of a vortex near a columnar pin now follow from standard results for a quantum particle in a cylindrical potential [22]. We assume for simplicity a cylindrical square well, i.e., $V_1(r) \equiv -U_0$, $r < b_0$, and $V_1(r) = 0$, $r > b_0$. The binding-free energy $U(T) = -E_0$ then takes the form [20]

$$U(T) = U_0 f(T/T^*) \qquad (2.7)$$

where $T^*$ is an important characteristic temperature defined by

$$T^* = \sqrt{\tilde{\epsilon}_1 U_o}\, b_0 . \qquad (2.8)$$

When $T \ll T^*$, the well depth is effectively infinite, and we find the usual particle-in-a-box result,

$$U(T) \approx U_0 - c_1 \frac{T^2}{2\tilde{\epsilon}_1 b_0^2} \qquad (2.9)$$

where $c_1$ is a constant related to the first zero of the Bessel function $J_0(x)$ which solves Eq. (2.4) in this limit. The localization length is then

$$\ell_\perp(T) \approx b_0[1 + \mathcal{O}(1/\kappa b_o)] , \qquad (2.10)$$

where $\kappa^{-1} \approx T/\sqrt{2\tilde{\epsilon}_1 U(T)} \ll b_0$ is the distance the "particle" penetrates into the classically forbidden region. The low-temperature correction in Eq. (2.9) represents the entropy lost each time a wandering flux line is reflected off the confining walls of the binding potential [23]. At the crossover temperature $T^*$, this "zero point energy" of confinement becomes comparable to the well depth.

When $T \gg T^*$, the flux line is only weakly bound, although a strictly localized ground state *always* exists in this effectively two-dimensional problem [22]. In the limit, one finds [15,20]

$$U(T) \approx \frac{1}{2} U_0 \left( \frac{T}{T^*} \right)^2 e^{-2(T/T^*)^2} , \tag{2.11}$$

and a localization length $\ell_\perp(T) \approx \kappa^{-1} = T/\sqrt{2\tilde{\epsilon}_1 U(T)}$, so that

$$\ell_\perp(T) \approx b_0 e^{(T/T^*)^2} . \tag{2.12}$$

The flux line now "diffuses" within a confining tube with radius of order $\ell_\perp(T)$ as it crosses the sample. The length along $\hat{z}$ required to "diffuse" across this tube is $\ell_z \approx \ell_\perp^2 / (T/\tilde{\epsilon}_1)$, because vortex line fluctuations are equivalent to "diffusion" along the $z$-axis with diffusion constant $D = T/\tilde{\epsilon}_1$ [17], i.e.,

$$\ell_z = (b_0^2 \tilde{\epsilon}_1 / T) e^{2(T/T^*)^2} . \tag{2.13}$$

These results imply a strong thermal renormalization of the critical current $J_c(T)$ [15]. $J_c(T)$ is the current necessary to produce a Lorentz force $f_c = J_c \phi_0 / c$ so strong that thermal activation is unnecessary to tear a flux line away from its columnar pin. At low temperatures, one would expect (ignoring constants of order unity) that $f_c \approx U_0/b_0$, i.e.,

$$J_c \approx c U_0 / \phi_0 b_0 . \tag{2.14}$$

Here we assume that the confining potential $V_1(r)$ does not really jump abruptly at $r = b_0$, but instead rises smoothly to zero in a distance of order $b_0$. Results such as (2.11–2.13) are in any case independent of the precise form of the microscopic potential when $T \gg T^*$ [22]. To account for line wandering, we should replace $U_0$ by $U(T)$ and $b_0$ by $\ell_\perp(T)$ in Eq. (2.14), which leads to

$$J_c(T) \approx J_c(0) e^{-3(T/T^*)^2} . \tag{2.15}$$

## 2.3. ONE FLUX LINE AND MANY COLUMNAR PINS

The localization length $\ell_\perp(T)$ cannot continue to grow indefinitely as in Eq. (2.12) when additional unoccupied columnar pins are nearby. For $T > T_{dp}$, where $T_{dp}$ is defined by

$$\ell_\perp(T_{dp}) = d , \tag{2.16}$$

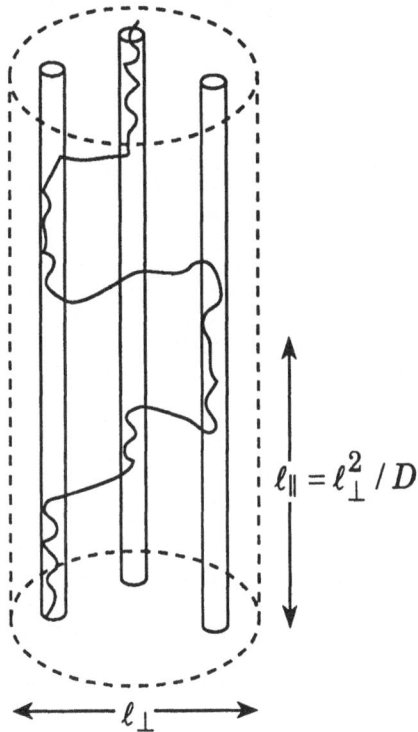

$$\ell_\parallel = \ell_\perp^2 / D$$

$$\ell_\perp$$

Figure 5. Flux trapped in a localization tube of radius $\ell_\perp(T)$ by a fluctuation in the density of columnar defects.

the localization length of one vortex line is determined by its interactions with *many* columnar pins. Thus, when $T > T_{dp}$, where

$$T_{dp} \approx T^* \ln^{1/2}(d/b_0) \,, \tag{2.17}$$

multiple rods in a transverse region of size $\ell_\perp(T) > d$ will now be incorporated into the ground state. See Figure 5. Energies should now be measured relative to $\bar{V}_D \approx -(b_0^2/d^2)U_0$, and we need to solve a Schröedinger equation like Eq. (2.10) with the random potential $\delta V_D(\vec{r}) \equiv V_D(\vec{r}) - \bar{V}_D$ (see Eq. (2.1b)) replacing $V_1(\vec{r})$. When the localization length encompases many columnar defects, we can model $\delta V_D(\vec{r})$ as a random variable satisfying [15]

$$\overline{\delta V_D(\vec{r})\delta V_D(\vec{r'})} = \Delta_1 \delta^{(2)}(\vec{r} - \vec{r'}) \,, \tag{2.18a}$$

with

$$\Delta_1 = (U_0^2 b_0^4/d^2)[1 + \mathcal{O}(b_0^2/d^2)] \,. \tag{2.18b}$$

We proceed variationally by minimizing

$$E_0(\psi) = \frac{T^2}{2\tilde{\epsilon}_1} \int d^2r (\vec{\nabla}\psi)^2 + \int d^2r \psi^2 \delta V_D(\vec{r}) , \qquad (2.19)$$

with a wave function $\psi_0(r)$ of spatial extent $\ell_\perp$ where $\psi_0(r) \sim 1/\ell_\perp$ for $r < \ell_\perp$ so that

$$\int d^2r \psi_0^2(r) = 1 . \qquad (2.20)$$

The localized state is produced by a favorable *fluctuation* in the local density of pins. To estimate the second term, we compute its mean square fluctuation using Eq. (2.18),

$$\overline{\left[ \int d^2r \delta V_D(\vec{r}) \psi_0^2(r) \right]^2} \approx \frac{\Delta_1}{\ell_\perp^2} \approx \frac{U_0^2 b_0^4}{\ell_\perp^2 d^2} . \qquad (2.21)$$

A typical contribution from the random potential in a favorable region is thus of order $-U_0 b_0^2 / \ell_\perp d$. A straightforward estimate of the first term in (2.19) then gives

$$E_0 \approx \frac{T^2}{\tilde{\epsilon}_1 \ell_\perp^2} - \frac{U_0 b_0^2}{\ell_\perp d} . \qquad (2.22)$$

Minimizing over $\ell_\perp$ now leads to our basic result,

$$\ell_\perp(T) \approx d(T/T^*)^2 \quad (T > T_{dp}) , \qquad (2.23)$$

and a binding energy $U(T) = -E_0$ which is approximately

$$U(T) \approx U_0 \left( \frac{b_0}{d} \right)^2 \left( \frac{T^*}{T} \right)^2 \quad (T > T_{dp}) . \qquad (2.24)$$

The distance along $z$ required for the vortex line to sample this localization tube is now $\ell_z(T) = \tilde{\epsilon}_1 \ell_\perp^2 / T$. We see that the localization length grows much more slowly with temperature than the isolated pin result (2.12). This slow growth arises because the spreading of the ground state wave function is restrained by the multiple pinning centers away from the central one. Upon substituting for $\ell_\perp(T)$ and $U(T)$ in Eq. (2.14), we derive an estimate for the critical current when $T > T_{dp}$ [15],

$$J_c(T) \approx J_c(0) \left( \frac{b_0}{d} \right)^3 \left( \frac{T^*}{T} \right)^4 \quad (T > T_{dp}) . \qquad (2.25)$$

## 3. Flux Melting and the Quantum Harmonic Oscillator

The analogy with quantum mechanics can also be used to estimate the melting temperature in the absence of columnar pins. Consider one representative fluxon in the confining potential "cage" provided by its surrounding vortices in a triangular lattice. The partition function for a fixed entry point $\vec{0}$ and exit point $\vec{r}_\perp$ in a sample of thickness $L$ is

$$Z_1(\vec{r}_\perp, \vec{0}; L) = \int_{\vec{r}(0)=\vec{0}}^{\vec{r}(L)=\vec{r}_\perp} D\vec{r}(z) \exp\left\{ -\frac{1}{T} \int_0^L \left[ \tfrac{1}{2}\tilde{\epsilon}_1 \left( \frac{d\vec{r}(z)}{dz} \right)^2 + V_1[\vec{r}(z)] \right] dz \right\},$$
(3.1)

where $V_1[\vec{r}]$ is now a one-body potential chosen to mimic the interactions in Eq. (2.1a). We assume clean samples and high temperatures so that both correlated and point disorder can be neglected.

Three important field regimes for fluctuations in vortex crystals are easily extracted from this simplified model. Following the approach in the previous section, we rewrite this imaginary time path integral as a quantum mechanical matrix element,

$$Z(\vec{r}_\perp, \vec{0}; L) = \langle \vec{r}_\perp | e^{-L\mathcal{H}/T} | \vec{0} \rangle ,$$
(3.2)

where $|\vec{0}\rangle$ is an initial state localized at $\vec{0}$, $\langle \vec{r}_\perp |$ is a final state localized at $\vec{r}_\perp$, and the "Hamiltonian" $\mathcal{H}$ is the operator which appears in Eq. (2.4),

$$\mathcal{H} = -\frac{T^2}{2\tilde{\epsilon}_1} \nabla_\perp^2 + V_1(\vec{r}) .$$
(3.3)

Recall that the probability of finding the flux line at transverse position $\vec{r}$ within the crystal is $\psi_0^2(\vec{r})$, where $\psi_0(\vec{r})$ is the normalized ground state eigenfunction of (3.3).

When $B \gg B_{c1} \equiv \phi_0/\lambda_{ab}^2$, the pair potential $V(r_{ij}) = 2\epsilon_0 K_0(r_{ij}/\lambda_{ab})$ is logarithmic, $K_0(x) \approx \ln x$, and we expand $V_1(\vec{r}_\perp)$ about its minimum at $\vec{r}_\perp = 0$ to find

$$\left[ -\frac{T^2}{2\tilde{\epsilon}_1} \nabla_\perp^2 + \frac{1}{2} kr_\perp^2 \right] \psi_0 = E_0 \psi_0$$
(3.4)

where (neglecting logarithmic corrections to $\epsilon_0$ and constants of order unity)

$$k \approx \frac{d^2 V}{dr^2}\bigg|_{r=a_0}$$
$$\approx \frac{\epsilon_0}{a_0^2},$$
(3.5)

and $a_0$ is the mean vortex spacing. Equation (3.4) is the Schröedinger equation for a two-dimensional quantum oscillator, with $\hbar \to T$ and mass $m \to \tilde{\epsilon}_1$. The ground state wave function is

$$\psi_0(r_\perp) = \frac{1}{\sqrt{2\pi}\, r_*}\, e^{-r^2/4r_*^2} \tag{3.6}$$

with spatial extent

$$r_* = \left(\frac{T^2 a_0^2}{\epsilon_0 \tilde{\epsilon}_1}\right)^{1/4}. \tag{3.7}$$

Melting occurs when $r_* = c_L a_0$, where $c_L$ is the Lindemann constant, so the melting temperature is

$$T_m = c_L^2 \sqrt{\epsilon_0 \tilde{\epsilon}_1}\, a_0\,, \qquad (B_{c1} < B \lesssim B_\times) \tag{3.8}$$

in agreement with other estimates [24]. Vortices in the crystalline phase will travel across their confining tube of radius $r_\perp^*$ in a "time" along the $z$ axis of order $\ell_z$, where

$$\ell_z \approx r_*^2/(T/\tilde{\epsilon}_1)$$
$$\approx \sqrt{\frac{\tilde{\epsilon}_1}{\epsilon_0}}\, a_0\,. \tag{3.9}$$

A new high field regime arises when $\ell_z \lesssim d_0$, where $d_0$ is the average spacing of the copper-oxide planes, i.e., for $B \gtrsim B_\times$, with decoupling field

$$B_\times \approx \frac{\tilde{\epsilon}_1}{\epsilon_0}\, \frac{\phi_0}{d_0^2}\,, \tag{3.10}$$

again in agreement with earlier work [8, 24]. Above this field, the planes are approximately decoupled, and $T_m$ may be estimated from the theory of two-dimensional dislocation mediated melting [8, 24]

$$T_m \approx \frac{\epsilon_0 d_0}{8\pi\sqrt{3}}\,, \qquad (B \gtrsim B_\times)\,. \tag{3.11}$$

The estimate (3.8) also breaks down at low fields $B \lesssim B_{c1}$ where the logarithmic interaction potential must be replaced by an exponential repulsion. The two-dimensional harmonic oscillator model again applies, with the replacement

$$k \to \frac{\epsilon_0}{\lambda_{ab}^2}\, e^{-a_0/\lambda_{ab}}. \tag{3.12}$$

Table 1. Estimates for the flux lattice melting temperature determined for the three regimes discussed in the text

| Regime | $T_m(B)$ | |
|---|---|---|
| $B_\times \lesssim B$ | $\epsilon_0 d_0/8\pi\sqrt{3}$ | |
| | | $B_\times \approx \frac{\tilde{\epsilon}_1}{\epsilon_0}\frac{\phi_0}{d_0^2}$ |
| $B_{c1} \lesssim B \lesssim B_\times$ | $c_L^2\sqrt{\epsilon_0\tilde{\epsilon}_1}(\phi_0/B)^{1/2}$ | |
| | | $B_{c1} \approx \phi_0/\lambda_{ab}^2$ |
| $B \lesssim B_{c1}$ | $c_L^2\sqrt{\epsilon_0\tilde{\epsilon}_1}\lambda_{ab}\left(\frac{B_{c1}}{B}\right)e^{-\frac{1}{2}(B_{c1}/B)^{1/2}}$ | |

The transverse wandering distance is now

$$r_* \approx \left(\frac{T^2\lambda_{ab}^2}{\epsilon_0\tilde{\epsilon}_1}\right)^{1/4} e^{a_0/4\lambda_{ab}}. \tag{3.13}$$

and takes place over a longitudinal distance

$$\ell_z^0 = \sqrt{\frac{\tilde{\epsilon}_1}{\epsilon_0}}\,\lambda_{ab}e^{a_0/2\lambda_{ab}}. \tag{3.14}$$

The low field melting temperature becomes

$$T_m \approx c_L^2\sqrt{\epsilon_0\tilde{\epsilon}_1}\,\frac{a_0^2}{\lambda_{ab}}e^{-a_0/2\lambda_{ab}} \qquad (B \lesssim B_{c1})\,, \tag{3.15}$$

consistent with earlier predictions [17, 8]. Although we have retained the distinction between $\tilde{\epsilon}_1$ and $\epsilon_0$ in these formulas, we note that $\tilde{\epsilon}_1 \approx \epsilon_0$ in this regime [8].

The predictions (3.8), (3.11) and (3.15) are combined to give the reentrant phase diagram for melting shown in Fig. 1. Analytic estimates and boundaries for melting in the various regimes are summarized in Table 1.

Although the quantum mechanical energy is quite useful, it is worth noting that similar results can be obtained by direct estimates of path integrals such as Eq. (2.3) [26]. In the harmonic oscillator approximation, vortex configurations are controlled by a free energy

$$F = \frac{1}{2}\tilde{\epsilon}_1\int_0^L \left(\frac{d\vec{r}}{dz}\right)^2 dz + \frac{1}{2}k\int_0^L r^2(z)\,dz\,, \tag{3.16}$$

where $k$ is given by Eqs. (3.5) or (3.12) above. Let us assume that a typical vortex trajectory wanders a perpendicular distance $r$, from one side of the parabolic confining potential to the other, and that it takes a "time" $\ell$ along the $z$-axis to do this. The free energy of the segment length $\ell$ is then approximately

$$F_\ell \approx \frac{1}{2} \left[ \frac{\tilde{\epsilon}_1}{\ell} + k\ell \right] r^2 . \tag{3.17}$$

Upon optimizing $\ell$, we find that the preferred segment length is

$$\ell^* \approx \sqrt{\tilde{\epsilon}_1 / k} \tag{3.18}$$

(compare Eq. (3.9)), while the corresponding free energy of the segment is

$$F_\ell^* \approx \sqrt{\tilde{\epsilon}_1 k} \, r^2 . \tag{3.19}$$

We now assume that successive segments of length $\ell$ along the vortex fluctuate independently, and apply the equipartition theorem to one such segment to obtain

$$\langle r^2 \rangle \approx \frac{T}{\sqrt{\tilde{\epsilon}_1 k}} . \tag{3.20}$$

Upon setting $\langle r^2 \rangle = c_L^2 a_0^2$, we recover the estimates of the low and intermediate field melting temperatures discussed earlier.

## 4. Landau Theory of the Freezing Transition

Unfortunately, crude estimates such as the Lindemann criterion tell us nothing about the nature of the freezing transition from the high temperature flux liquid to the crystalline state. We conclude with a brief review of theoretical arguments predicting first-order freezing from a flux liquid into a triangular vortex lattice [6]. We also argue that this transition can be continuous for freezing into a smectic-like state which forms upon cooling with flux lines at approximately right angles to an array of columnar defects [15]. From mean field theory one expects that the Abrikosov flux lattice appears in clean systems exactly at $H_{c2}$ via a continuous phase transition [1, 27]. A renormalization group analysis in $6 - \epsilon$ dimensions [6], however, shows that thermal fluctuations reduce the ordering temperature and force this transition to become first order below six dimensions. Although this effect is probably unobservable in most conventional superconductors, it should become more pronounced in the strongly fluctuating high-$T_c$ materials [17]. In dimension $d = 3$, we can appeal directly to the following physical argument [6]: Once fluctuations drive the ordering transition significantly below $H_{c2}$, the system acquires a nonzero local condensate density,

$$\rho(\vec{r}_\perp, z) = \langle |\Psi_{\text{BCS}}(\vec{r}_\perp, z)|^2 \rangle \quad , \tag{4.1}$$

where $\Psi_{\text{BCS}}(\vec{r}_\perp, z)$ is the BCS order parameter and the field is assumed to be along $\hat{z}$. In a flux liquid, the zeroes of $\Psi_{\text{BCS}}(\vec{r}_\perp, z)$ (i.e., embryonic vortices wandering along the $z$ axis) are disordered, so the thermal average in Eq. (4.1) is independent of position. This quantity becomes modulated in directions perpendicular to $\vec{H}$, however, once these zeroes crystallize. We can then follow the classic analysis of Landau [28] and expand $\rho$ in the Fourier components of the triangular flux lattice:

$$\rho(\vec{r}_\perp) = \rho_0 + \sum_{\vec{G} \neq 0} \rho_{\vec{G}} e^{i\vec{G} \cdot \vec{r}_\perp} , \qquad (4.2)$$

where the $\{\vec{G}\}$ are reciprocal lattice vectors lying in a plane perpendicular to $\hat{z} \| \vec{H}$. The free-energy density difference $\delta\mathcal{F}$ between the liquid and crystalline phases can then be expressed as a Taylor series in the $\{\rho_G\}$, which are order parameters for the freezing transition,

$$\delta\mathcal{F} = \frac{1}{2} r \sum_{j=1}^{6} |\rho_{\vec{G}_j}|^2 + w \sum_{\vec{G}_i + \vec{G}_j + \vec{G}_k = 0} \rho_{\vec{G}_i} \rho_{\vec{G}_j} \rho_{\vec{G}_k} + \cdots . \qquad (4.3)$$

The $\vec{G}$-vectors included here come from the first ring of six around the origin. The crucial element is the third-order term allowed by the symmetry of a triangular lattice, which leads to a first-order phase transition when $r = r(T)$ decreases with decreasing temperature. Toner has discussed how to incorporate hexatic bond orientational order and quenched random disorder into this picture [29].

A recent computer simulation of a lattice gauge glass model by Hetzel *et al.* does indeed find a first-order transition [30]. The flux array studied in Ref. 30 is somewhat artificial, because it is exactly commensurate with the underlying mesh of points which defines the model. In addition, the star of reciprocal lattice vectors which defines the order parameter points in a discrete rather than continuous set of directions. These artifacts should not alter the basic prediction of a first-order freezing transition, however. The Landau argument correctly predicts a first-order freezing transition for liquids of point particles in three dimensions, and is likely to be correct for three-dimensional line liquids as well.

When columnar pins are present, one expects a continuous Bose glass transition when the vortices are nearly parallel to the pin direction [15]. The Landau theory of freezing becomes relevant, however, when flux lines are tipped at large angles relative to this preferred direction. Consider first a magnetic field oriented in the $ab$-plane exactly perpendicular to columnar pins aligned with the $c$ axis. We again neglect point disorder, which is assumed to affect the physics only at length scales much larger than the vortex lattice spacing. Vortex lines then run at right angles through the defect "forest" on predominantly "horizontal" trajectories. Although fluxons may prefer to deviate to the right or left to pass through particularly favorable "thickets" of columnar pins, their vertical deflections are unconstrained. A low temperature crystalline phase periodic in two directions is impossible because of the disorder [31]. The vortices can order, however, in smectic-like sheets which are periodic in $z$, along the $c$-axis. This crystal should freeze from a flux liquid in a way somewhat reminiscent of the nematic to smectic-A transition in liquid crystals [32].

The condensate density (4.1) is now modulated in the $z$-direction below the freezing transition, and the order parameter expansion (4.2) becomes

$$\rho(z) \approx \rho_0[1 + Re\{\Phi_{q_0} e^{iq_0 z}\}] \tag{4.4}$$

where $\Phi_{q_0}$ is a complex order parameter describing the amplitude and phase of a single density wave with wave vector $\vec{q}_0 = q_0 \hat{z} \approx (2\pi/a_0)\hat{z}$. Although integer multiples of $\vec{q}_0$ could also appear in the expansion, wave vectors with components perpendicular to $\hat{z}$ are excluded by the disorder. As in the de Gennes theory of the nematic to smectic-A transition [32], we expand the free energy density difference between the liquid and crystal phases to quartic order in $\Phi_{q_0}$,

$$\delta\mathcal{F} = \frac{1}{2} r|\Phi_{q_0}|^2 + u|\Phi_{q_0}|^2 \tag{4.5}$$

with $r \propto (T - T_m)$. Unlike Eq. (4.3), a third order term is not permitted by symmetry, so the transition can be continuous. Of course, a first order transition caused by a negative value of $u$ cannot be ruled out, however. To treat thermal fluctuations more accurately, we should allow $\Phi_{q_0}$ to have spatial dependence and include gradient terms like $|\vec{\nabla}\Phi_{q_0}(\vec{r}_\perp, z)|^2$. The continuous phase transition which results should be in the universality class of the three dimensional $XY$ model. At low temperatures, the lattice constant of the smectic sheets will lock into an integral multiple of the lattice constant of the $CuO_2$ planes. Genuine long range transitional order will survive in this "commensurate smectic" phase even after the imposition of point disorder, because long wavelength phonons are now massive and behave like Einstein instead of Debye oscillators. The commensurate smectic state (accessible via a *continuous* freezing transition) should be observable even in the *absence* of columnar pins provided the field direction is carefully aligned with the *ab* planes [33].

Suppose the field is tilted at some angle away from the *ab*-plane, but not so close to the *c*-axis as to produce a lock-in to a Bose glass phase (see Ref. 15). Fluxon trajectories will now consist of approximately horizontal vortex segments (kinks) which connect vertical portions running along the columnar pins. The low temperature phase will now consist of vortex sheets tilted on average which will remain periodic along the *c*-axis. Because the periodicity is still only possible in one direction, the analysis of smectic freezing sketched above for perpendicular field orientations should remain valid. A low temperature lock-in of the translational periodicity is no longer expected, however.

### Acknowledgements

This research was supported by the National Science Foundation, through Grant No. DMR 91-15491 and through the Harvard Materials Research Laboratory. Much of this work is the result of a stimulating collaboration with V.M. Vinokur. See Ref. 15 for a more complete treatment of correlated pinning. Discussions with D. Bishop, R. Budhani, L. Civale, G. Crabtree, D.S. Fisher, P.L. Gammel, T. Hwa, P. Le Doussal and M.C. Marchetti, are also gratefully acknowledged.

## Appendix A: Transfer Matrix Representation of the Partition Function

We review here how path integrals like those represented in Eq. (2.3) can be rewritten in terms of a transfer matrix, which is the exponential of the Schröedinger operator which appears in elementary quantum mechanics [16].

We first consider

$$
Z(\vec{r}_\perp, \vec{0}; \ell) = \int_{\vec{r}(0)=\vec{0}}^{\vec{r}(\ell)=\vec{r}_\perp} \mathcal{D}\vec{r}(z) \exp\left[ -\frac{\tilde{\epsilon}_1}{2T} \int_0^\ell \left(\frac{d\vec{r}}{dz}\right)^2 dz - \frac{1}{T} \int_0^\ell V_1[\vec{r}(z)]\, dz \right]
$$

$$(A1)$$

and discretize this path integral as indicated in Figure 6, where the planes of constant $z$ are separated by a small parameter $\delta$. This is precisely the situation which arises in the high $T_c$ superconductors with field parallel to the $c$ axis, provided $\delta$ represents the mean spacing between $CuO_2$ planes. The discretized path integral reads

$$
Z(\vec{r}_\perp, \vec{0}; \ell) \approx \left( \prod_{j=2}^N \frac{\tilde{\epsilon}_1}{2\pi T\delta} \int d^2 r_j \right) \exp\left[ \frac{-\tilde{\epsilon}_1}{2T\delta} \sum_{j=2}^{N+1} (\vec{r}_j - \vec{r}_{j-1})^2 - \frac{\delta}{T} \sum_{j=1}^{N+1} V_1(\vec{r}_j) \right],
$$

$$(A2)$$

where it is understood that $\vec{r}_1 = \vec{0}$, $\vec{r}_{N+1} = \vec{r}_\perp$, and the normalization of the integrals is chosen so that $Z(r_\perp, 0; \ell) = 1$ when $V_1(\vec{r}) = 0$.

As usual in statistical mechanics, the effect of adding one copper oxide plane to the system can be represented in terms of a transfer matrix

$$
Z(\vec{r}, \vec{0}; \ell + \delta) = \int d^2 r' T(\vec{r}, \vec{r}')\, Z(\vec{r}', \vec{0}; \ell) \tag{A3}
$$

where (neglecting small edge effects at the top and bottom of the sample)

$$
T(r, \vec{r}') = \frac{\tilde{\epsilon}_1}{2\pi T\delta} \exp\left\{ \frac{-\tilde{\epsilon}_1}{2T\delta} |\vec{r} - \vec{r}'|^2 - \frac{\delta}{2T} [V_1(\vec{r}) + V_1(\vec{r}')] \right\} . \tag{A4}
$$

Ryu *et al.* have studied the spectrum of $T(\vec{r}, \vec{r}')$ for finite $\delta$ [32]. Here, we shall instead consider the limit of small $\delta$. We can then expand the potential term and derive a differential equation for $Z$,

$$
T\partial_\ell Z(\vec{r}, \vec{0}; \ell) = \left[ \frac{T^2}{2\tilde{\epsilon}_1} \nabla_\perp^2 - V_1(\vec{r}) \right] Z(\vec{r}, 0; \ell) . \tag{A5}
$$

A formal expression for the partition function results from integrating Eq. (A5) across a sample thickness $L$ with initial state $|0\rangle$ and final state $|\vec{r}_\perp\rangle$,

$$
Z(\vec{r}_\perp, 0; \ell) = \langle \vec{r}_\perp | e^{-\mathcal{H}L/T} | \vec{0}\rangle , \tag{A6}
$$

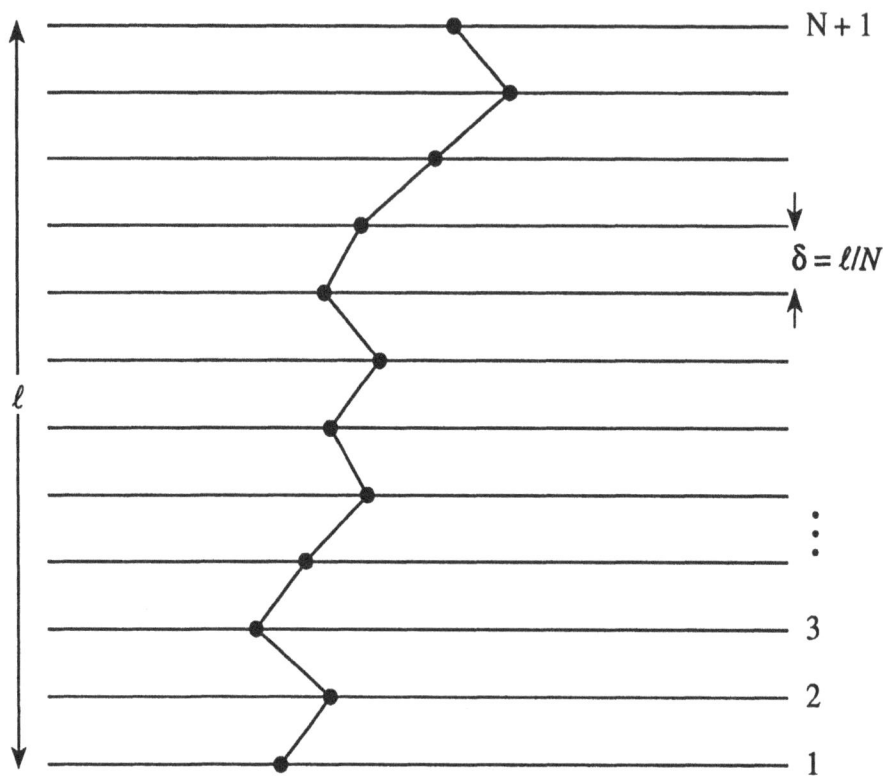

Figure 6. Discretized path integral representation of vortex line passing through $N+1$ CuO$_2$ planes with average spacing $\delta = \ell/N$.

where $\mathcal{H} = \frac{-T^2}{2\tilde{\epsilon}_1} \nabla^2_\perp + V(\vec{r})$. The partition function $Z(\vec{r}_\perp, \vec{0}; L)$ has been defined to be the ratio of the path integrals which appear in Eq. (2.3). Upon inserting a complete set of normalized eigenfunctions $\{\psi_n(\vec{r})\}$ of $\mathcal{H}$ with energies $\{E_n\}$ into (A6), we obtain

$$
\begin{aligned}
e^{U(T)L/T} &= Z(\vec{r}_\perp, 0; L) \\
&= \sum_n \psi_n(\vec{r}_\perp)\psi_n(\vec{0})e^{-E_n L/T}
\end{aligned}
\tag{A7}
$$

which leads when $L \to \infty$ to the result

$$
U(T) = -E_0(T)
\tag{A8}
$$

quoted in the text.

## Appendix B: Vortex Probability Distributions

We show here that the probability of finding an individual vortex line at height $z$ with position $\vec{r}_\perp$ in an arbitrary binding potential $V_1(\vec{r}_\perp)$ is related to the ground state wave function of the corresponding Schröedinger equation. The probability distribution at a free surface is proportional to the wave function itself, while the probability far from the surface is proportional to the wave function squared.

Consider first a fluxon which starts at the origin $\vec{0}$ and wanders across a sample of thickness $L$ to position $\vec{r}_\perp$. As discussed in Appendix A, the partition function associated with this constrained path integral may be written as a quantum mechanical matrix element

$$
\mathcal{Z}(\vec{r}_\perp, \vec{0}; L) = \int_{\vec{r}(0)=\vec{0}}^{\vec{r}(L)=\vec{r}_\perp} \mathcal{D}r(z) \exp\left[ -\frac{\tilde{\epsilon}_1}{2T} \int_0^L \left|\frac{d\vec{r}}{dz}\right|^2 dz - \frac{1}{T} \int_0^L V_1[\vec{r}(z)]dz \right]
$$
$$
\equiv \langle \vec{r}_\perp | e^{-L\mathcal{H}/T} | 0 \rangle \tag{B1}
$$

where $|\vec{0}\rangle$ is an initial state localized at $\vec{0}$ while $\langle \vec{r}_\perp |$ is a final state localized at $\vec{r}_\perp$. The "Hamiltonian" $\mathcal{H}$ appearing in (B1) is the Schröedinger operator, Eq. (3.3). The probability distribution $\mathcal{P}(\vec{r}_\perp)$ for the vortex tip position at the upper surface is then

$$
\mathcal{P}(\vec{r}_\perp) = \mathcal{Z}(\vec{r}_\perp, \vec{0}; L) \bigg/ \int d^2 r_\perp \mathcal{Z}(\vec{r}_\perp, \vec{0}; L) . \tag{B2}
$$

Upon inserting a complete set of (real) energy eigenstates $|n\rangle$ with eigenvalues $E_n$ into Eq. (B1), we have

$$
\mathcal{P}(\vec{r}_\perp) = \frac{\sum_n \psi_n(\vec{0})\psi_n(\vec{r}_\perp)e^{-E_n L/T}}{\sum_n \psi_n(0) \int d^2 r_\perp \psi_n(\vec{r}_\perp)e^{-E_n L/T}} . \tag{B3}
$$

In the limit $L \to \infty$ the ground state dominates, the probability $\mathcal{P}(\vec{r}_\perp)$ becomes

$$
\mathcal{P}(\vec{r}_\perp) \approx \frac{\psi_0(\vec{r}_\perp)}{\int d^2 r_\perp \psi_0(\vec{r}_\perp)} \left[ 1 + \mathcal{O}\left( e^{-(E_1-E_0)L/T} \right) \right] , \tag{B4}
$$

where $E_1$ is the energy of the first excited state. Because the ground wave function is nodeless [21], $\mathcal{P}(r_\perp)$ is always positive and well defined.

Consider now a more general problem of a vortex which enters the sample at $\vec{r}_i$, exits at $\vec{r}_f$, and passes through $\vec{r}$ at a height $z$ which is far from the boundaries. The normalized probability distribution is now

$$
\tilde{\mathcal{P}}(\vec{r}; L) = \tilde{\mathcal{Z}}(\vec{r}; L) / \int d^2 r \tilde{\mathcal{Z}}(\vec{r}; L) \tag{B5}
$$

where

$$
\tilde{\mathcal{Z}}(\vec{r}; L) = \int d^2 r_i \int d^2 r_f \mathcal{Z}(\vec{r}_f, \vec{r}; L - z)\mathcal{Z}(\vec{r}, \vec{r}_i; z) \tag{B6}
$$

and $Z(\vec{r}_2, \vec{r}_1; L)$ is given by Eq. (B1). Upon inserting complete sets of states as before, we find that

$$\tilde{\mathcal{P}}(\vec{r}; L) = \frac{\psi_0^2(r)}{\int d^2 r_\perp \psi_0^2(r)} \left[ 1 + \mathcal{O}\left( e^{-L(E_1 - E_0)/2T} \right) \right] , \tag{B7}$$

where the correction assumes $\vec{r}$ is at the midplane of the sample.

## References

1. A.A. Abrikosov, *Zh. Eksperim. i Theor. Fiz.* **32**, 1442 (1957) [*Sov. Phys. JETP* **5**, 1174 (1957)].
2. M. Charalmabous *et al.*, *Phys. Rev.* **B45**, 45 (1992).
3. W.K. Kwok, S. Fleshler, U. Welp, V.M. Vinokur, J. Downey, G.W. Crabtree and M.M. Miller, *Phys. Rev. Lett.* **69**, 3370 (1992).
4. D.E. Farrell, J.P. Rice and D.M. Ginsberg, *Phys. Rev. Lett.* **67**, 1165 (1991).
5. H. Safar, P.L. Gammel, D.A. Huse, D.J. Bishop, J.P. Rice and D.M. Ginsberg, *Phys. Rev. Lett.* **69**, 824 (1992).
6. E. Brezin, D.R. Nelson and A. Thiaville, *Phys. Rev.* **B31**, 7124 (1985).
7. D.R. Nelson and P. Le Doussal, *Phys. Rev.* **B42**, 10112 (1990).
8. D.S. Fisher, M.P.A. Fisher and D.A. Huse, *Phys. Rev.* **B43**, 130 (1991).
9. R.C. Budhani, M. Suenaga and H.S. Liou, *Phys. Rev. Lett.* **69**, 3816 (1992).
10. M. Konczykowski *et al.*, *Phys. Rev.* **B44**, 7167 (1991).
11. L. Civale, A.D. Marwich, T.K. Worthington, M.A. Kirk, J.R. Thompson, L. Krusin-Elbaum, Y. Sun, J.R. Clem and F. Holtzberg, *Phys. Rev. Lett.* **67**, 648 (1991).
12. W. Gerhauser *et al.*, *Phys. Rev. Lett.* **68**, 879 (1992).
13. V. Hardy *et al.*, *Nucl. Instr. and Meth.* **B54**, 472 (1991).
14. A.P. Malozemoff, T.K. Worthington, Y. Yeshurun and F. Holtzberg, *Phys. Rev.* **B38**, 7203 (1988).
15. D.R. Nelson and V.M. Vinokur, *Phys. Rev. Lett.* **68**, 2392 (1992); and *Phys. Rev.* **B** (in press).
16. R.P. Feynman and A.R. Hibbs, *Quantum Mechanics and Path Integrals* (McGraw-Hill, New York, 1965); R.P. Feynman, *Statistical Mechanics* (Benjamin, Reading, MA, 1972).
17. D.R. Nelson, *Phys. Rev. Lett.* **60**, 1973 (1988); D.R. Nelson and S. Seung, *Phys. Rev.* **B39**, 9153 (1989).
18. S.F. Edwards, *Proc. Phys. Soc.* **85**, 613 (1965); P.G. de Gennes, *Rep. Prog. Phys.* **32**, 187 (1969).
19. M.P.A. Fisher, P.B. Weichman, G. Grinstein and D.S. Fisher, *Phys. Rev.* **B40**, 546 (1989), and references therein.
20. D.R. Nelson, in *Phenomenology and Applications of High Temperature Semiconductors*, edited by K. Bedell, M. Inui, D. Meltzer, J.R. Schrieffer, and S. Doniach (Addison-Wesley, New York, 1991).
21. D.R. Nelson, in *Phase Transitions in Systems with Competing Energy Scales*, edited by T. Riste and D. Sherrington (Kluwer, Boston, to be published).

22. L.D. Landau and E.M. Lifshitz, *Quantum Mechanics*, 2nd. Edition (Pergammon, New York, 1965).
23. See, e.g., D.R. Nelson, *J. Stat. Phys.* **57**, 511 (1989).
24. See, e.g., L.I. Glazman and A.E. Koshelev, *Phys. Rev.* **B43**, 2835 (1991).
25. D.S. Fisher, *Phys. Rev.* **B22**, 1190 (1980).
26. D.S. Fisher has also constructed an argument along these lines (private communication).
27. A.A. Abrikosov, Zh. Eksperim. i Teor. Fiz. **32**, 1442 (1957) [Sov. Phys. JETP **5**, 1174 (1957)]; S. Fetter and P.C. Hohenberg, in *Superconductivity*, edited by R.D. Parks, Vol. II (Dekker, New York, 1969).
28. L.D. Landau, *Phys. Z.* **11**, 26 (1937); *The Collected Papers of L.D. Landau*, edited by D. ter Haar (Gordon and Breach-Pergamon, New York, 1965).
29. J. Toner, *Phys. Rev. Lett.* **66**, 2523 (1991).
30. R.E. Hetzel, A. Sudbo and D.A. Huse, *Phys. Rev. Lett.* **69**, 518 (1992).
31. A.I. Larkin and Y.M. Ovchinnikov, *J. Low Temp. Phys.* **34**, 409 (1979).
32. P.G. de Gennes, *Solid. State Commun.* **10**, 753 (1972).
33. W. Kwok *et al.*, Argonne National Laboratories, preprint.
34. S. Ryu, A. Kapitulnik and S. Doniach, Stanford University preprint.

# TWO-DIMENSIONAL SUPERFLUIDS

A. P. Young

*Physics Department,*
*University of California Santa Cruz,*
*Santa Cruz, CA 95064.*

ABSTRACT. These lectures describe the behaviour of two-dimensional superfluids, which differ from three-dimensional systems in having phase transitions driven by *thermally excited* vortices. Firstly, I will discuss the Kosterlitz-Thouless-Berezinskii (KTB) theory of the superfluid transition in helium films, which is due to unbinding of vortex-antivortex pairs. Next, I will describe the modifications of this transition due to screening effects at the zero field transition in superconductors. The transition of a clean superconducting film in a perpendicular magnetic field, caused by melting of the flux lattice, will be discussed next. This may also be of the KTB type, involving unbinding of dislocations (with a second transition due to disclination unbinding first discussed by Halperin and Nelson) or may be first order. Finally, the case of a dirty two-dimensional superconductor in a field will be discussed. This has a vortex glass transition (analogous to the spin glass transition in magnetic systems) at zero temperature. Results of recent experiments and simulations which provide evidence for vortex glass behavior will be discussed.

# 1    Introduction

In bulk superfluids, vortices are lines and have an energy which is proportional to their length (neglecting logarithmic factors coming from long range interactions between different segments of the vortex). The probability of thermally exciting a vortex is therefore very small because its energy is large compared with $k_B T$. Vortices have to be forced into the system, either by applying a field in the case of superconductors [1, 2], or rotating the system in the case of helium [3]. By contrast, in two-dimensional systems, vortices are point defects so the energy to create them is finite (more precisely the energy to create a pair of opposite vortices is finite) and not necessarily very much greater than $k_B T$ so thermal excitation of vortices is quite probable. In fact, as pointed out by Berezinskii [4] and Kosterlitz and Thouless [5] the superfluid transition in two dimensional systems is actually *controlled* by the unbinding of vortex-antivortex pairs.

In these lectures I will discuss the KTB theory as applied to the superfluid transition in $^4$He, the zero field transition in superconductors, and the transition in a perpendicular magnetic field for clean superconductors. There are several good reviews on the KTB theory, see e.g. [6, 7], and the KTB theory is also discussed at this school by Mooij [8]. My lectures conclude with some recent developments on the transition in a field for dirty superconductors. The disorder destroys both translational [9] and orientational [10] order in the flux lattice, so melting is not expected. Recently, a new type of transition in such systems has been proposed [11] in which

*N. Bontemps et al. (eds.), The Vortex State, 63–83.*

the low-$T$ state is a "vortex glass", similar to the spin glass state [12] in magnetic systems, *i.e.* there is no order at any wavevector but the system has a stiffness associated with it so that if it is perturbed in one region, then the state is altered out to infinite distance. Above the transition temperature, $T_c$, the vortex glass correlation length, the distance over which a local perturbation affects the system, is finite and diverges as $T$ approaches $T_c$. The necessary conditions for spin glass behavior are randomness and frustration [12] and these indeed occur in disordered type-II superconductors in a magnetic field. In two-dimensions, it appears that the vortex glass transition occurs at $T = 0$ [13, 14], but, as we shall see, the temperature dependence of the correlation length, which diverges as $T \to 0$, leads to observable effects at finite temperature.

Before going into details, it is worth discussing when a film can be considered two-dimensional. If the film thickness is $d$ and the *bulk* correlation length is $\xi$ (for anisotropic systems this would be the correlation length perpendicular to the surface of the film) then the film is effectively two-dimensional when $d < \xi$ and three-dimensional when $d > \xi$. Since the correlation length is temperature dependent and diverges at $T_c$, a film which is many atoms thick (for a superconductor one would replace the atomic dimension by the Cooper pair size) shows bulk behavior for $T$ significantly above $T_c$ and then crosses over to two-dimensional behavior close to $T_c$. For really thick films, the 2-$d$ region would be unobservably narrow. On the other hand, films which are only a few atomic layers thick would not show bulk behavior in any region of $T$ since, by the time the bulk correlation length is large enough for the system to be in the critical region, it is already larger than $d$.

# 2 The KTB Phase Transition in $^4$He films

Superfluidity in $^4$He is believed to be due to Bose-Einstein condensation. The order parameter is therefore the amplitude of the condensate, which has a magnitude and a phase. Long wavelength fluctuations in the phase cost very little energy and so are expected to play an important role. For the moment, let us *only* include such fluctuations. They can be described by the spin-wave Hamiltonian

$$\mathcal{H}_{sw} = \frac{J}{2} \int (\nabla \theta)^2 \, d^2r \quad , \tag{1}$$

where $J$ denotes the interaction strength and $\theta$ is the local phase of the condensate. Neglecting the periodicity of $\theta$ we can integrate from $-\infty$ to $\infty$, so the partition function is

$$Z = \int_{-\infty}^{\infty} D[\theta] \exp\left(-\frac{\mathcal{H}}{k_B T}\right) \quad . \tag{2}$$

The spin-wave Hamiltonian is a simplification of a lattice model for the lambda transition in $^4$He,

$$\mathcal{H}_{XY} = -J \sum_{\langle i,j \rangle} \cos(\theta_i - \theta_j) \quad , \tag{3}$$

called the $XY$ model. Unlike the spin-wave Hamiltonian, the $XY$ model includes vortices and spinwave-spinwave interactions and so is much more realistic. In fact, the lambda transition in $^4$He is believed to be in the same universality class as the $XY$ model.

For the spin-wave Hamiltonian, correlation functions can be evaluated straightforwardly because only Gaussian integrals are involved. One finds

$$
\begin{aligned}
C(r) &= \langle \exp i[\theta(r) - \theta(0)] \rangle \\
&= \exp\left[-\frac{1}{2}\left\langle (\theta(r) - \theta(0))^2 \right\rangle\right] \quad,
\end{aligned}
\tag{4}
$$

where

$$
\begin{aligned}
\left\langle (\theta(r) - \theta(0))^2 \right\rangle &= \frac{2k_BT}{J} \int \frac{d^2k}{(2\pi)^2}\left[1 - e^{i\mathbf{k}\cdot\mathbf{r}}\right] \\
&\simeq \frac{k_BT}{\pi J} \ln(r/a_0) \quad,
\end{aligned}
\tag{5}
$$

so

$$
C(r) \sim \frac{1}{r^{\eta(T)}} \quad,
\tag{6}
$$

where $a_0$ is a short distance cutoff of order the atomic size and

$$
\eta(T) = \frac{k_BT}{2\pi J} \quad.
\tag{7}
$$

In other words, correlation functions decay with a power law, in which the power varies continuously with $T$. Since power law decay is generally associated with a system at its critical point, this behavior is like a *line* of critical points, called the spin-wave line.

The power law decay in Eq. (7) implies that there is no long range order at finite temperature, a result which can be proved rigorously [15, 16]. On the other hand, a power law decay of the correlations, while reasonable at low temperature, is certainly wrong at high $T$ where the correlations should decay exponentially. One expects, then, that inclusion of additional terms in the Hamiltonian, neglected in the spin-wave approximation, would terminate the spinwave line at a certain temperature $T_c$, which marks a transition where the decay of the correlations changes from power law to exponential. It turns out that spin-wave interactions do not induce this transition, but rather it is the vortices that do it.

We shall explain the vortex driven transition shortly, but first let us discuss another property of the spin-wave line. Suppose we apply a twist to the phase by suitable boundary conditions, so that there is an *externally imposed* phase gradient $\nabla\Theta$. The extra free energy can be written as

$$
\delta F = \frac{S}{2} \int (\nabla\Theta)^2 \, d^2r
\tag{8}
$$

where $S$, the stiffness coefficient, is related to the superfluid density, $\rho_s$, by

$$S = \left(\frac{\hbar}{m}\right)^2 \rho_s \quad , \tag{9}$$

where $m$ is the mass of helium atom. In the spinwave approximation it is easy to see that

$$S = J \quad . \tag{10}$$

Thus, even though there is no long range order, the spin-wave approximation gives a finite superfluid density, so the system is actually a superfluid. Note that it is $\rho_s$, rather than the order parameter (condensate fraction), $\psi$, which determines superfluid properties. In mean field theory, these are related by

$$S = J\psi^2 \qquad \text{MFT} \quad , \tag{11}$$

but in 2-$d$, fluctuation effects are so big that mean field theory is completely invalid and we have $S > 0$ (below $T_c$) even though $\psi = 0$ (at all $T$).

As stated above, $T_c$ is determined by vortices. These are configurations of the local order parameter, $\psi(\mathbf{r}) = |\psi(\mathbf{r})| \exp i\theta(\mathbf{r})$ such that the phase changes by a multiple of $2\pi$ if one follows a closed path around a vortex, *i.e.*

$$\oint \nabla\theta \cdot d\mathbf{l} = 2\pi n \quad , \tag{12}$$

where $n$ is an integer. Vortices with $|n| > 1$ have a high energy and can decay into two or more vortices with $|n| = 1$, so we will only consider the cases $n = \pm 1$. At $T = 0$, far away from an isolated vortex the amplitude is constant and we shall set it to be unity by absorbing factors into the definition of $J$ in Eq. (1). However, the amplitude must vanish at the center of the vortex otherwise the vortex energy would be infinite (note that the gradient of the phase is infinite there). The distance over which the amplitude is depressed is called the core size; it is of order the atomic size and we shall denote it by $a_0$. We shall not need to discuss in detail what happens in the core, which is useful because the core is complicated and not very well understood. According to spin-wave theory, Eq. (1), the energy of a single vortex is

$$E_{\text{1-vortex}} = \pi J \ln(L/a_0) \quad , \tag{13}$$

where $L$ is the linear size of the system. A single vortex therefore costs an infinite energy in the thermodynamic limit. A pair of opposite vortices separated by a distance $r$ has finite energy however:

$$E_{\text{pair}} = 2\pi J \ln(r/a_0) \quad . \tag{14}$$

Because the energy of a pair diverges with $r$ we expect that at low-$T$ all vortices form bound pairs. However, at high-$T$, the pairs can unbind because of entropy effects. Consider, then, a single vortex. Its entropy is ($k_B$ times) the log of the number of points that the vortex can be put, *i.e.* $k_B \ln(L/a_0)^2$. The *free* energy of a single free vortex is then given by

$$F_{\text{1-vortex}} = (\pi J - 2k_B T) \ln(L/a_0) \quad , \tag{15}$$

which is negative for $k_B T > \pi J/2$. We interpret the temperature at which $F_{\text{1-vortex}}$ changes sign to be $T_c$, and so, according to this simple 1-vortex argument,

$$k_B T_c = \frac{\pi J}{2} \quad . \tag{16}$$

Despite the simplicity of the argument, we shall see that it contains the essential physics of the problem and gives the transition temperature exactly (including vortex-vortex and spinwave-vortex interactions which have been neglected here) provided the factor of $J$ in Eq. (16) is replaced by the fully renormalized stiffness, $S$. This is equal to $J$ in the spin-wave approximation, see Eq. (10), but is less than $J$ when screening due to vortices (and also spinwave effects) are included. When the vortices unbind we shall see that $S = 0$ and, as a result, the correlations fall off exponentially with distance.

To describe the KTB transition we need to go beyond the simple case, studied above, of one vortex and treat the many body problem of a finite density of vortices. The vortex Hamiltonian is

$$-\beta \mathcal{H}_V = \frac{2\pi J}{k_B T} \sum_{(i,j)} n_i n_j \ln(r_{ij}/a_0) + \ln y \sum_i n_i^2 \tag{17}$$

where, $n_i = \pm 1$,

$$y = e^{-E_0/k_B T} \tag{18}$$

is a fugacity for vortices, $E_0$ is the core energy, and the total vorticity must be zero, *i.e.*

$$\sum_i n_i = 0 \quad , \tag{19}$$

otherwise the energy diverges. The vortex Hamiltonian, Eq. (17), can be derived microscopically if the cosine in the $XY$ model, Eq. (3), is replaced by the form in the periodic Gaussian or Villain [17, 18] model. (The precise form of the function should be irrelevant). The parameter $J$ in Eq. (17) now includes effects of spin-wave vortex interactions which exist in real $^4$He films and the $XY$ model, but which vanish for the Villain model. To represent the core, no two vortices can approach closer than the core size $a_0$.

The vortex Hamiltonian is equivalent to the 2-$d$ Coulomb gas

$$\mathcal{H}_{CG} = -\sum_{(i,j)} q_i q_j \ln(r_{ij}) \tag{20}$$

with

$$q_i = (2\pi J)^{1/2} n_i \tag{21}$$

the $i$-th charge, (note that $q_i^2 \equiv q^2 = 2\pi J$) and so the scalar potential, $\phi$, satisfies

$$\nabla^2 \phi = -2\pi\rho \tag{22}$$

with

$$\rho(r) = \sum_i q_i \delta(r - r_i) \quad . \tag{23}$$

In this representation, the system is an insulator at low-$T$ because the charges form neutral "molecules", so the dielectric constant, $\epsilon$, is finite. As we shall see, the stiffness is given by

$$S = \frac{J}{\epsilon} \tag{24}$$

which is finite, so the system is a superfluid. At high temperature there are free charges and so the dielectric constant diverges. Consequently the renormalized stiffness vanishes and the system is in the normal state.

The KTB theory considers the screening effect that *smaller* pairs have on larger pairs. Consider a pair separated by a distance $r$. The force between them is reduced from $2\pi J/r$ to $2\pi J/(r\epsilon(r))$, where $\epsilon(r)$ is a *scale dependent* dielectric constant representing the screening effect of pairs with separation $r'$ less than $r$. One determines $\epsilon(r)$ in the usual way from electrostatics,

$$\epsilon(r) = 1 + 2\pi\chi(r) \quad , \tag{25}$$

where the susceptibility, $\chi(r)$, is related to the polarizability of a pair, $\alpha(r)$, and the density of pairs, $n(r)$, by

$$\chi(r) = \int_{a_0}^{r} n(r')\alpha(r')dr' \quad . \tag{26}$$

The polarizability is given by the standard expression

$$\alpha(r) = \frac{(qr)^2}{2k_B T} \tag{27}$$

where the factor of $1/2$ comes from integrating over the direction of the dipole in two-dimensions. The vortex density is given by

$$n(r) = 2\pi r \frac{y^2}{a_0^4} e^{-V(r)/k_B T} \tag{28}$$

where $V(r)$, the energy between two vortices, is obtained by integrating up the force, *i.e.*

$$V(r) = q^2 \int_{\ln a_0}^{\ln r} \frac{d\ln r'}{\epsilon(r')} \quad . \tag{29}$$

These equations enable one to determine $\epsilon(r)$ self-consistently. To simplify the analysis let us define a logarithmic scale, $l$, by $r = a_0 e^l$ and then define

$$K(l) = \frac{J}{k_B T \epsilon(r = a_0 e^l)} \tag{30}$$

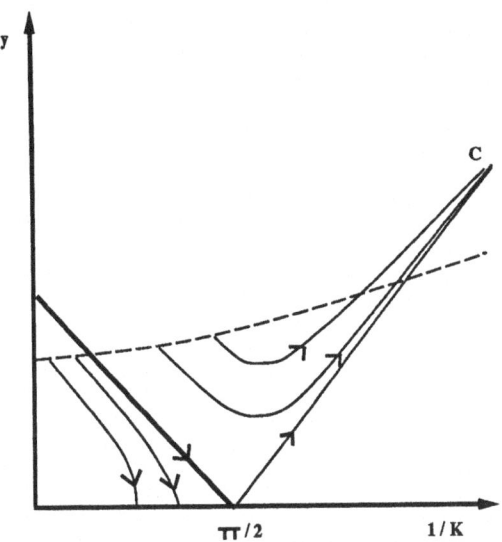

Figure 1: Sketch of solutions of the Kosterlitz equations, Eqs. (32). The dashed line is a locus of physical values of the parameters, on which integration of the equations is started. The solid line terminating at $y = 0, K^{-1} = \pi/2$ is the phase boundary separating the superfluid phase to the left from the normal phase to the right. In the superfluid region the equations flow to the spinwave line, $y = 0$. In the normal region the equations flow to large $y$.

$$y^2(l) = y^2 \left( \frac{r}{a_0} \right)^4 e^{-V(r)/k_B T} \quad , \tag{31}$$

where $K(l)$ is the effective stiffness (divided by $k_B T$) on scale $l$ and $y(l)$ is an effective scale dependent fugacity. Equations (25)-(29) can then be reexpressed as

$$\frac{dK^{-1}(l)}{dl} = 4\pi^3 y^2(l)$$
$$\frac{dy(l)}{dl} = [2 - \pi K(l)] y(l) \tag{32}$$

which were first obtained by Kosterlitz [19], though the original Kosterlitz-Thouless theory [5] is equivalent to them, see [20]. Eqs. (32) have to be solved subject to the boundary conditions

$$K(l = 0) = \frac{J}{k_B T}$$
$$y(l = 0) = y \quad . \tag{33}$$

The right hand side of Eqs. (32) is just the first term in an expansion in powers of $y(l)$, so the density has been assumed to be small.

Eqs. (32) can be integrated analytically for $[2 - \pi K(l)] \ll 1$ and the result is shown in Fig. 1. The thick line, which terminates at $y = 0, K^{-1} = \pi/2$ is the locus of critical points. If the physical values of $K$ and $y$, *i.e.* $K(l = 0)$ and $y(l = 0)$, lie to the left of this line, then the trajectory flows to the line $y = 0$, which corresponds to there being no vortices on long length scales. This is expected since all the vortices form bound pairs in this region. In other words the long distance physics is described by the spin-wave line but with the effective value of $J$ reduced from its bare value, *i.e.*

$$J \longrightarrow S = \frac{J}{\epsilon} \tag{34}$$

where

$$\epsilon \equiv \epsilon(r = \infty) = \frac{K(l = 0)}{K(l = \infty)} \quad . \tag{35}$$

Thus the spin-wave theory can be taken over but with $J$ replaced by the stiffness $S$. Since $\epsilon$ is finite to the left of the solid line in Fig. 1, the stiffness is finite so this region is a superfluid.

By contrast, if the starting values of $y$ and $K^{-1}$ lie to the right of the thick line in Fig. 1, then the trajectories flow to large values of $y$, presumably indicating the presence of free vortices, where the equations are inapplicable. The strategy then is to integrate the equations not up to $l = \infty$ but only to a point (C in Fig. 1) where $y(l)$ is still less than unity but the system is far from the critical point. One can then calculate physical quantities from some other theory, such as Debye-Huckel theory, which should be fairly reliable in this region. Eqs. (32) are then used to relate the parameters in the Debye-Huckel theory to the original parameters in the Hamiltonian, see *e.g.* [21].

According to Debye-Huckel theory,

$$\epsilon(\mathbf{k}) = 1 + \frac{2\pi n_f q^2}{k^2} \tag{36}$$

for small $\mathbf{k}$, where $\mathbf{k}$ is the wavevector, $q$ the magnitude of the charge, and $n_f$ the density of free charges. The separation between the free charges must be comparable with the correlation length above $T_c$, *i.e.*

$$n_f \sim \xi_+^{-2} \quad , \tag{37}$$

and integration of Eqs. (32) gives [19]

$$\xi_+ \sim e^{C/t^{1/2}} \quad , \tag{38}$$

where $t = (T - T_c)/T_c$ and $C$ is a non-universal constant.

From Fig. 1 it is apparent that the critical temperature, $T_c$, is where

$$\frac{S(T_c^-)}{k_B T_c} = K(l = \infty) = \frac{2}{\pi} \quad , \tag{39}$$

which is to be compared with the simple energy-entropy argument above for a single vortex, Eq. (16), which can be expressed in a similar manner as $K(l = 0) = 2/\pi$. From Eqs. (39) and (9) it follows that [22]

$$\left(\frac{\hbar}{m}\right)^2 \frac{\rho_s(T_c^-)}{k_B T_c} = \frac{2}{\pi} \quad, \tag{40}$$

so that $\rho_s$ has a universal jump at $T_c$ (remember that $\rho_s = 0$ for $T > T_c$). This prediction has been verified in experiments on helium films [23], though to make a precise comparison one has to generalize the KTB theory to dynamics [24].

Note that the KTB theory is able to make *exact* predictions for the critical behavior because this is determined by the spin-wave line ($y = 0$) where the physics is simple. If the initial value of $y$ is small, then the renormalized fugacity $y(l)$ is always small (at least at and below $T_c$) so the theory is reliable. Suppose, though, that the initial value of $y$ is not small. Then there are two possibilities. The first is that, below $T_c$, the fugacity decreases with increasing $l$ down to the region where the theory is applicable and the behavior is still given by the KTB theory. The other possibility is that some other physics may intervene while $y(l)$ is still large. One recent interesting suggestion [25, 26] is that there may be a vortex-antivortex solid in a region of the phase diagram with $y$ large. However, at present, the phase diagram of the Coulomb gas for large fugacity is uncertain.

# 3  Superconducting Films in Zero Field

Several authors [27, 28, 29] have applied the KTB theory to the superconducting transition in thin films. The main difference compared with helium films is the effect of screening, as a result of which the interaction between vortices only varies as $\ln(r)$ for $r$ less than the screening length, and vanishes at larger $r$. For bulk superconductors, the screening length is just the London penetration length, $\lambda_L$, [1, 2]. However, for films whose thickness, $d$, is less than $\lambda_L$, the effective screening length is [1, 30]

$$\Lambda_{\text{eff}}(T) = \frac{2\lambda_L^2}{d} \quad, \tag{41}$$

which is much larger. This larger screening length arises because the magnetic field, responsible for the screening, is largely *outside* the film. In principle, then, there are always some free vortices so the KTB transition is rounded. However, we shall see that the value of $\Lambda_{\text{eff}}(T_c^-)$ is enormous so the effects of rounding are probably too small to be observed. For practical purposes, therefore, the transition will be sharp.

According to Pearl [30] the interaction between two vortices is

$$U(r) = \frac{\pi}{2} \frac{n_s \hbar^2}{m} \ln(r) + \text{const.} \qquad (r \ll \Lambda) \quad, \tag{42}$$

where $m$ is the electron mass and $n_s$ is the density of superconducting electrons per unit *area*. For short distances one should use the bare value of $n_s$ in this expression but at long distances one should put the fully renormalized value, including screening effects of vortices. According to Eq. (39), the KTB theory predicts that $2\pi S$, the coefficient of $\ln r$ at distances large compared with microscopic lengths but small compared with $\Lambda_{\text{eff}}$, is $4k_B T_c$. Hence the analogue of the universal jump in Eq. (40) is

$$\frac{n_s(T_c^-)}{k_B T_c} \frac{\hbar^2}{m} = \frac{8}{\pi} \quad . \tag{43}$$

This can be checked experimentally from the low frequency form of the conductivity:

$$\text{Im}\,\sigma(\omega) = \frac{e^2 n_s}{m\omega} \quad . \tag{44}$$

Eq. (43) does not appear to be satisfied in all systems studied so far [26, 31, 8].

The universal jump in $n_s$ at $T_c$ can also be related to the screening length, $\Lambda_{\text{eff}}$. Noting that the London penetration depth is related to the *bulk* density of superconducting electrons, $n_s/d$, by

$$\lambda_L^2 = \frac{mc^2}{4\pi(n_s/d)e^2} \tag{45}$$

we have

$$\Lambda_{\text{eff}}(T) = \frac{mc^2}{2\pi n_s e^2} \tag{46}$$

and so

$$\Lambda_{\text{eff}}(T_c^-) = \Lambda_{T_c} \tag{47}$$

where the "fluctuation length", $\Lambda_T$, is given by

$$\Lambda_T = \frac{\phi_0^2}{16\pi^2} \frac{1}{k_B T} \quad , \tag{48}$$

with $\phi_0 = hc/2e$, the flux quantum. Numerically [32]

$$\Lambda_T = \frac{2}{T} \tag{49}$$

where $\Lambda_T$ is measured in centimeters and $T$ in Kelvin. Thus typically $\Lambda_{\text{eff}}(T_c^-)$ is of order 1 cm, a macroscopic length which might be even larger than the sample size. Thus, any rounding due to the screening length being finite should be unobservably small.

It is important to discuss the size of the critical region, $\Delta T$, *i.e.* the range of temperature over which one can observe the KTB behavior. For clean samples, *i.e.* those where the Cooper pair size, $\xi_0$, is much smaller than the mean free path $l$, it turns out to be very small [27, 28]. One way to see this is to assume that the critical region is roughly the difference between $T_c$ and the mean field transition temperature, $T_c^{MF}$. Using the mean field divergence of $\lambda_L$ one has roughly

$$\frac{\Lambda_{\text{eff}}(T)}{\Lambda_{\text{eff}}(T=0)} \sim \frac{T_c^{MF}}{T_c^{MF} - T} \tag{50}$$

so

$$\frac{\Delta T}{T_c} \sim \frac{T_c^{MF} - T_c}{T_c} \sim \frac{\Lambda_{\text{eff}}(0)}{\Lambda_{\text{eff}}(T_c^-)} \tag{51}$$

which is very small even for films only a few angstroms thick. It is larger, however, for high-$T_c$ systems, because $\Lambda(T_c^-)$ is smaller, see Eqs. (47) and (49). For dirty films, $\Lambda(0)$ is increased and hence so is the critical region. Beasley *et al.* have shown that, in this limit,

$$\frac{\Delta T}{T_c} = 0.17 \frac{e^2}{\hbar} \frac{1}{\sigma_n} \quad , \tag{52}$$

where $\sigma_n$ is the normal state conductivity. Thus the critical region is significant for films sufficiently dirty that the $\sigma_n$ is of order the quantum of conductance $e^2/\hbar$.

Next we discuss, following Halperin and Nelson [27], how the dc conductivity diverges as $T \to T_c^+$. The voltage is related to the rate of change of the phase difference, $\Delta\theta$, across the sample by the Josephson relation

$$V = \frac{\hbar}{2e} \frac{d\Delta\theta}{dt} \quad . \tag{53}$$

The change of the phase difference is due to the flow of vortices, *i.e.*

$$\frac{d\Delta\theta}{dt} = 2\pi L n_f v_{\text{vortex}} \tag{54}$$

where $n_f \sim \xi_+^{-2}$ is the density of free vortices, $L$ is the sample size, and the mean vortex velocity, $v_{\text{vortex}}$, is related to the Lorentz force,

$$f_L = \frac{\phi_0}{c} J \quad , \tag{55}$$

by

$$v_{\text{vortex}} = \mu f_L \tag{56}$$

where $J$ is the current density and $\mu$ the vortex mobility. Hence we have

$$E = \frac{\hbar^2 \pi^2}{e^2} n_f \mu J \tag{57}$$

and so

$$\sigma = \frac{e^2}{\hbar^2 \pi^2} \frac{1}{n_f \mu} \quad . \tag{58}$$

According to the dynamical theory [24] $\mu$ is a constant near $T_c$ and so

$$\sigma \sim \frac{1}{n_f} \sim \xi_+^2 \sim e^{2C/t^{1/2}} \quad . \tag{59}$$

Somewhat further away from $T_c$ one expects [27] that the relation $\sigma \sim \xi_+^2$ will still hold but with $\xi_+$ replaced by its mean field form, $\xi \sim (T - T_c^{MF})^{-1/2}$, which gives, for

74

the resistance, $R \sim T - T_c^{MF}$. This result is in agreement with the Aslamazov-Larkin theory [33] which calculates the *leading* fluctuation correction as $T$ is reduced towards $T_c$.

At and below $T_c$ the linear resistance vanishes. In this region, the electric field varies as $J$ to a power greater than unity [27] *i.e.*

$$E \sim J^{x(T)} \tag{60}$$

where $x(T_c^-) = 3$ and $x$ increases as $T$ decreases below $T_c$. Because of this power law behavior, different from the exponential dependence expected below $T_c$ in 3-$d$, see *e.g.* [34], the critical current density, *i.e.* the value of $J$ where the dissipation rapidly increases, is zero in 2-$d$.

We conclude this section by noting that disorder is not expected to change the universal behavior at the KTB transition in superconducting films at zero magnetic field. This is because, according to the Harris [35] criterion, weak disorder is irrelevant if $\nu_{\text{pure}} > 2/D$ where $D$ is the space dimension and $\nu_{\text{pure}}$ is the correlation length exponent of the pure system. At the KTB transition, the correlation length diverges exponentially, see Eq. (38), which corresponds to $\nu_{\text{pure}} = \infty$, so disorder is strongly irrelevant. We shall see, however, that disorder *does* strongly affect the transition in a magnetic field discussed below. This is because disorder acts like a random *field* on the vortex lattice and this is always relevant at the critical point [36]. For $D < 4$, disorder is also relevant at the $T = 0$ fixed point describing the vortex lattice phase and so destroys the vortex lattice [9] in two and three dimensions.

# 4 Clean Superconducting Films in a Field

In a clean sample of *bulk* type-II material, a flux lattice forms for magnetic field, $H$, between the critical fields $H_{c_1}$ and $H_{c_2}$ [1, 2]. For the case of two-dimensional thin films, the flux always penetrates and so $H_{c_1} = 0$ [37]. Furthermore, because fluctuations are more important in 2-$d$ systems than in the bulk, there is no sharp transition at the mean field $H_{c_2}$ but this is rather a crossover region where superconducting short range order starts to develop. Just below and to the left of this region, it makes sense to talk about the instantaneous positions of vortices but there is no long range order, so the system can be described as a "vortex liquid", see Fig. 2. At lower temperatures and fields the vortices form a solid triangular array. The transition in a field therefore corresponds to a melting of this "vortex solid" [37, 38].

The properties of this solid phase in two dimensions are very analogous to the properties of the superfluid films discussed above. Firstly, there is no true long-range translational order but density-density correlations decay with a power of the distance. Consequently the 2-$d$ solid is not a crystal. Nonetheless, it is a solid since there is a finite "stiffness", which in this case is just the shear modulus. As noted by Kosterlitz and Thouless [5] melting of the solid might take place by unbinding of dislocation pairs, in a very similar manner to unbinding of vortex-antivortex pairs in the superfluid transition. A major advance in the theory of 2-$d$ melting was made

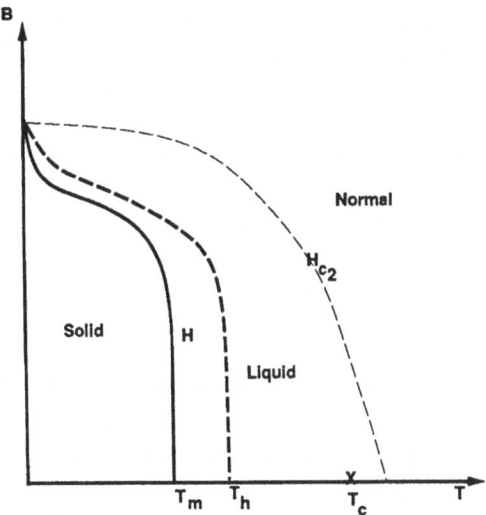

Figure 2: Phase diagram of a superconducting film in a perpendicular field within the KTBHN theory, according to ref. [37]. The phases are normal, vortex liquid, hexatic labelled H, and vortex solid. The line marked $H_{c_2}$ is the mean field transition line, which, including fluctuation effects, is no longer a sharp transition but rather a crossover *region* between the vortex liquid and normal regions. The transition from solid to hexatic is mediated by unbinding of dislocation pairs, and that from hexatic to vortex liquid is mediated by unbinding of disclination pairs. Alternatively, the transition from solid to liquid might take place through a single first order transition, at a temperature below that of the solid hexatic transition in the KTBHN theory.

by Halperin and Nelson [39], who noted that solid phase has true long-range *orientational* order and that the phase just above the dislocation unbinding transition is not a liquid, because it has power-law, rather than exponential, decay of orientational correlation functions. They called this new state a "hexatic" because it has 6-fold orientational correlations analogous to the 2-fold orientational correlations in a nematic liquid crystal. According the the KTBHN theory there is then a second KTB type transition from hexatic to liquid involving the unbinding of another type of topological defect, the disclination. In addition to the scenario just presented, where melting proceeds by two KTB transitions, it is also possible that melting proceeds by a single first order transition, direct from the solid to the liquid. Continuous melting has been observed in liquid crystal films, but simulations of atoms in 2-*d* have generally indicated a first order transition, though with much larger fluctuations than occur in three-dimensional melting. In particular, for the melting of vortices, recent Monte Carlo simulations [40] indicates a first order transition, though earlier work [41, 42] also appeared consistent with the KTBHN scenario, and it has even been claimed that that there is *no* transition [43].

Assuming that the vortex solid melting transition is of the KTBHN type, the phase diagram has been discussed in refs. [37, 38]. Figure 2 shows a sketch based on the results of Fisher [37]. The different phases are the normal state, which crosses over smoothly to the vortex liquid at around the mean field $H_{c_2}$, the hexatic phase and the vortex solid phase. The hexatic phase would be difficult to detect because its transport properties are essentially the same as those in the vortex liquid region. At the melting line there is a universal jump in a combination of elastic constants, analogous to the universal jump in $\rho_s$ in helium films. To be precise one has

$$\frac{\mu B}{k_B T_m(\mu + B)} a_0^2 = 4\pi \quad , \qquad (61)$$

where $\mu$ is the shear modulus, $B$ the bulk modulus, $T_m$ the melting temperature, and $a_0$ is the lattice spacing. For the vortex lattice, $B$ is in general much larger than $\mu$ in which case Eq. (61) shows that $\mu(T_m)/T_m$ is universal. Eq. (61) has been used [37] to determine the form of the melting line in Fig. 2. Experimental evidence for melting of the vortex solid close to the point expected from the KTBHN theory has been found [44, 45].

# 5 The Vortex Glass

As first shown by Larkin [9] disorder destroys translationsl order in the flux lattice for dimension, $D$, less than four, on scales larger than the "Larkin length", $L_P$, which depends on the strength of the disorder and tends to infinity in the clean limit. Orientational order is also destroyed by disorder [10], though the orientational correlation length can be much longer than the translational correlation length, see for example [46]. Thus there can be no hexatic or vortex solid phase in the presence of disorder.

According to conventional flux creep theory [47, 48, 34], the system is then not a superconductor because vortices will move under the action of the Lorentz force arising from the current see Eq. (55), and the flow of vortices will give a voltage, see Eq. (53). The vortices are pinned by the disorder, so vortex motion is activated, and therefore very slow at low-$T$. Nonetheless, the linear resistivity is finite at all temperatures. Furthermore, according to conventional flux creep theory, the current will affect fluctuations in the system, and hence lead to non-linearities in the current-voltage characteristics, (*i.e.* deviations from Ohms law) when $f_L L_P \sim k_B T$ where $f_L$ is the Lorentz force in Eq, (55) and $L_P$ is the Larkin length, the range of vortex lattice translational order. In other words, deviations from Ohm's law will occur for $J \simeq J_{nl}$ where $J_{nl} \propto T$.

More recently, another picture has been proposed [11]. It is argued that, even though there is no ordering at any wavevector due to the effects of frustration and disorder, the system has a stiffness associated with it such that a perturbation applied in a localized region will affect the system out to a distance, $\xi$, called the vortex glass correlation length, which is presumably something like the Larkin length, $L_P$, at high temperatures but which can be much larger at low-$T$. The vortex glass correlation length might even diverge at some temperature $T_c$, below the system is in a vortex glass state, analogous to the spin glass state in magnetic systems [12], with vanishing linear resistivity. There is both experimental [49] and theoretical [50, 51] evidence that a vortex glass transition occurs at a finite $T_c$ in three-dimensional systems. However, in two-dimensional systems, the vortex glass transition takes place at $T = 0$ [13, 14], just as in spin glasses. Nonetheless, there are experimental consequences of the $T = 0$ vortex glass transition since the correlation length diverges in this limit as $T^{-\nu}$. Hence deviations from Ohm's law set in when $f_L \xi \simeq k_B T$, which, from Eq. (55), gives the current scale for this to occur as

$$J_{nl} \propto T^{1+\nu} \quad , \tag{62}$$

different from the conventional flux creep prediction, $J_{nl} \propto T$. Recent experiments [14] on very thin (16Å) films of YBCO show that $J_{nl} \propto T^3$, implying a vortex glass correlation length exponent, $\nu \simeq 2.0$, in agreement with Monte Carlo simulations [13].

We conclude this lecture by describing how Monte Carlo simulations combined with finite-size scaling can be used to study vortex glass behavior in a simple two-dimensional model. The model that we study, known as the "gauge glass", has the following Hamiltonian:

$$\mathcal{H} = - \sum_{<i,j>} \cos(\phi_i - \phi_j - A_{ij}) \quad . \tag{63}$$

The phase, $\phi_i$, is defined on each site of a square lattice, with periodic boundary conditions. The sum is over all nearest neighbor pairs on the lattice. The effects of the magnetic field and disorder are represented by the quenched vector potentials, $A_{ij}$, which we take to be independent random variables with a uniform distribution between 0 and $2\pi$. This model seems to be the simplest one with the correct ingredients of randomness, frustration and order parameter symmetry. It does, however, ignore screening, which should be an excellent approximation in two-dimensions since the effective screening length, $\Lambda_{\text{eff}}$ is very large, as discussed above.

If the $A_{ij}$ are restricted to the values 0 and $\pi$, the model becomes the $XY$ spin glass[12], for which the lower critical dimension is believed [52] to be 4. However, earlier work [51, 13, 50], has shown that the gauge glass is in a different universality class from the $XY$ spin glass, presumably because it does not have the the "reflection" symmetry, $\phi_i \rightarrow -\phi_i \ \forall i$ [51]. As discussed in refs. [50, 13], it is useful to consider the change in free energy $\Delta F$ when one imposes a twist $\Theta$ along one of the space directions, $x$ say. More precisely, the periodic boundary conditions, $\phi_i = \phi_{i+L\hat{x}}$ are replaced by the twisted boundary conditions, $\phi_i = \phi_{i+L\hat{x}} + \Theta$. By a simple redefinition of the phases $\phi_i$ one can replace this situation by a system with periodic boundary conditions and an extra contribution, $\Theta/L$, to the vector potential on bonds in the $x$-direction.

78

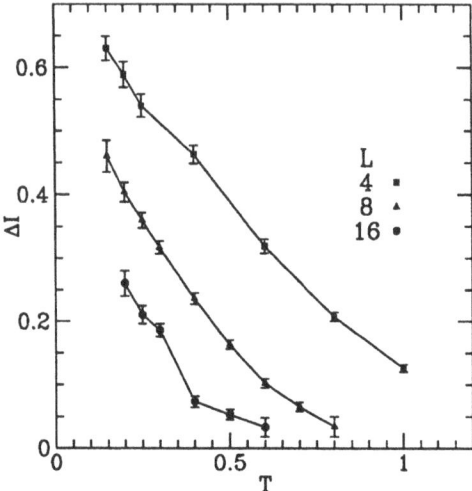

Figure 3: The r.m.s. current, $\Delta I = [I^2]_{av}^{1/2}$, for $d = 2$ determined by Monte Carlo simulations for different sizes and temperatures. The curves for different sizes do not come together, even at the lowest temperature. This behavior indicates a transition at $T = 0$.

By Monte Carlo methods one can calculate derivatives of the free energy w.r.t. $\Theta$, so, for a single sample, we define a current, $I$, and a stiffness, $Y$, by

$$I \equiv \frac{\partial F}{\partial \Theta} = \frac{1}{L} \sum_i \langle \sin \Delta_i \rangle_T \quad , \tag{64}$$

$$Y \equiv \frac{\partial^2 F}{\partial \Theta^2} = \frac{1}{L^2} \left\{ \sum_i \langle \cos \Delta_i \rangle_T - \frac{1}{T} \sum_{i,j} \left[ \langle \sin \Delta_i \sin \Delta_j \rangle_T - \langle \sin \Delta_i \rangle_T \langle \sin \Delta_j \rangle_T \right] \right\} \quad , \tag{65}$$

where $\Delta_i = \phi_i - \phi_{i+\hat{x}} - A_{i,i+\hat{x}}$, $F$ is the total free energy and $i + \hat{x}$ refers to the nearest neighbor site in the $x$-direction from $i$. Note that both $I$ and $Y$ are gauge invariant so they are still useful even if one includes fluctuating gauge fields.

Above $T_c$, $\Delta F$, and hence both $I$ and $Y$, go to zero rapidly with increasing system size because the system is insensitive to boundary conditions when $L$ is much greater than the vortex glass correlation length $\xi$. If $T_c$ is finite, then, below $T_c$, $I$ and $Y$ vary with $L$ as $L^\theta$ where $\theta$ ($> 0$), is an exponent describing the low temperature phase. In other words, $I$ and $Y$ *increase* with increasing $L$ below $T_c$, the opposite of what happens above $T_c$. Precisely at $T_c$, both $I$ and $Y$ are independent of size. Hence if $T_c$ is finite, $I$ and $Y$ should come together at $T_c$ and splay out again at lower temperatures. By contrast, if $T_c = 0$, then, at $T = 0$, $I$ and $Y$ vary as $L^\theta$ but with $\theta < 0$. Consequently, $I$ and $Y$ decrease with $L$ even at $T = 0$.

In a disordered system, it is necessary to perform an average over different realizations of the disorder, which we indicate by $[\cdots]_{av}$. For the gauge glass, the average values of $I$ and $Y$ are both zero, i.e.

$$[Y]_{av} = [I]_{av} = 0 \quad , \tag{66}$$

because the configuration in which the vector potentials in the $x$ direction have been increased by $\Theta/L$ has the same weight in the configurational average as the original choice of vector potentials. One is therefore interested in the root mean square fluctutation between samples. This means that many samples must be averaged over, typically several thousand. If $T_c$ is finite, the finite size scaling form for the r.m.s. current, $\Delta I$ is therefore

$$\Delta I \equiv [I^2]_{av}^{1/2} = \tilde{I}(L^{1/\nu}(T - T_c)) \quad (T_c > 0) \quad , \tag{67}$$

where $\nu$ is the correlation length exponent. We shall concentrate on the r.m.s. current in what follows, rather than the stiffness, because sample to sample flucutations in the stiffness have an asymmetric distribution with a long tail, which makes it difficult to get good statistics [50]. If $T_c = 0$ then $\Delta I$ decreases with size even at $T = 0$, i.e. $\Delta I \sim L^\theta$ where $\theta$ is negative and related to the exponent $\nu$ giving the divergence of the correlation length as $T \to 0$ by $-\theta = 1/\nu$. The finite size scaling form is then

$$L^{1/\nu}\Delta I = \tilde{I}(L^{1/\nu}T) \quad (T_c = 0) \quad . \tag{68}$$

Tests to ensure equilibration were carried out as described elsewhere [53].

The results for $\Delta I$ are shown in Fig. 3. Notice that they even at the lowest temperature, $\Delta I$ decreases with increasing size. This is precisely what is expected at a zero temperature transition, and the scaling plot in Fig. 4 corresponding to Eq. (68) works very well. From the fit we estimate

$$T_c = 0, \quad \nu = 2.2 \pm 0.2 \quad (d = 2) \quad . \tag{69}$$

The value for $\nu$ agrees with earlier work [13], in which a different finite-size-scaling technique was used, and with recent experiments [14] on very thin (16Å) films of YBCO, in which current-voltage characteristics were determined.

# 6   Conclusions

In these lectures we have seen that two dimensional superfluids can have a finite-temperature phase transition, controlled by thermally excited vortices, with no condensate fraction in the low-$T$ phase (*i.e.* there is no order parameter) but with a finite superfluid density. The KTB transition determines the nature of the transition in $^4$He films and very thin superconducting films in the absence of a magnetic field. It may also describe the melting of the flux lattice in clean superconducting films subjected a perpendicular field, though it is also possible that this transition may be

80

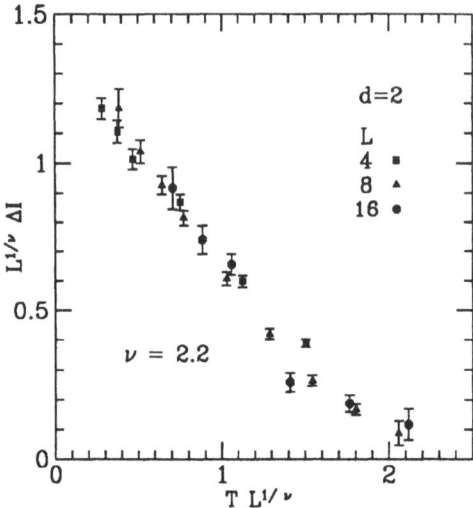

Figure 4: The same data as in Figure 3 but in a finite size scaling plot, with $T_c = 0$ and $\nu = 2.2$.

first order. Disordered superconducting films in a field show a vortex glass transition at $T = 0$, which has been seen experimentally.

**Acknowledgments:** This work was supported in part by the NSF grant No. DMR 91-11576.

# References

[1] A. A. Abrikosov, lectures at this school.

[2] A. L. Fetter and P. C. Hohenberg, in *Superconductivity*, edited by R. D. Parks (Dekker, New York, 1969), Vol. 2.

[3] V. W. Vinen, lectures at this school.

[4] V. L. Berezinskii, *Zh. Eksp. Teor. Fiz.* **61**, 1144 (1971) [*Sov. Phys. JETP* **34**, 610 (1972)].

[5] J. M. Kosterlitz and D. J. Thouless, *J. Phys. C* **5**, L124 (1972); *J. Phys. C* **6**, 1181 (1972).

[6] B. I. Halperin, in Proceedings of Kyoto Summer Institute 1979, *Physics of Low Dimensional Systems* (Publication Office, Progress of Theoretical Physics), p. 54 (1979).

[7] D. R. Nelson in *Phase Transitions*, Vol. 7, Academic Press (London), p. 1 (1983).

[8] J. E. Mooij, lectures at this school.

[9] A. I. Larkin, *Sov. Phys. JETP* **31**, 784 (1970); A. I. Larkin and Yu. N. Ovchinikov, *J. Low. Temp. Phys.* **34**, 409 (1979).

[10] J. Toner, *Phys. Rev. Lett.* **66**, 2523 (1991).

[11] M. P. A. Fisher, *Phys. Rev. Lett.* **62**, 1415 (1989).

[12] K. Binder and A. P. Young, *Rev. Mod. Phys.* **58**, 801 (1986).

[13] M. P. A. Fisher, T. A. Tokuyasu and A. P. Young, *Phys. Rev. Lett.* **66**, 2931 (1991).

[14] C. Dekker, P. J. M. Wöltgens, R. H. Koch, B. W. Hussey and A. Gupta, *Phys. Rev. Lett.* **69**, 2717 (1992).

[15] N. D. Mermin and H. Wagner, *Phys. Rev. Lett.* **17**, 1133 (1966).

[16] P. C. Hohenberg, *Phys. Rev. B* **158**, 383 (1967).

[17] J. Villain, *J. de Phys. (Paris)* **36**, 581 (1975).

[18] J. V. José, L. P. Kadanoff, S. Kirkpatrick and D. R. Nelson, *Phys. Rev. B* **16**, 1217 (1977).

[19] J. M. Kosterlitz, *J. Phys. C* **7**, 1046 (1974).

[20] A. P. Young, *J. Phys. C* **11**, L453 (1978).

[21] A. P. Young and T. Bohr, *J. Phys. C* **14**, 2713 (1981).

[22] D. R. Nelson and J. M. Kosterlitz, *Phys. Rev. Lett.* **39**, 1201 (1977).

[23] D. J. Bishop and J. Reppy, *Phys. Rev. Lett.* **40**, 1727 (1978).

[24] V. Ambegaokar, B. I. Halperin, D. R. Nelson and E. D. Siggia, *Phys. Rev. Lett.* **40**, 783 (1978); *Phys. Rev. B* **21**, 1806 (1980).

[25] S. Zhang (preprint).

[26] M. Gabay and A. Kapitulnik (preprint).

[27] B. I. Halperin and D. R. Nelson, *J. Low Temp. Phys* **36**, 599 (1979).

[28] M. R. Beasley, J. E. Mooij and T. P. Orlando, *Phys. Rev. Lett.* **42**, 1165 (1979).

[29] S. Doniach and B. A. Huberman, *Phys. Rev. Lett.* **42**, 1169 (1979).

[30] J. Pearl, *Appl. Phys. Lett.* **5**, 65 (1964); in *Low Temperature Physics, LT9*, J. G. Gaunt, D. O. Edwards, F. J. Milford and M. Yaqub, eds. (Plenum Press, New York, 1965), p.566.

[31] A. Yazdani, M. R. Hahn, W. White, P. Lerch, M. R. Beasley and A. Kapitulnik, preprint.

[32] To see why $\Lambda_T$ is so large note that it is roughly (neglecting numerical factors) $(a_B/\alpha^2)(R_H/k_BT)$ where $a_B$ is the Bohr radius, $\alpha \simeq 1/137$ is the fine structure constant, and $R_H$ is the Rydberg, *i.e.*, to sufficient accuracy, an electronic energy scale of order a few ev.

[33] L. G. Aslamazov and A. I. Larkin, *Phys. Lett.* **26A**, 238 (1968).

[34] D. S. Fisher, M. P. A. Fisher and D. A. Huse, *Phys. Rev. B* **43**, 130 (1991).

[35] A. B. Harris, *J. Phys. C* **7**, 1671 (1974).

[36] See *e.g.* A. Aharony, *Europhys. Lett.* **1**, 617 (1986) and references therein.

[37] D. S. Fisher, *Phys. Rev. B* **22**, 1190 (1980).

[38] B. A. Huberman and S. Doniach, *Phys. Rev. Lett.* **43**, 950 (1979).

[39] B. I. Halperin and D. R. Nelson, *Phys. Rev. Lett.* **41**, 121 (1978); **41**, 519(E); and D. R. Nelson and B. I. Halperin, *Phys. Rev. B* **19**, 2457 (1979).

[40] Y. Kato and N. Nagaosa, *Phys. Rev. B* **47**, 2932 (1993); preprint.

[41] Z. Tešanović, *Phys. Rev. B* **44**, 12635 (1991); Z. Tešanović and L. Ting, *Phys. Rev. Lett.* **67**, 2729 (1991).

[42] S. Hikami, A. Fujita and A. I. Larkin, *Phys. Rev. B* **44**, 10400 (1991).

[43] J. A. O'Neill and M. A. Moore, *Phys. Rev. Lett.* **69**, 2582 (1992); preprint (1992); N. K. Wilkin and M. A. Moore, *Phys. Rev. B* **47**, 957 (1993).

[44] P. Berghuis, A. L. F. van der Slot and P. H. Kes, *Phys. Rev. Lett.* **65**, 2583 (1990).

[45] A. Yazdani, W. R. White, M. R. Hahn, M. Gabay, M. R. Beasley and A. Kapitulnik, *Phys. Rev. Lett.* **70**, 505 (1993).

[46] D. J. Bishop, lectures at this school.

[47] P. W. Anderson and Y. B. Kim, *Rev. Mod. Phys.* **36**, 39 (1064).

[48] Y. B. Kim and M. J. Stephen, in *Superconductivity*, edited by R. D. Parks (Dekker, New York, 1969), Vol. 2, Chapter 19.

[49] R. H. Koch, V. Foglietti, W. J. Gallagher, G. Koren, A. Gupta and M. P. A. Fisher, *Phys. Rev. Lett.* **63**, 1511 (1989); P. L. Gammel, L. F. Schneemener and D. J. Bishop, *Phys. Rev. Lett.* **66**, 953 (1991); H. K. Olsson, R. H. Koch, W. Eidelloth and R. P. Robertazzi, *Phys. Rev. Lett.* **66**, 2661 (1991).

[50] J. D. Reger, T. A. Tokuyasu, A. P. Young and M. P. A. Fisher, *Phys. Rev. B* **44**, 7147 (1991).

[51] D. A. Huse and H. S. Seung, *Phys. Rev. B* **42**, 1059 (1990).

[52] S. Jain and A. P. Young, *J. Phys. C* **19**, 3913 (1986); B. W. Morris, S. G. Colborne, M. A. Moore, A. J. Bray and J. Canisius, *J. Phys. C* **19**, 1157 (1986); J. R. Banavar and M. Cieplak, *Phys. Rev. Lett.* **48**, 832 (1982).

[53] R. N. Bhatt and A. P. Young, *Phys. Rev. B* **37**, 5606 (1988); R. N. Bhatt and A. P. Young , *Phys. Rev. Lett.* **54**, 924 (1985).

# SHORT COHERENCE LENGTH SUPERCONDUCTORS

G. DEUTSCHER

*School of Physics and Astronomy*

*Tel Aviv University*

*Ramat Aviv, Tel Aviv*

*Israel*

ABSTRACT. We first briefly review the phenomenological Ginzburg Landau and the microscopic BCS concepts of the coherence length. We compare the BCS and the experimental values, as determined from measurements of the nucleation field. It appears that the BCS expression is in excellent agreement with experiment for the low Tc superconductors, but that for all known superconducting compounds having a critical temperature higher than 10K, the coherence length is anomalously short. This is also the case for Heavy Fermion superconductors, which suggests the existence of strong electron correlation effects in the oxides. The screening problem in short coherence length superconductors is discussed. The special properties due to and the problems posed by the short coherence length are introduced.

## 1. The Ginzburg Landau coherence length.

### 1.1 THE GINZBURG LANDAU FUNCTIONAL

While the London penetration depth $\lambda_L$ was already introduced in the 30's and gave a quantitative explanation to the Meissner effect, the second characteristic length scale of the superconducting state – called somewhat unfortunately the coherence length – was introduced only in the early 50's by Ginzburg and Landau (GL) in a phenomenological way. While in the London theory the "superconducting wave function is rigid" in the GL theory the behavior of the complex order parameter $\psi(r)$ is determined by the minimisation of the free energy using the GL density functional:

$$F_s = F_n + a|\psi|^2 + \frac{b}{2}|\psi|^4 + \frac{1}{2m}|(-i\hbar\nabla - \frac{2\,eA}{c})\psi|^2 + \frac{h^2}{8\pi} \quad (1)$$

where $a = \bar{a}(t-1)$, t being the reduced temperature $t = (T/Tc)$, b is a

85

*N. Bontemps et al. (eds.), The Vortex State, 85–97.*

constant, A is the vector potential and h is the local magnetic field
at point r.

Minimisation of $\int F(r)dr$ with respect to a variation $\delta\psi(r)$ of the GL
order parameter leads to the well known equation for the order
parameter:

$$a|\psi| + b|\psi|^2\psi + \frac{1}{2m} (-i\hbar\nabla - \frac{2e}{c}A)^2\psi = 0 \qquad (2)$$

and to the introduction of the new length scale $\xi$ defined by:

$$\xi^2 = \frac{\hbar^2}{2m|a|} \qquad (3)$$

$\xi(t)$ is the natural length scale for the spatial variation of $\psi(r)$.
It should really have been called correlation length rather than
coherence length, the coherence of the superconducting state having
more to do with the phase than with the amplitude of the order
parameter.

## 1.2 EXAMPLES OF SITUATIONS WHERE THE ORDER PARAMETER IS POSITION DEPENDENT

1.2.1. *Proximity effect.* To illustrate the meaning of $\xi$ we set in one
dimension $\psi(x)=0$ at point x=0. The solution of Eq.2 is then:

$$\psi(x) = \psi_0 \tanh\frac{|x|}{\sqrt{2}\xi(t)} \qquad (4)$$

where $\psi_0$ is the equilibrium value of $\psi$ ($\psi_0^2=-(a/b)$).

At a boundary between an insulator and a superconductor, GL assume:

$$\frac{d\psi}{dx} = 0 \qquad (5)$$

while at a boundary with a normal metal one must use the more general
de Gennes condition (1):

$$\frac{1}{\psi}\left|\frac{d\psi}{dx}\right| = \frac{1}{b} \qquad (6)$$

where b is called the "extrapolation length" for the order parameter.

Above Tc, Eq.3 defines a "decay length" for $\psi$, i.e. if we set $\psi(x=0)$
finite, the solution for $\psi(x)$ (neglecting the $|\psi|^2\psi$ in Eq.2) is given

by :

$$\psi(x) = \psi(x=0)\exp{-\frac{|x|}{\xi(t)}}$$ (7)

At T > Tc the condition $\psi(x=0)$ finite can be achieved if x=0 is an interface with a (higher critical temperature) superconductor S'. Because in an S'/S/S' sandwich the transverse critical current Ic is directly related to the smallest value of $\psi$ in the system, proximity effect experiments where Ic is measured as a function of the thickness d of S can be used to obtain $\xi$:

$$Ic = Ico.\exp(-d/2\xi)$$ (8)

1.2.2. *Nucleation field.* The most common way to determine $\xi$ is through a measurement of the nucleation field:

$$H_n = \frac{\phi_o}{2\pi\xi^2}$$ (9)

$H_n$ is a solution of Eq.2 in its linearised form, assuming that the field is uniform in the sample. These conditions are fulfilled when the nucleation occurs, since $\psi$ is then infinitely small. As is well kmowm, if $\kappa = (\lambda/\xi) > 1/\sqrt{2}$, $H_n$ is the upper critical field $H_{c2}$. If $\kappa < 1/\sqrt{2}$, $H_n$ is the supercooling field $H_{sc}$. Eq.9 holds when the magnetic field is applied perpendicular to the surface of the sample. In the parallel orientation, because of the boundary condition Eq.5:

$$H_n = H_{c3} = 1.69\frac{\phi_o}{2\pi\xi^2}$$ (10)

In that case, nucleation of the superconducting state starts at the surface of the specimen over a depth $\xi(t)$.

Eq.1 through 10 summarize the main GL results relative to the coherence length that we shall need in this lecture.

## 2. The BCS coherence length

As introduced by GL, $\xi$ is a purely phenomenological length scale . Eq.3 can be rewritten as:

$$\xi(t) = \xi(0).(1-t)^{-1/2} \qquad (11)$$

The value of $\xi(0)$ is deeply related to the structure of the superconducting state. The superconducting wave function is constructed from states lying within an energy range $(-\Delta, +\Delta)$ around the Fermi level. Hence, they comprise a range of wave vectors:

$$\delta k \cong \left(\frac{d\varepsilon}{dk}\right)^{-1}_{\varepsilon=0} \cdot \Delta \qquad (12)$$

where $\varepsilon$ is the energy measured from the Fermi level. Since the Fermi velocity is given by:

$$v_F = \frac{1}{\hbar}\left(\frac{d\varepsilon}{dk}\right)_{\varepsilon=0} \qquad (13)$$

we have:

$$\delta k \cong \frac{\Delta}{\hbar v_F} \qquad (14)$$

This relation defines a fundamental length scale $\delta k^{-1}$. The exact BCS relation for the coherence length is:

$$\xi_o = \frac{\hbar v_F}{\pi \Delta} \qquad (15)$$

This relation is valid at T=0 and for a clean superconductor. The BCS and the GL coherence lengths are related by:

$$\xi(t) = 0,74\xi_o(1-t)^{-1/2} \qquad (16)$$

in the clean limit and:

$$\xi(t) = 0.85(\xi_o l)^{1/2}(1-t)^{-1/2} \qquad (17)$$

in the dirty limit $l < \xi_o$, where $l$ is the normal state mean free path.

Eq.15 is valid for a spherical Fermi surface. For an anisotropic

Fermi surface, $v_F$ must be appropriately averaged. For instance, in the case of a van Hove singularity as may be relevant to the high $T_c$ oxides (2), the Fermi velocity goes to zero at the singular points. For this case, Force has calculated that the effective velocity to be used is half its maximum value (3).

## 2.1. EFFECT ON THE COHERENCE LENGTH OF MANY BODY INTERACTIONS

The value of $v_F$ to be used in Eq.16 is the quasiparticle Fermi velocity, as renormalized by many body interactions. These include the electron-electron and electron-phonon interactions. $v_F$ is given by (4):

$$v_F = \bar{v}_F \cdot z \tag{18}$$

where z is the mass renormalisation factor and $\bar{v}_F$ is an effective Fermi velocity that differs from that of the bare Bloch electron $v_{FB}$ if the self energy correction $\Sigma$ depends on k, and not only on the frequency $\omega$. If $(\partial\Sigma/\partial k)=0$ and if the electron-electron interaction can be neglected, $v_F=v_{FB} \cdot z$ with $z=(1+\lambda_{ep})^{-1}$ where $\lambda_{ep}$ is the McMillan electron-phonon interaction parameter. The BCS coherence length is then given by:

$$\xi_o = \frac{\hbar v_{FB}}{\pi(1+\lambda_{ep})\Delta} \tag{19}$$

where $v_{FB}$ is the band value.

## 2.2. COMPARISON BETWEEN THE BCS AND EXPERIMENTAL VALUES OF THE COHERENCE LENGTH

For usual metals with a broad band, $v_{FB}$ is roughly a constant, i.e. it does not vary much in k space. Its absolute value does not vary much either from one metal to another, since in a free electron model it changes only as the cubic root of the electron density. Hence, in the weak coupling limit $\lambda_{ep}\to 0$, where the ratio of $\Delta$ to Tc is a constant $(2\Delta=3.5k_B Tc)$ we expect that for simple metals the product $\xi_o \cdot Tc$ should be roughly a constant. Because of the intimate link between $\xi_o$ and the structure of the superconducting state, a test of this BCS prediction is of interest.

As seen in Table I, $\xi_o \cdot T_c$ is indeed remarkably constant for conventional low Tc superconductors, for which it converges towards a value close to 1.5 μm.K in the limit $T_c$ ->0. This value is in good agreement with $v_{FB} \cong 1.10^8$ cm/sec as expected from free electron theory for a typical electronic density of the order of 5 $10^{22}$/cm$^3$.

We also note a trend towards lower values of $\xi_o$.Tc for higher Tc's. This trend is expected within the framework of the BCS theory: i)higher Tc superconductors tend to have a larger electron-phonon coupling; ii)for strong coupling superconductors $2\Delta > 3,5 k_B T_c$ and iii)stronger electron-phonon coupling may result from a higher density of states. According to the weak coupling BCS expression for Tc:

$$k_B Tc = 1.13 \hbar \omega_D \exp(-1/\lambda) \qquad (20)$$

where $\omega_D$ is the Debye frequency and $\lambda = N(0)V$, $N(0)$ being the density of atates at the Fermi level and V the net attractive electron-electron interaction. Neglecting the repulsive Coulomb interaction, a point to which we come back later, $V=V_{ep}$. A higher $N(0)$, which enhances $\lambda$ and hence Tc (if V stays constant, which is not necessarily the case), would normally be associated with a narrow band. This implies a lower $v_{FB}$, hence a shorter $\xi_o$.

Thus the gemeral trend towards lower $\xi_o$.Tc products at higher Tc's is expected within the BCS theory. Actually up to Pb (Tc = 7.2K) the quantitative agreement with Eq.19 is impressive. Using known values for $v_{FB}$ (5), $\Delta$ and $\xi_o$ (measured), we calculate $\lambda_{ep} = 1.3$, as obtained from other determinations of the mass renormalisation factor (such as low temperature heat capaacity measurements (5), and tunneling measurements (6)). In effect, Eq.19 is a simple and effective way to obtain $\lambda_{ep}$ for conventional low Tc superconductors.

## 2.3. STRONG CORRELATION EFFECTS

However, when we look up Table I at higher Tc compounds (Tc > 10K), we immediatly see that the smooth and monotonous correlation between Tc and $\xi_o$.Tc breaks down. All of these compounds are characterized by anomalously small $\xi_o$.Tc products; we call them "Short Coherence Length Superconductors". In some of them such as the A15 compounds, the Fermi level does fall in a narrow band with a high density of states, giving a smaller Fermi velocity than in usual metals. For them (7), as well as for the high $T_c$ oxides(2), van Hove singularity effects should be taken into account. Yet, unrealistically high values of $\lambda_{ep}$ would still be necessary to explain the small $\xi_o$.$T_c$ values. How can we understand them?

A clue to the answer is the very small value of $\xi_o$.$T_c$ seen in Heavy

TABLE I

| | Tc (K) | $\xi$ ($\mu$m) | $\xi$Tc ($\mu$m.K) |
|---|---|---|---|
| Aluminium (1) | 1,19 | 1,20 | 1,4 |
| Indium (1) | 3,40 | 0,33 | 1,1 |
| Tin (1) | 3,72 | 0,26 | 0,97 |
| Gallium (1) | 5,90 | 0,16 | 0,94 |
| Lead (1) | 7,20 | 0,080 | 0,58 |
| Niobium (1) | 9,25 | 0,035 | 0,32 |
| PbMoS$_8$ (2) | 15 | 0,0025 | 0,04 |
| Nb$_3$Sn (1) | 17 | 0,0040 | 0,07 |
| C$_{60}$K$_3$ (3) | 19 | 0,0030 | 0,06 |
| C$_{60}$Rb$_3$ (3) | 31 | 0,0023 | 0,07 |
| Y$_{0.6}$Pr$_{0.4}$Ba$_2$Cu$_3$O$_7$ (4) | 40 | 0,007 | 0,3 |
| YBa$_2$Cu$_3$O$_7$ (1) | 93 | 0.0015 | 0.14 |

(1)     G. Deutscher, in "Concise Encyclopedia of Magnetic and Superconducting Materials" Ed Jan Evetts, Pergamon Press

(2)     J. Cors, PhD thesis, University of Geneva, 1990

(3)     see Ref. 3

(4)     see Ref. 7

Fermion (low $T_c$) superconductors. For instance in UPt$_3$ ($T_c$=0.5, $\xi_o \cong 100$A) it is 200 times smaller than in Al! Clearly, Eq. 19 is not applicable to this compound. This does not come as a surprise. As is well known, mass renormalisation in Heavy Fermions is not due primarily to the electron-phonon interaction, but to strong electron correlation effects. In general, $z^{-1} = \lambda_{ep} + \lambda_c$ where $\lambda_c$ describes the renormalisation due to electron-electron interactions. Table I suggests that $\lambda_c$ can be neglected (as done in Eq. 19) in the low $T_c$ elements, but not in the higher $T_c$ compounds. In that sense, they are also "Heavy Fermions".

## 3. THE SCREENING PROBLEM IN SHORT COHERENCE LENGTH SUPERCONDUCTORS

Morel and Anderson pointed out many years ago (8) that a high DOS in a narrow band (small $v_F$) cannot by itself give a high Tc, contrary to what the simple BCS expression Eq. 20 would indicate. This is because not only N(0) but also the effective strength of the screened repulsive Coulomb interaction $\mu^*$ are functions of the bandwidth:

$$\mu^* = \frac{\mu}{1 + \mu \ln(\frac{E_F}{\hbar\omega_D})} \qquad (21)$$

where $\mu$ is the unscreened Coulomb interaction. In a narrow band, $E_F \cong \hbar\omega_D$, $\mu^* \cong \mu$ and the effective interaction parameter $(\lambda_{ep}-\mu^*)$ will in fact be reduced (typically $\lambda \cong \mu$). Thus Tc will go down instead of going up. Note that according to this argument, Tc should go through a maximum as a function of $\xi_0$. Tc. Anderson argued that this maximum should be of the order of 10K, and indeed this is the impression one gets from Table I if one limits oneself to the pure elements.

The short coherence length in the cuprates raises this screening problem. Effective screening clearly requires:

$$\xi_0 > k_F^{-1} \qquad (22)$$

or by using the BCS expression Eq. 15:

$$\frac{mv_F^2}{\pi\Delta} > 1 \qquad (23)$$

but for practical $\Delta$ values, $\Delta \leq \hbar\omega_D$, hence condition (23) can only be fulfilled if:

$$\frac{mv_F^2}{\pi\hbar\omega_D} > 1 \qquad (24)$$

which is essentially the condition for $\mu^* <\mu$ according to Eq. (21).

In the cuprates, $k_F \cong (\pi/2a)$, $k_F^{-1} \cong 3A$ and with $\xi_{ab} \cong 15A$, we get $\xi k_F \cong 5$. This implies $(2E_F/\pi\Delta) \cong 5$, and with the value for YBCO $\Delta=20$meV, we obtain $E_F \cong 150$meV. This is only about 3 times the typical phonon frequency, which is insufficient to obtain a small $\mu^*$.

The short $\xi$ thus seems incompatible with the high Tc, at least within the standard framework of the phonon mediated theory of superconductivity.

There are two possible approaches to this problem. One is to conclude that superconductivity in the cuprates is not phonon mediated, but is due to an electronic interaction (such as antiferromagnetic fluctuations). Screening then becomes unecessary.

Another approach, proposed by Bok and Force (9) is based on a DOS that comprises both a narrow and a broad band, with the Fermi level falling in the narrow band (such a situation is realised for instance when the Fermi level falls at or near a van Hove singularity). The idea is then that the effective Fermi velocity that determines the coherence legth is dominated by the contribution of the narrow small

velocity band (hence the small effective $E_F$ calculated above), while the broad band background provides the screening.
    Which approach is the correct one remains to be established.


## 4. Experimental determination of the coherence length in the HTSC

### 4.1. STANDARD METHODS

The standard way to obtain $\xi(t)$ is through a measurement of $H_{c2}$ Eq.9. In principle, $H_{c2}$ can be determined in many different ways: magnetization measurement M(H), heat capacity C(H), density of states measurements N(E,H), resistivity $\rho$(H). All of these have been used successfully in the LTSC with excellent consistency and good agreement with theory:

Magnetization:

$$M = -\frac{1}{4\pi} \frac{H_{c2}-H}{(2\kappa^2-1)\beta_A} \qquad (25)$$

where $\beta_A=1.16$.


Heat capacity jump at $H_{c2}$:

$$\Delta C = \frac{T}{4\pi} (\frac{dH_{c2}}{dT})^2 [(2\kappa^2-1)\beta_A]^{-1} \qquad (26)$$

    The density of states is gapless and returns progressively to that in the normal state linearly, similarly to M(H).

    In the pure flux flow regime the resistivity returns linearly to its normal state value $\rho_n$:

$$\rho(B) = \frac{B}{H_{c2}} \rho_n \qquad (27)$$

    In the (common for low $T_c$ alloys) case of strong pinning effective up to $H_{c2}$, the resitivity goes from zero to $\rho_n$ in a narrow range of fields near $H_{c2}$.

None of the above methods really works well for the HTSC. Because $\kappa$ is very large ($\kappa \cong 100$), $(dM/dH)$ is very small near $H_{c2}$, and difficult to distinguish from background effects. $C(H)$ does not show a jump at $\overline{H_{c2}}$. Good DOS measurements are not yet available on the HTSC, but would in any case have the same problem as $M(H)$. And finally the resistive transition is considerably broadened under applied fields. These difficulties are compounded by the large $(dH_{c2}/dT)$ of the order of of $-2T/K$ so that using standard superconducting magnets $H_{c2}$ can only be measured within 5 to 10K from Tc. Extremely high quality samples are then necessary to distinguish between intrinsic and extrinsic broadening effects (the transition width must be well below 1K).

## 4.2. THE RESISTIVE TRANSITION

Let us first make a simple remark concerning the resistive transition. If we neglect all fluctuation effects, and since $B \cong H$ at high fields (of order $H_{c2}$), according to Eq.27 the middle of the resistive transition occurs at $H \cong (H_{c2}/2)$. Any pinning will tend to make the transition narrower. Hence, at a given temperature:

$$H_{c2} < 2H(\rho = \rho_n/2) \qquad (28)$$

In the case of a so called "ideal" resistive transition with pinning remaining effective up to $H_{c2}$ (such as is commonly the case in the LTSC), the mid-point of the transition is a fair estimate of $H_{c2}$. Thus as rough bounds for $H_{c2}$ we can use:

$$H(\rho = \rho_n/2) < H_{c2} < 2H(\rho = \rho_n/2) \qquad (29)$$

Ullah and Dorsey (10) have proposed an extension of the Doniach model to high magnetic fields to calculate the effect of thermodynamic fluctuations on the resistive transition, neglecting all pinning effects. The basic idea is that in a strong applied field fluctuations become one dimensional in nature, and therefore quite large, because the basic transverse length scale:

$$L = (\phi_o/H)^{1/2} \qquad (30)$$

becomes small. The proposed expression for the fluctuation conductivity - i.e. the actual conductivity at temperature T and applied field H minus the normal state conductivity at the same temperature- is:

$$\sigma \propto (T^2/H^{1/3}) \ F \ [A(\frac{T-Tc(H)}{(TH)^{2/3}})] \qquad (31)$$

where F is a scaling function and A a constant.

Data obtained on high quality single crystals and thin films (11)(transition widths of less than 1K) have been successfully fitted to Eq.31, taking Tc(H) as the adjustable parameter. Tc(H) is the mean field critical temperature corresponding to the applied field H, i.e. the temperature at which superconductivity would nucleate in the absence of fluctuation effects or in other words the temperature at which the applied field is equal to $H_{c2}(T)$.

Results on YBCO single crystals and thin films (6)give a linear temperature dependence for $H_{c2}(T)$, with a slope $(dH_{c2}/dT) = -1.9T/K$ for the field applied along the c axis and -10T/K for the field applied along the (ab) planes (thin film values). These slopes are in good agreement with values obtained from magnetization measurements on single crystals. Racah et al.(11) also remark that the values obtained from the fit to the fluctuation theory are close to the mid-points of the resistive transitions, which also follow a linear temperature dependence. This linear dependence is only seen in high quality samples. For films with a transition width somewhat larger than 1K, the mid-point field does not vary linearly with temperature, but with some larger power of $(T_c-T)$. Such a behavior should not be considered as an intrinsic property ao the samples.

## 4.3. THE COHERENCE LENGTH IN DOPED SAMPLES

Using the clean limit expression for $\xi(t)$ Eq.16, one calculates from the above determinations of $H_{c2}$: $\xi_{ab}=18A$, $\xi_c=3,5A$. The dirty limit expression would give $\xi_{ab}=15A$ and $\xi_c=3A$, in round figures. In fact, the correct temperature dependence of $H_{c2}(t)$ is not known, thus some uncertainty is attached to the above values.

These small values of $\xi$ have provided the basis for the above discussion of the coherence length. Some significantly larger values have been obtained in Pr doped YBCO samples. For the $Pr_{0.4}Y_{0.6}$ composition, (Tc=40K) it has been reported that the resistive transition has a more conventional shape, i.e. the transition is shifted rather than broadened by an applied magnetic field. Taking then the mid point of the transition as $H_{c2}$, Racah et al.(12) get $\xi_{ab}=50-60A$, $\xi_c=10-12A$.

This result is quite interesting. Because the Pr doping results in a significantly lower conductivity, one would have expected the alloy to be in the dirty limit and to have a **shorter** $\xi$, not a larger one (Eq. 17). The increase of $\xi$ by a factor of about 4 is even larger than what one calculates in the clean limit, considering the decrease in Tc by a factor of 2.3. As seen in Table I, $\xi_o . T_c$ is now closer to that of the low $T_c$ superconductors.

This return to a more conventional behavior may be due to band effects, as proposed by Racah et al.(12) (for instance removal of the Fermi level from the van Hove peak in the DOS due to doping), or to a reduction in the mass renormalisation factor (reduction of the electron-phonon and/or electron-electron interaction). Additional experiments (for instance tunneling experiments) are necessary to fully understand this interesting behavior.

## 5. Conclusions

We have in this lecture concentrated our attention on the origin of the short coherence length, rather than on its consequences for the behavior of the HTSC, which is the main topic of this School. We have only discussed the in plane coherence length, leaving out the question of its anisotropy, for which we have really no microscopic model at the momemt.

The coherence length and its anisotropy provide the basis for a Ginzburg Landau description of the oxides. Special properties of the oxides involving the short $\xi$ and its large anisotropy include large thermodynamic fluctuations of the order parameter, weak pinning and vortex decoupling. They also include special boundary conditions. The Ginzburg Landau theory and its extensions seems to be reasonably successfull in explaining the observed critical fluctuations, broadened transtions under applied fields, and vortex pinning behavior as described in detail during at this School. Yet, we have not discussed the possible influence of strong correlation effects on these properties. Such a discussion might turn out in the future to be necessary: as we have seen, the short coherence length probably implies that the High $T_c$ oxides have some Heavy Fermion character.

I wish to acknowledge enlightening remarks by Philippe Nozieres during the final redaction of this lecture. This work has been supported in part by NEDO and by the Oren Family Chair of Experimental Solid State Physics.

## References

1) P. G. de Gennes, Rev. Mod. Phys. 36, 225(1964).
2) J. Labbe and J. Bok, Europhys. Lett. 3, 1225(1987)
3) L. Force, PhD Thesis, Paris 1993.
4) P. Nozieres and G. Deutscher, preprint.
5) C. Kittel, Introduction to Solid State Physics, John Wiley and Sons, New York (1971).
6) W. L. McMillan and J. M. Rowell, in "Superconductivity", ed. Parks, Marcel Dekker (New York, 1969), p. 561.
7) J. Labbe and J. Friedel, J. Phys. et le Radium 27, 153(1966).

8)P.Morel and P.W.Anderson, Phys.Rev.125,1263(1962).
9)J.Bok and L.Force, Solid State Comm.85,11(1993).
10)S Ullah and A.Dorsey, Phys.Rev.B44,262(1991)
11)D.Racah et al., preprint.
12)D.Racah, U.Dai and G.Deutscher, Phys.Rev.B46,14915(1992)

# Magnetic Decoration Studies of Flux Line Lattices in the Cuprate Superconductors

David J. Bishop, Peter L. Gammel and Cherry A. Murray
*AT&T Bell Laboratories*
*Murray Hill, New Jersey 07974*

**Abstract:** In this article we will discuss a variety of experiments in which the magnetic decoration technique is used to image the magnetic flux line lattice in the oxide superconductors. We will first briefly describe the technique and then various studies we have undertaken over the last six years using magnetic decoration to image the flux lattices. We will describe measurements of the magnetic flux quantum, the observation of hexatic order, the order-disorder transition, vortex chains in YBCO and BSCCO, tilt-induced hexatic order in $NbSe_2$, vortex lattices near sawtooth twins, twin boundary pinning, and thermal fluctuations of the vortices as the temperature is raised.

## Introduction

The behavior of superconductors in the presence of a magnetic field has been the subject of much scientific as well as practical interest over the past few decades. As has been emphasized in other parts of this book, superconductors can be divided into two classes: Type I superconductors which do not remain superconducting after an applied field penetrates the sample and Type II superconductors which do remain superconducting until the applied field exceeds a critical value for the destruction of superconductivity. The pioneering work of Abrikosov has shown us how this happens.

As reviewed by Abrikosov elsewhere in this book, for Type II superconductors, when the applied magnetic field exceeds the lower critical field, $H_{c1}$ but is below the upper critical field, $H_{c2}$, the field penetrates the sample in the form of quantized flux lines, each carrying exactly one quantum of flux, $\phi_0 = hc/2e$. In the absence of disorder, these lines of magnetic flux form an hexagonal array in the sample. The phase diagram for a conventional Type II superconductor is shown in Fig. 1a). At low fields and low temperatures, the field is expelled from the superconductor, the Meissner state. At intermediate fields, the magnetic field

99

*N. Bontemps et al. (eds.), The Vortex State*, 99–123.
© 1994 *Kluwer Academic Publishers.*

penetrates and forms a magnetic flux line lattice. This is the mixed state and imaging this lattice is the subject of this article. Finally at high fields and temperatures the superconducting state is destroyed and the normal state is obtained.

Soon after the discovery of the oxide superconductors, it became clear that this phase diagram needed to be modified. It is now clear that because of the increased importance of thermal fluctuations in these materials that there is a large vortex liquid regime and that the phase diagram needs to be modified as shown in Fig. 1b). The thermal fluctuations in these materials are much larger than for conventional Type II superconductors because of the high transition temperatures, short coherence lengths, long penetration lengths and the strong anisotropies[1]. These same effects also produce a number of novel static flux line lattice structures. The study of these using the magnetic decoration technique is the subject of this article.

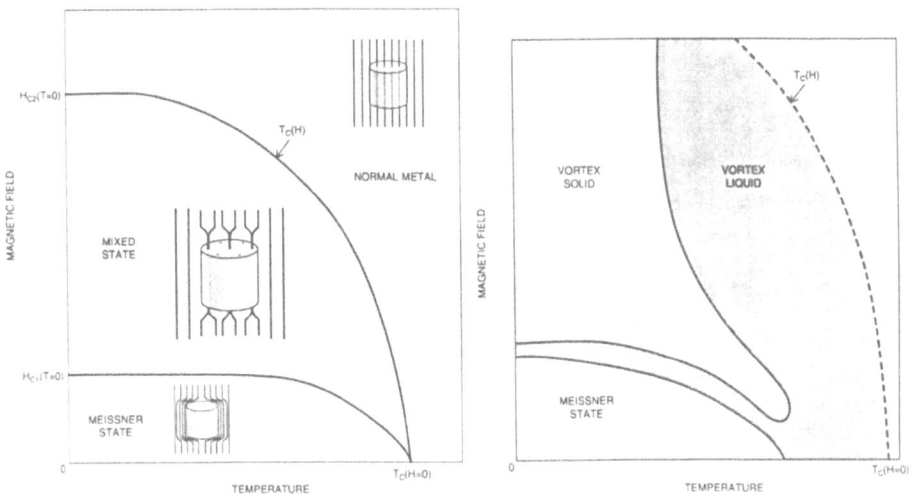

**Fig.1** (left) The phase diagram is shown for conventional type two superconductors. (right) The modified phase diagram is shown for the high Tc superconductors. The differences are due to the increased importance of thermal fluctuations.

## The Magnetic Decoration Technique

Direct information on the ordering of the magnetic vortices in the mixed state of the high Tc superconductors can be obtained through the use of the Bitter imaging technique in which samples are cooled in an applied magnetic field and subsequently exposed to a smoke of ferromagnetic particles formed by evaporation into a helium buffer gas. The apparatus is shown schematically in Fig. 2. The technique was pioneered by Trauble and Essmann[2], and Sarma[3] to study individual vortices. The ferromagnetic particles travel down magnetic field lines outside the surface of the superconducting sample and form clusters on the surface, which decorate the locations of the vortices. The particles stick to the surface with van der Waals forces. The applied field is then removed, the sample is warmed to room temperature, and the clusters of particles are viewed with an electron microscope. The van der Waals forces are sufficiently strong that the magnetic particles do not move after they become stuck onto the surface. These forces act like "atomic glue" for the magnetic particles.

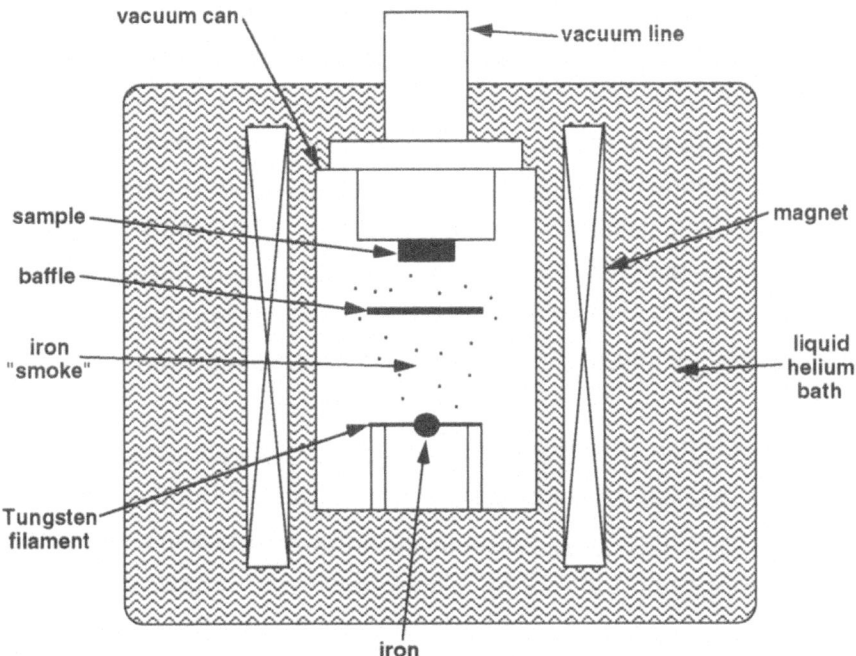

**Fig. 2** Shown is a schematic diagram of the magnetic decoration apparatus.

Direct, real-space imaging of the arrangement of individual vortices provides information on both the translational and bond orientational order of a two-dimensional slice of the vortex lattice as it pierces the sample surface. Present experiments have been limited for the most part to field-cooled samples at T=4.2K, subsequently viewed by scanning electron microscopy, for which sufficient contrast is obtained when vortices are separated by roughly $a_o > 0.3$ μm (H < ~200 G). Below this separation (or above this field), the ferromagnetic particles have a tendency to form strings by dipole-dipole interactions and the decorated images become difficult to interpret. One must take into account the demagnetizing factor of the sample to determine the actual magnetic field in the bulk of the sample. Most of the samples studied to date are thin slabs of ~ l mm extent along the a, b axes and ~5 to 30 μm thickness along c. The a and b axes span the copper oxide layers of these materials, while c is normal to the layers. For H along c, vortices penetrate the sample at H ~ 0.5G, rather than the measured $H_{c1}$ of ~150G, owing to this demagnetizing effect. The Bitter decoration technique is limited to a static snapshot of the vortex lattice arrangement in the sample averaged over the time required to decorate—about 1 second. Also, because there is pinning and possibly entanglement of the vortices, the actual temperature during the sample cool-down at which the microscopic arrangement of vortices goes out of equilibrium and freezes into the resulting Bitter pattern is currently unknown. This temperature must lie somewhere between the temperature at which the bulk DC magnetization goes out of equilibrium, $T_{irr}$, which for these low fields is quite close to Tc, and the lowest temperature obtained in the decoration experiment, which is 4.2 K.

## Static Flux-Lattice Structures for Low Temperatures and HIIc

The Bitter decoration experiments have established the following points about the mixed state of the high Tc superconductors for H parallel to c: (i) the vortices exist as hexagonally correlated, singly quantized vortices with one flux quantum, hc/2e, per vortex[4] as in a conventional Type II superconductor; (ii) the vortices undergo pinning[5] at twin boundaries, at crystal defects, and individually at other intrinsic lattice sites in ostensibly defect-free regions; (iii) in twin-free samples, instead of the long-range translational order expected in the crystalline state of the Abrikosov vortex lattice, the vortex arrangement has short-range translational order with correlation lengths on the order of a few nearest neighbor spacings, but long-range bond-orientational (hexatic) order[6] that extends over ten to several hundred nearest neighbor spacings; (iv) a rather sharp transition with applied field is observed in BSCCO between isotropic disorder in the vortex arrangement for H < 20 G and hexatic order at higher fields[7]; this transition occurs at a considerably lower field (H ~ 8 G) for samples that have been annealed in oxygen; for both types of samples the translational and bond orientational order in the hexatic arrangement increases monotonically with field up to 100 G; (v) motion of individual vortices comparable to their separation within the one second decoration time appears to occur[8] at 15 K in BSCCO, presumably because of thermal motion of the vortices; (vi) the vortex lattice exhibits[9] the expected ~20% anisotropy from a perfect hexagonal structure in the a-b plane in YBCO due to the in-plane effective mass anisotropy of the electrons but a smaller anisotropy of ~3-10% rather than the nearly 50% expected in BSCCO from a-b mass anisotropy[6,7]; (vii) twins in YBCO pin[5, 10] the vortex lattice such that the flux lattice tends to run parallel to the twin; (viii) for vortices near sawtooth twins in YBCO one finds evidence[11] for magnetic pinning of vortices; (ix) for H not parallel to c, a variety of novel structures have been observed, including oval vortices[9] and flux-line chains of several types[12,13]; and finally (x) for vortices in $NbSe_2$ we find evidence for tilt induced orientational order[14] as found for smectic liquid crystals.

The advantage of using BSCCO in the decoration experiments is that excellent quality, untwinned single crystals can be obtained that can be cleaved to expose a clean surface layer for decoration. In addition, the Ginsburg-Landau parameter $\kappa = \lambda/\xi \sim 200$ and the a-c carrier mass anisotropy parameter $\Gamma \sim 55$ for BSCCO make it a rather exotic superconductor. For comparison, YBCO has $\kappa$ ~100 and $\Gamma \sim 5$, and a conventional Type II .superconductor such as $Nb_3Sn$ has

104

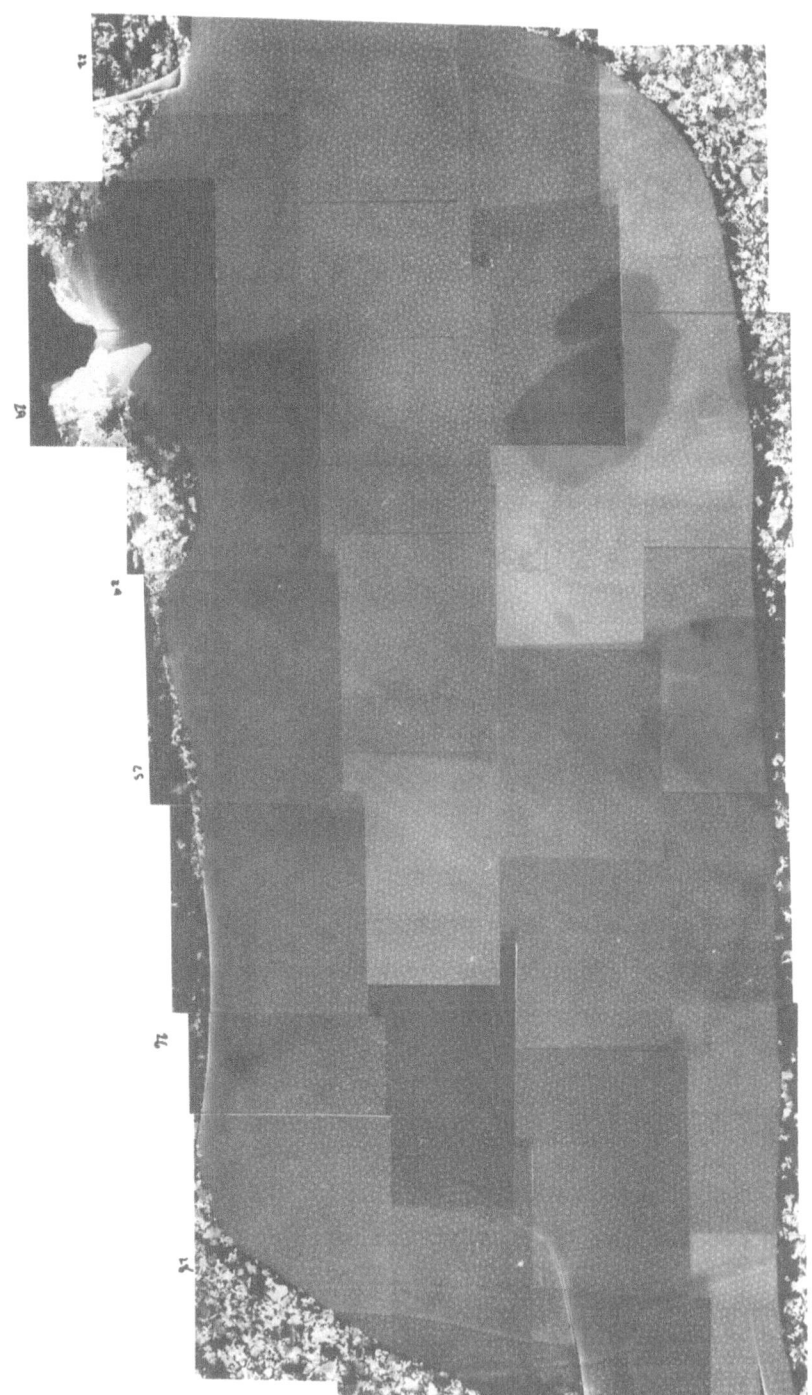

**Fig.3** Shown is a collage of the magnetic flux lines in a superconducting sample of BSCCO taken at 8 Gauss, H||c and 4.2 Kelvin. The individual magnetic flux lines are the white spots in the photo and are roughly 1.7 µm apart.

$\kappa \sim 20$ and $\Gamma = 1$.

A decoration collage of a BSCCO sample at H = 8G (H||c) is shown in Fig. 3. In that picture one can observe nearly all of the points mentioned above for H||c. Individual vortices of size hc/2e show up as white clusters of particles. They are strongly pinned along stripe defects in the center of the sample (possibly borders between domains in which the a and b axes are interchanged[7]) and there are other obvious sample defects such as surface steps and tears caused by the cleaving procedure. There are dark regions of the sample that exhibit no apparent magnetization or vortices; and vortices show some tendency to align with some sample edges, but not all edges. The sample shows two large uniform interior regions of ~100 by 100 vortices each that are free of obvious sample defects. In our study of the applied field dependence of the order of the vortex arrangement, we have analyzed digitized images from two to three such regions of size ~4,000 vortices for each of 20 decorated samples. Some studies have examined regions of as many as 15,000 vortices.

In Fig. 4, a) through d), are shown defect maps known as Delaunay triangulations of the arrangement of roughly 4000 vortices from four different BSCCO samples cooled to 4 K in fields parallel to c of 69, 23, 11 and 8 G respectively[7]. These samples had been previously annealed at 600°C for ~24 hours in 1 atm oxygen and then quenched to room temperature. Indications are that the oxygen annealing process probably does not greatly affect either $\kappa$ or $\Gamma$ when compared to those of the as-made samples but does reduce the concentration of oxygen vacancies, which could serve as pinning centers for vortices. In the defect maps shown, each vortex center is represented as a vertex of nearest neighbor bonds. Non-sixfold-coordinated centers, defects in a perfect hexagonal array, are shaded in the figure. The vortices are quite disordered at 8 G and 11 G whereas much less so at 23 G, and no topological defects in the vortex lattice are visible in the field of view at 69 G. The translational correlation length $\xi_G$, as determined from exponential fits to the correlation function of the translational order parameter of the vortex lattice

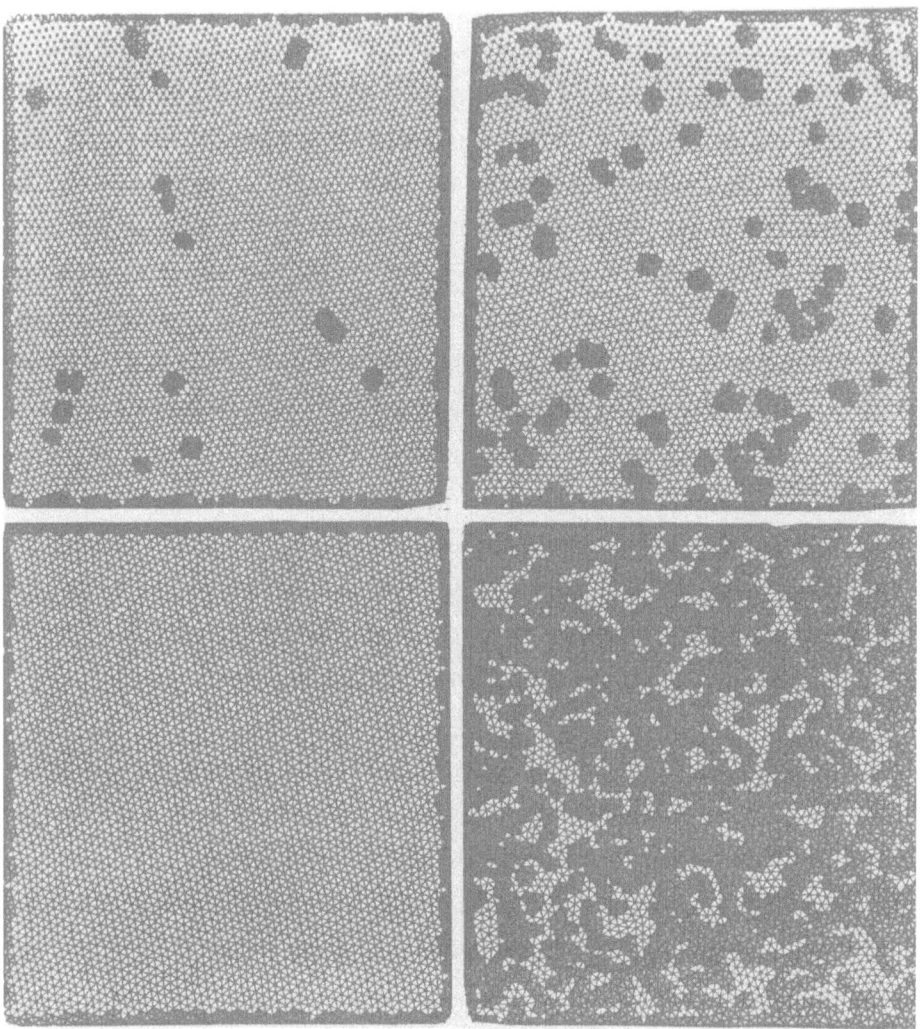

**Fig.4** Shown are Delaunay triangulations for image processed scanning micrographs of Bitter decorated BSCCO crystals at 4.2 Kelvin in fields of 69, 23, 11 and 8 Gauss respectively (starting with the lower left and going clockwise). The shaded triangles join vertices that are not six-fold coordinated and are the topological defects in the flux lattice.

$\Psi_G(r) = \exp(iG \bullet r)$, where r is a vortex position, is shown versus H in Fig. 5a) for both annealed and as-made (unannealed) samples. A monotonic increase of $\xi_G$ from ~2 $a_o$ at 5 G to ~20 $a_o$ is observed for the annealed samples, whereas the as-made samples have $\xi_G$ roughly half that value at each field.

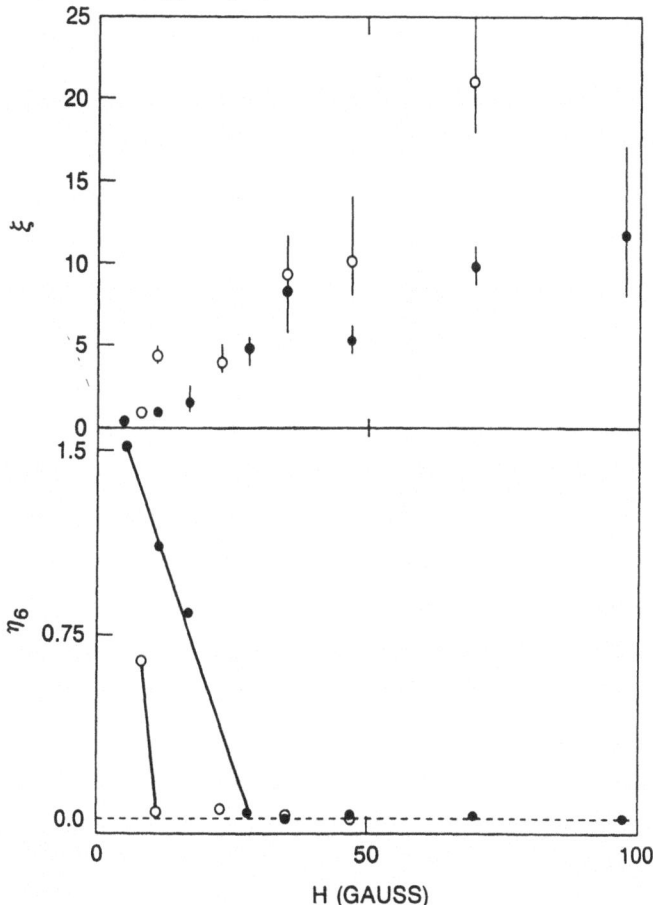

**Fig. 5** (upper) The translational correlation lengths $\xi$ for annealed and as-made samples in units of nearest neighbor spacings. (lower) The bond orientational correlation exponents $\eta_6$. The line at $\eta_6$~0.06 is our limit of experimental resolution.

In the 69-G and 23-G defect maps (Fig. 4, a) and b)) one can easily sight down rows of vortices, despite the relatively small value of $\xi_G$ compared to the size of the image. This is the signature of an hexatic, which exhibits short-range translational order and long range bond-orientational order[15-17]. The bond-orientational order of the vortex lattice is characterized by an order parameter $\Psi_6(r) = \exp[i6\theta(r)]$. The correlation function of this bond-orientational order parameter measures the correlation of a bond angle $\theta(r)$ at r (modulo $2\pi/6$) with that at the origin. Assuming a power law dependence for the decay of correlations of the form $\langle \Psi_6^*(0) \cdot \Psi_6(r) \rangle \sim r^{-\eta_6}$, one can extract a correlation exponent, $\eta_6$, from fits to computed bond-orientational correlation functions from the measured position of vortex centers[18]. The exponents are shown for the same series of decorations in Fig. 5b). Immediately obvious from the figure is the abrupt change of the fitted $\eta_6$ from a relatively large value of 0.8, a rather steep decay of the bond orientational order, to the limits of the experimental resolution ~ 0.06, where it does not decay at all to within our experimental resolution. The change occurs rapidly in a change in the applied field of only a few gauss, but at different fields for the as-made and annealed samples, presumably reflecting the change in the concentration of intrinsic pinning sites in the two types of samples. This abrupt change in the bond-orientational order with field is difficult to reconcile with theories that include only a range of pinning energies but no phase transition. These data are consistent with the predictions of an isotropic vortex fluid to hexatic vortex glass or hexatic vortex fluid phase transition or a transition between a strongly pinned disordered glassy phase to a less strongly pinned hexatic near $H_{c1}$. Experimental data on vortex mobility versus temperature and the microscopic irreversibility temperature versus H in this field range are needed to discriminate among the various possibilities.

Shown in Fig. 6 are three views of a sawtooth twin in YBCO and the novel vortex structure that forms near it[11]. In Fig 6a) is shown a Nomarski micrograph where the different twin domains show up as the light and dark regions. The sawtooth twin is easily visible in this micrograph while the vortices show up as the small bumps. Twins in YBCO more typically grow as straight lines but on occasion they grow in a sawtooth fashion as shown. In Fig. 6b) is shown an SEM micrograph of the vortex structure near the twin. The vortices show up as

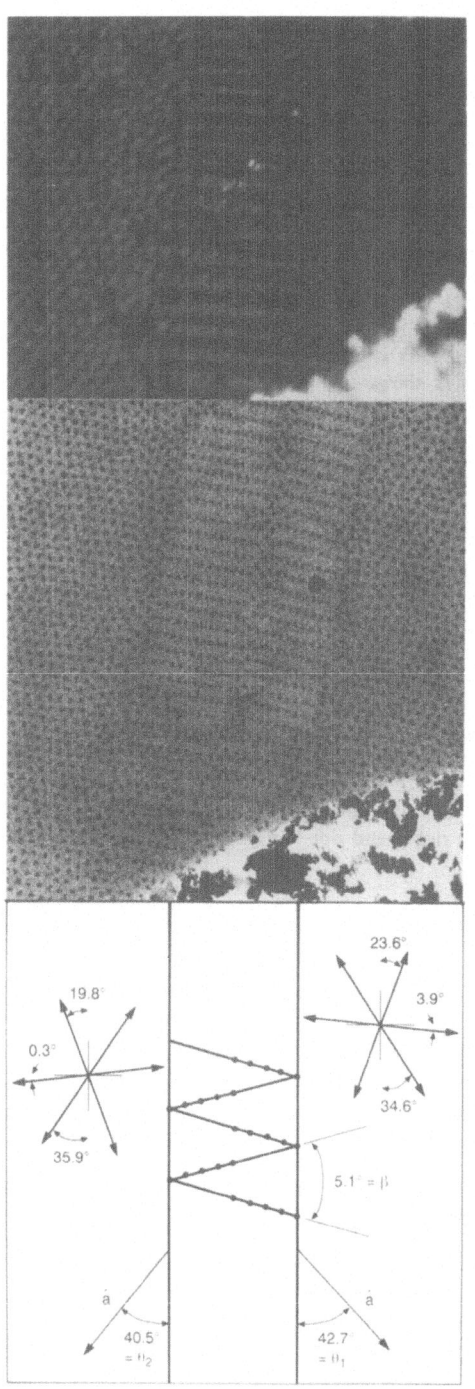

**Fig. 6** Shown are three views of a sawtooth twin in single crystal YBCO and the novel vortex structure which forms there.

(top) A Nomarski micrograph where the different twin domains show up as light and dark. The sawtooth twin is easily visible in this micrograph while the vortices show up as the small bumps.

(middle) An SEM micrograph of the vortex structure near the twin. The vortices show up as the black dots while the twin domains cannot be seen with this type of imaging.

(lower) A drawing which combines the information found in the upper two images. The vortices which decorate the sawtooth twin are shown as the solid dots in the central panel. Note the reduced symmetry of the vortex structure relative to the sawtooth twin itself. The upper diagrams show the basis vectors for the flux lattice well away from the sawtooth twin and at the bottom of the figure are shown the a directions of the crystal as determined by using flux lattice crystallography as described in reference 11.

the black dots while the twin domains cannot be seen with this type of imaging. In Fig. 6c) is shown a drawing which combines the information found in the upper two images. The vortices which decorate the sawtooth twin are shown as the black dots in the central panel. Note the reduced symmetry of the vortex structure relative to the sawtooth twin itself. The upper diagrams show the basis vectors for the flux lattice well away from the sawtooth twin and at the bottom of the figure are shown the a directions of the crystal as determined by using flux lattice crystallography as described in reference 11. This unusual pattern only occurs for sawtooth twins for which the pitch of the sawtooth is roughly comparable to the intervortex spacing and the penetration depth.

The origin of this pattern is a direct result of the magnetic interaction which occurs between a vortex and the twin boundary at an asymmetric twin. For a twin which runs in a low symmetry direction, there is a bound state which forms

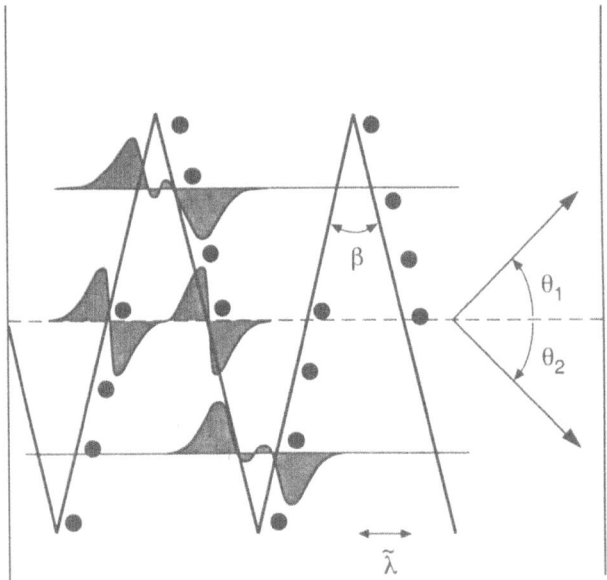

**Fig.7** Shown is our model for the formation of the novel vortex structure near a sawtooth twin. The angles $\theta_1$ and $\theta_2$ are the angles that the a directions form with respect to the average twin boundary which is shown as the dotted line. $\beta$ is the opening angle of the sawtooth twin and $\lambda$ is the average penetration depth. The sawtooth twin is shown as the heavy, solid line and the solid circles are the vortices. The shaded regions represent schematically the interaction potentials as discussed in reference 11.

on one side of the twin and a repulsive interaction on the other side. Near the twin, the vortices prefer to sit in this bound state. At the ends of the sawtooth, this bound statein one arm gets filled in by the peak in pinning energy from the other arm of the sawtooth. This is drawn schematically in Fig. 7. This reduced symmetry vortex structure which results from these interactions is the first observation of a structure which is caused by the magnetic interaction with a crystalline defect and not the much more usual core pinning which typically drives the vortex interactions near crystalline defects.

In the last part of this section, we present another novel vortex structure found in YBCO near twins. Shown in Fig. 8 is a decoration image taken in lightly twinned YBCO. In that material the twins meet at angles of 90°. The crystal twins are drawn and labeled on the figure. The interesting issue is how a six

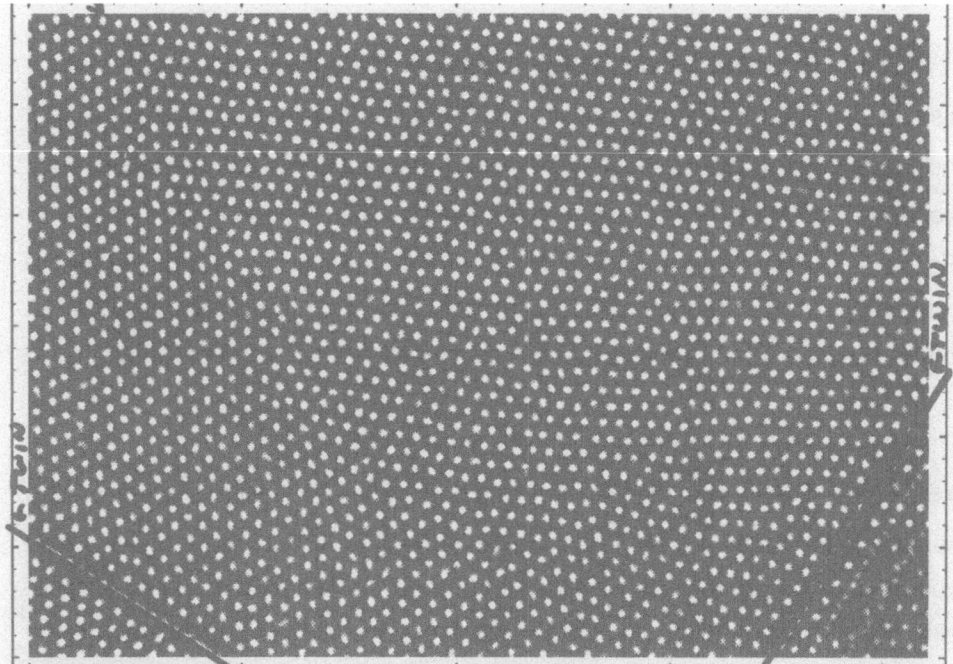

**Fig.8** Shown is an image of the flux lattice in YBCO near the intersection of two twins. The twins are labeled with the solid black lines. Note how the flux lattice tends to line up parallel to the twins which meet at an angle of roughly 90°. Note that roughly bisecting the twins is a large angle grain boundary in the flux lattice. This large angle grain boundary is labeled with an arrow.

-sided flux lattice chooses to fit into a square corner. The figure shows how it does this. Shown in the decoration is a large angle grain boundary that the flux lattice forms. Near each twin, the flux lattice lines up with the twin. Eventually a defect must form in order to fit. This defect is labeled and is the large grain boundary shown.

## Flux-Line Lattice Structures for Other Field Orientations

Decoration experiments can also be performed with the magnetic field applied at an angle with respect to the c axis. In the experiments to be discussed here, all the samples were mounted at a fixed angle $\theta$ with respect to the field and field-cooled with all decorations having been done at 4.2 K. For these orientations a variety of novel structures have been seen, including flux-line chains, oval vortices and tilt angle induced hexatics.

In Bi:2212 there have been a variety of results. The first measurements we will discuss were of the average density of vortices on the cleaved face ($\perp$ to c) of the crystal as a function of angle[12]. For all angles up to 85° the average density was found to follow a $(B/\phi_o)\cos\theta$ dependence. This dependence is shown in Fig. 9. In other words, the vortex lattice on the surface is only induced by the component of the magnetic field B parallel to c. This dependence had been inferred by Kes *et al*[19] through an analysis of torque magnetometer data.

The vortex lattice structures seen on the surface of Bi:2212 fall into two regimes, $\theta<60°$ and $\theta>60°$. For the smaller angles, the vortex lattice present on the surface of the sample is isotropic, to within the 5 to 10% distortions discussed above. In addition, there is no apparent preferred orientation of the flux lattice, either with respect to the sample's crystallographic axis or with respect to the magnetic field's tilt axis. Interestingly enough, this measured isotropy implies an extreme anisotropy in the penetration depth or effective mass. To clarify this point, consider the case of an isotropic system. In the absence of surface effects, a hexagonal vortex lattice will form with the vortex lines parallel to the applied field. When the field is not perpendicular to the decorated surface, the resulting pattern seen is a distorted lattice, with a distortion factor $\rho=1/\cos\theta$. In the limit of large anisotropy $\Gamma\to\infty$ , however, the distortion of the hexagonal vortex lattice in the bulk predicted from the anisotropic London equations[20] is likewise $1/\cos\theta$. This exactly cancels the distortion in the surface pattern due to the tilted field, and the observed vortex lattice will then be isotropic. Assuming a maximum experimentally measured distortion of 5% at 60°, we find that a limit of $\Gamma> 8$ can be placed on the effective mass ratio using the full form of the London equations. Although this is certainly consistent with torque magnetometry[21] and resistivity, which give $\Gamma\sim$ 60-200, this analysis is insensitive to the value of $\Gamma$ when the anisotropy is large.

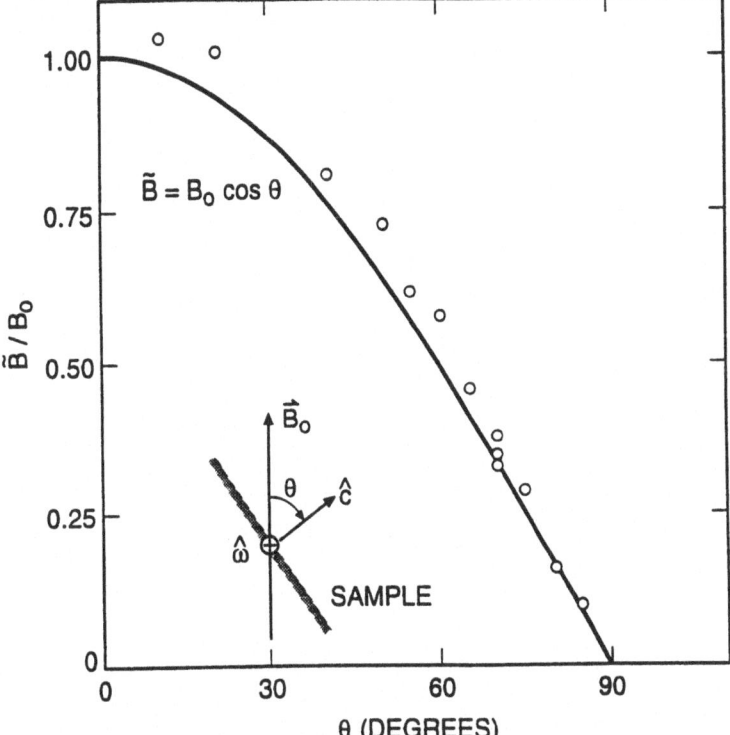

**Fig.9** Shown is the field $B=n\phi_0$ defined from the vortex density  n divided by the applied field $B_0$ vs the tilt angle $\theta$ for BSCCO. At low angles the density of vortices is what would be expected from the normal component of $B_0$ and *not* the total applied field.

114

**Fig.10** Shown are vortex chains in BSCCO. The field of 35 Gauss was applied at an angle of 70° with respect to the c axis. The vortices are the dark spots in this picture. The chains run approximately perpendicular to the rotation axis and are roughly 1.4 μm apart.

For angles $\theta > 60°$, a dramatic new structure emerges (Fig. 10). An array of flux chains lies in the plane spanned by B and c. The chains have an increased line density of vortices with respect to the background lattice. The chains also orient the intervening lattice so that one of the lattice vectors is parallel to the chain direction and perpendicular to the tilt axis. This is very different from the case for $\theta < 60°$ where no preferred orientation of the lattice is seen. Independent of the crystallographic a axis direction, this is the direction selected by London theory , although without having included or predicted the chains that are so prominent in the picture. This orientation was not defined before the appearance of the chains. It has been widely reported that neutron scattering experiments are consistent with a lattice rotated 90° from this. Actually, this is based on three results, Technetium[22], YBCO[23] and UPt$_3$ [24]. In all three cases, the field orientation corresponded to $\theta=90°$, in which special case the two lattice orientations are again degenerate within the anisotropic London equations. The decoration results presented here represent the first true test of this prediction of the flux lattice orientation as a complete function of angle.

We can further analyze the photo shown in Fig. 10. In the plane of the photograph, we can let the spacing between vortices along the chain be D and that between chains be C. Although C has large variations due the chain wandering which can be seen in Fig. 10, both D and C are found to scale as B$^{-1/2}$ at a fixed angle. Because this is the same scaling as the vortex lattice constant, the picture at a fixed angle is field independent in the sense that the field only sets the overall magnification, not the structure we see. For all fields and angles studied, we find DC $\sim \phi_o/B$. If the chain structure is interpreted as a superlattice modulation, this scaling is equivalent to incorporating one extra flux quantum per superlattice unit cell. A necklace of dislocation pairs, with zero net Burgers vector, dresses the chains to accommodate the extra flux line. These dislocations can be identified in the Delaunay triangulation of the overall structure shown in Fig. 11.

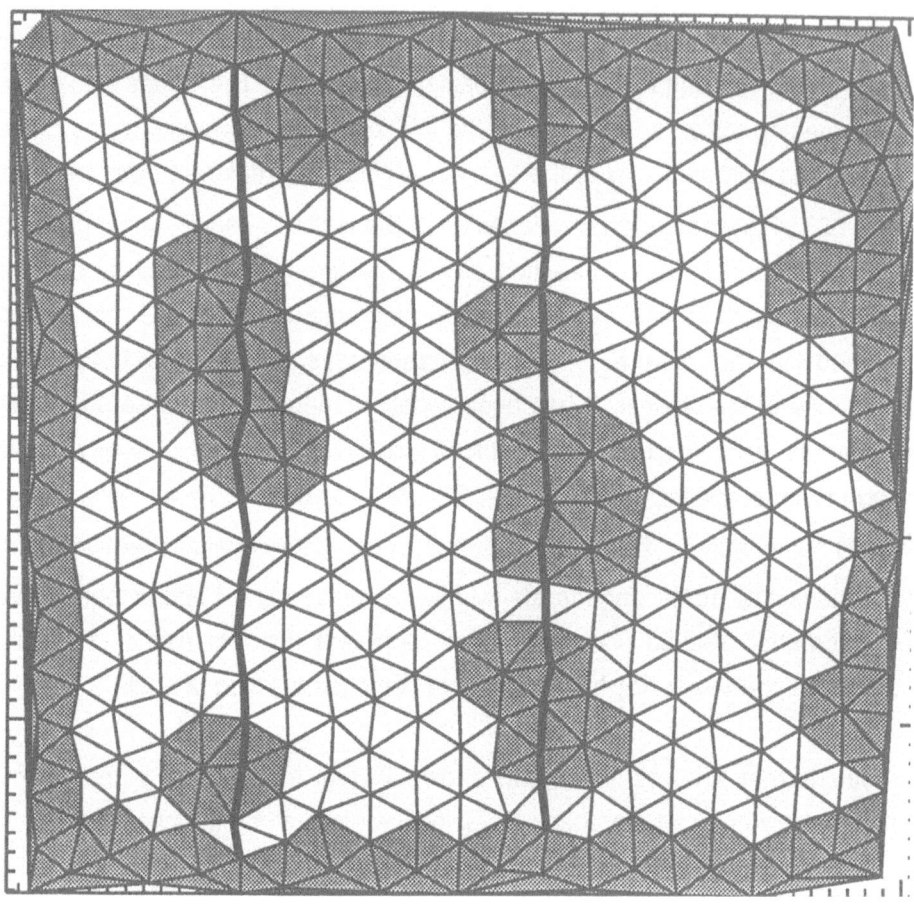

**Fig.11** Shown is a triangulation of vortex chains in BSCCO taken at 97 Gauss and at an angle of 70°. To accomodate the increased vortex density along the chain, pairs of dislocations (shaded) are formed.

As the tilt angle increases toward 90°, the number of Abrikosov vortices between chains decreases. Owing to the large fluctuations in C, the exact from of this reduction is difficult to to state quantitatively. One form, consistent with the data, arises from considerations of an anisotropic lattice. In this case, $D=0.75(\phi_o/Bcos\theta)^{1/2}$. Experimentally, the number of lattice constants between chains varies from 11 at 60° to 2 at 85°.

Flux-line chain structures were anticipated theoretically[25,26] although different in detail from our results. It is clear experimentally that the chains are formed by at least a weakening of the repulsive vortex-vortex interaction in the plane formed by B and c. Considering the enormous anisotropy, this is presumably related to the current paths which tend to stay in the a-b plane, independent of field orientation. Calculations have centered on the effective mass approximation. Such theories suggest that, near $H_{c1}$, a vortex-vortex attraction develops for extreme anisotropy in tilted fields. This is a result of the current paths remaining in the planes and the simple observation that two dipoles attract when their axes are both parallel and along the line connecting them. Extensions of these theories have shown that this state should consist of chains only. The spacing of vortices along the chains is estimated to be $D \sim (\lambda_a \lambda_b \lambda_c)^{1/3}$ and independent of field. Our observations in BSCCO contradict this[12]. However we have also seen chains in YBCO[13] and they do agree with these theoretical predictions. Shown in Fig. 12 is a pinstripe array of vortex chains seen in YBCO for an applied field of 24.8 Gauss applied at an angle of 70°. This pattern is clearly very different from that seen in BSCCO. In YBCO one sees only chains, there are no Abrikosov vortices in between. In addition, one finds that as a function of field that the density of vortices in the chains remains constant. As the field changes, the distance between chains changes in order to conserve flux. In YBCO we find a constant D and C ~ 1/B. These vortex chain structures in the moderately anisotropic YBCO agree in detail with theoretical predictions. Therefore the structures in BSCCO appear to be a result of the extreme anisotropy in that system.

**Fig.12** Shown is an SEM micrograph of decorated vortex chains in YBCO taken at a field of 24.8 Gauss and an angle of 70°. The chains run perpendicular to the rotation axis and are independent of the a axis of the crystal which runs roughly 45° from the chains. The bar is 10 microns.

We now believe that the structure in BSCCO actually represents two sublattices of vortices running essentially orthogonal to one another. One sublattice is the Abrikosov lattice which dominates the picture. The second sublattice of Josephson vortices causes the superlattice modulation as the vortices attract each other when orthogonal. Hence the chains in that case are actually an image of vortices running parallel to the a-b plane. That two such sublattices should appear is implicit in the flux-line free energy calculated by Sudbo and Brandt[27], which shows two pronounced minima as a function of angle for certain applied fields.

In a separate set of experiments Dolan *et al*[9] were able to decorate the surface perpendicular to the a-b plane in YBCO(123). As seen in Fig. 13, a dramatically distorted lattice was seen as well as oval vortices with the major axis aligned parallel to the a-b planes. Their results can be explained by assuming a mass ratio of $\Gamma^2 = 30$. This mass ratio is in rough agreement with other estimates. The oval structure of the individual vortices can be similarly explained. Their pattern is strongly influenced by the presence of a twin domain boundary (marked with the arrow) that aligns one chain but then rapidly degrades. This degradation of the order may be due to the weak shear modulus[28] in this direction, reduced by a factor $\Gamma^4 \sim 900$.

Finally in this section we present results on tilt-induced orientational order as seen in a conventional but anisotropic superconductor $NbSe_2$ using magnetic decoration[14]. For this material, for low applied fields and with the field applied parallel to the c axis of the crystal, we find a disordered flux lattice with short ranged positional and orientational order. This is shown in Fig. 14 where is shown the Fourier transform of the flux line lattice for 47.1 Gauss applied field for three different angles. For the bottom frame $\theta = 30°$ and the transform only shows a ring of scattering indicative of short ranged positional and orientational order. As the tilt angle of the applied magnetic field increases, we find that the orientational order in the flux lattice grows dramatically. At high tilt angles, we

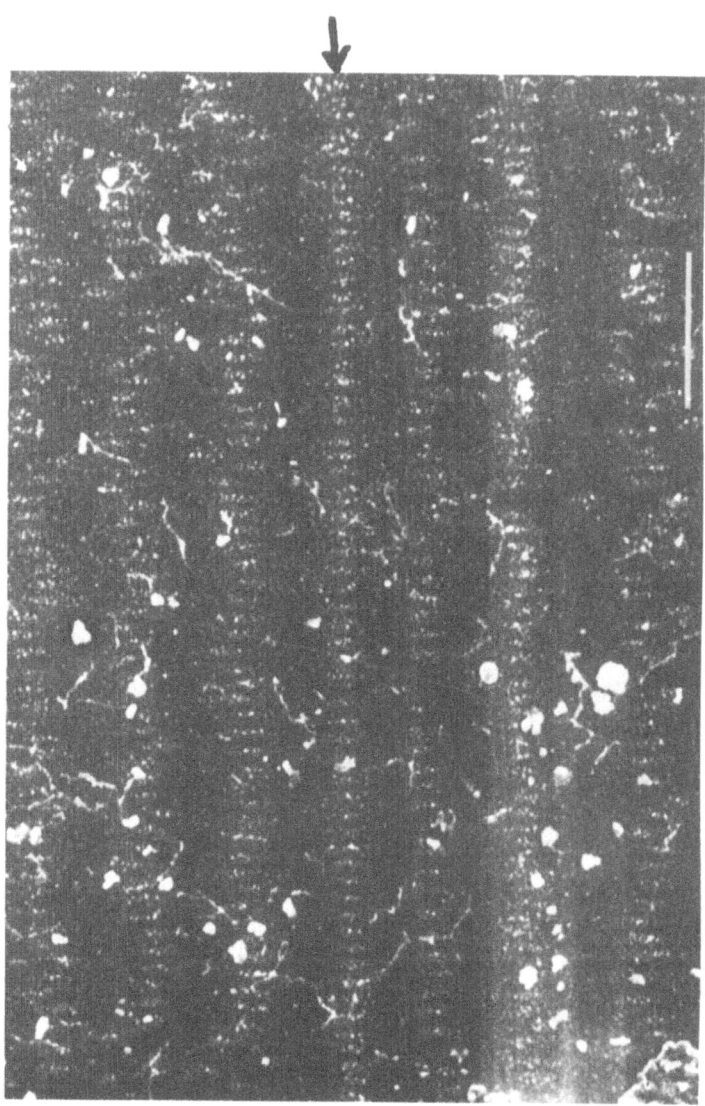

**Fig.13** Shown is a decoration of a face of YBCO parallel to the c axis at a field of 8 Gauss. The arrow marks a twin boundary and is also the c direction. Note the oval vortices. This data is from reference 9.

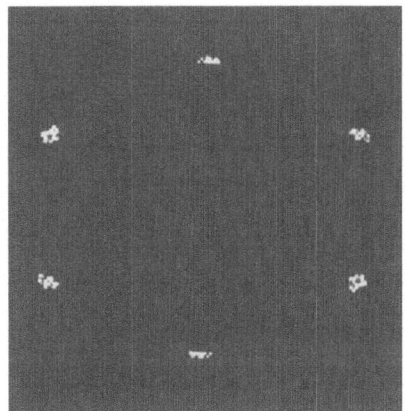

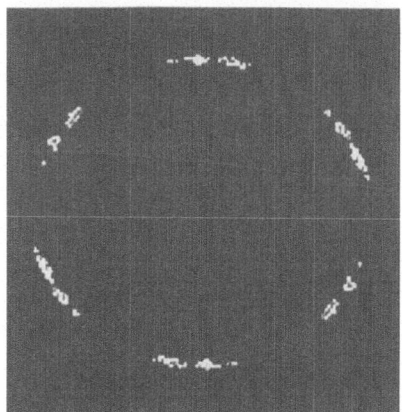

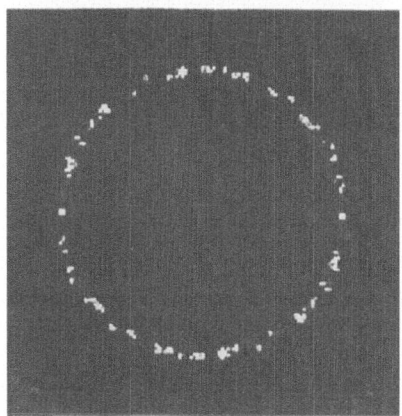

**Fig. 14** Angular dependence of the Fourier transform for the flux line lattice for 47.1 Gauss applied field and various tilt angles in NbSe$_2$. The upper figure is for $\theta=70°$, the middle for $\theta=50°$ and the lower for $\theta=30°$. Note that as the tilt angle is increased how **the** azimuthal modulation developes **in**dicative of developing hexatic order in the **flux** line lattice in this system.

find a hexatic solid with long ranged orientational order but with only short ranged positional order. This can be seen in the middle and upper frames of Fig. 14. As the tilt angle is increased, one sees an azimuthal modulation developing (center frame) and at the highest tilt angle one sees six well defined spots indicative of long ranged hexatic order (upper frame). This is the first observation of a flux lattice in which the orientational order is stabilized by changing the tilt angle of the applied magnetic field in an anisotropic superconductor. These results show that flux lattices behave in a way very similar to smectic liquid crystals where it is found that a spontaneous tilt of the liquid crystal molecules can stabilize long-ranged orientational order.

## Conclusions

We conclude from the above, that the mixed state of extremely anisotropic superconductors can produce a wide variety of unusual vortex lattice structures. Magnetic decoration images such as those we have presented here are one road to the goal of understanding how and why these structures form and ultimately how they can be pinned.

The authors thank David Nelson, Daniel Fisher and Christian Bolle for many helpful discussions.

## References

1. For a review, see D. J. Bishop *et al.*, Science **255**, 165 (1992).
2. H. Trauble and U. Essmann, J. Appl. Phys. **39**, 4052 (1968).
3. N. V. Sarma, Philos. Mag. **17**, 1233 (1968).
4. P. L. Gammel *et al.*, Phys. Rev. Lett. **59**, 2952 (1987).
5. G. J. Dolan *et al.*, Phys. Rev. Lett. **62**, 827 (1989).
6. C. A. Murray *et al.*, Phys. Rev. Lett. **64**, 2312 (1990).
7. D. G. Grier *et al.*, Phys. Rev. Lett. **66**, 2270 (1991).
8. R. N. Kleiman *et al.*, Phys. Rev. Lett. **62**, 2331 (1989).
9. G. J. Dolan *et al.*, Phys. Rev. Lett. **62**, 2184 (1989).
10. P. L. Gammel, private communication.
11. P. L. Gammel *et al.*, Phys. Rev. Lett. **69**, 3808 (1992).
12. C. A. Bolle *et al.*, Phys. Rev. Lett. **60**, 112 (1991).
13. P. L. Gammel *et al.*, Phys. Rev. Lett. **68**, 3343 (1992).
14. C. A. Bolle *et al.*, Phys. Rev. Lett. **71**, 4039 (1993).
15. B. I. Halperin and D. R. Nelson, Phys. Rev. Lett. **41**, 121 (1978).
16. D. R. Nelson *et al.*, Philos. Mag. **A46**, 105 (1982).
17. E. M. Chudnovsky, Phys. Rev. **B40**, 11355 (1989).
18. D. R. Nelson, in "Phase Transitions and Critical Phenomena", C. Domb and J. L. Lebowitz, Eds. (Academic Press, London, 1983), vol. 7, pp. 1-99.
19. P. H. Kes *et al.*, Phys. Rev. Lett. **64**, 1063 (1990).
20. L. J. Campbell *et al.*, Phys. Rev. **B38**, 2439 (1988).
21. D. E. Ferrell *et al.*, Phys. Rev. Lett. **63**, 782 (1989).
22. J. Schelten *et al.*, J. Low Temp. Phys. **14**, 213 (1974).
23. E. M. Forgan *et al.*, Nature **343**, 735 (1990).
24. R. N. Kleiman *et al.*, Phys. Rev. Lett. **69**, 3120 (1992).
25. A. M. Grishin *et al.*, Sov. Phys. JETP **70**, 1089 (1990).
26. A. J. Buzdin and A. Yu Simonov, JETP Lett. **51**, 191 (1990).
27. A. Sudbo and E. H. Brandt, Phys. Rev. Lett. **67**, 3176 (1991).

# FLUX CREEP AND AC MEASUREMENTS IN SUPERCONDUCTORS

ERNST HELMUT BRANDT
*Max-Planck-Institut für Metallforschung, Institut für Physik*
*D-70506 Stuttgart*
*FR Germany*

## 1. Introduction

## 2. Critical Currents

## 3. Thermally Activated Depinning

## 4. Further Topics Related to Depinning

## 5. Linear AC Response of the Pinned Flux-Line Lattice

## 6. Depinning Lines

*N. Bontemps et al. (eds.), The Vortex State*, 125–157.
© 1994 *Kluwer Academic Publishers.*

ABSTRACT: As discovered by Abrikosov, a magnetic field $B$ can penetrate a type-II superconductor in form of flux lines or vortices which carry a quantum of flux each and arrange to a more or less regular triangular lattice. The flux-line lattice has interesting elastic and fluctuation properties, in particular in the highly anisotropic high-$T_c$ superconductors (HTSC) with layered structure. Under the action of an electric current density $J > J_c$ the flux lines move and dissipate energy, but for $J < J_c$ they are pinned by material inhomogeneites. In HTSC thermally activated depinning causes a finite resistivity $\rho$ even at current densities $J < J_c$. At sufficiently high temperature $T$ ohmic resistivity $\rho(T, B)$ is observed down to $J \to 0$. This indicates that the flux lines are in a "liquid state" with no shear stiffness and with small depinning energy. At lower $T$, $\rho(T, B, J)$ is non-linear since the pinning energy of an elastic vortex lattice or "vortex glass" increases with decreasing $J$. In the extremely anisotropic Bi- and Tl-based HTSC short vortex segments ("pancake vortices" in the CuO layers) can depin individually with very small activation energy.

# 1. Introduction

## 1.1. THE FLUX-LINE LATTICE

The occurrence in superconductors of a lattice of tiny current vortices each carrying a quantum of magnetic flux $\Phi_0 = h/2e = 2.07 \times 10^{-15}\,\mathrm{Tm^2}$ was predicted by Alexei Abrikosov [1] when he discovered a periodic solution of the Ginzburg-Landau (GL) equations. This flux-line lattice (FLL) was observed by neutron scattering [2,3] and by a Bitter decoration technique [4-7]. The FLL occurs when the GL parameter $\kappa = \lambda/\xi$ of the material is $\kappa > 1/\sqrt{2}$ (type-II superconductor) and the applied magnetic field $B_a$ is in the range $B_{c1} < B_a < B_{c2}$ where $B_{c1} \approx (\Phi_0/4\pi\lambda^2)(\ln\kappa + 0.5)$ and $B_{c2} = \Phi_0/2\pi\xi^2$ are the lower and upper critical fields and $\lambda$ and $\xi$ the magnetic penetration depth and the coherence length of the GL theory, respectively. Approximate temperature dependences of these quantities are $\lambda \sim \xi \sim (1 - T/T_c)^{-1/2}$ (near $T_c$) and $B_{c1} \sim B_{c2} \sim 1 - T^2/T_c^2$, where $T_c$ is the superconducting transition temperature [8-10]. Each flux line has a core of radius $\approx \sqrt{2}\xi$ in which the superconducting order parameter $\psi(\mathbf{r})$ is suppressed; the line $\psi = 0$ defines the vortex position. The magnetic field $\mathbf{B}(\mathbf{r})$ of a flux line is concentrated in a tube of radius $\lambda$. For an isolated straight flux line along $z$ one has for $\kappa \gg 1$ [11, 12]

$$\psi(\mathbf{r}) \approx \frac{x + iy}{r}[1 - \exp(-\frac{r^2}{2\xi^2})]^{1/2} \tag{1}$$

$$B(\mathbf{r}) \approx (\phi_0/2\pi\lambda^2)K_0[(r^2 + 2\xi^2)^{1/2}/\lambda] \tag{2}$$

where $r^2 = x^2 + y^2$ and $K_0(r)$ is a modified Bessel function with $K_0(r) \approx -\ln r$ for $r \ll 1$ and $K_0(r) \approx (\pi/2r)^{1/2}\exp(-r)$ for $r \gg 1$. Approximate solutions for curved flux lines and for arbitrary arrangements of flux lines may be found in [13-16] and in the review papers [12, 17, 18].

When the applied field $B_a$ is increased, more flux lines penetrate the superconductor and form a more or less regular triangular FLL with vortex spacing $a = (2B/\sqrt{3}\Phi_0)^{1/2}$. The vortex *fields* start to overlap when the average induction (flux density) $B = \langle B(\mathbf{r})\rangle$ exeeds the value $B_{c2}/2\kappa^2 = \Phi_0/4\pi\lambda^2$. The vortex *cores* overlap when $B > 0.25B_{c2}$. At lower $B$ the vortex cores are well separated; the order parameter $|\psi|$ is then practically constant outside the small core regions and the magnetic field, current, and energy of a vortex arrangement follow from London theory. The London equation, a linear differential equation for $\mathbf{B}(\mathbf{r})$, is particularly useful for superconductors with large $\kappa \gg 1$ and large $B_{c2}$ and it is easily generalized to anisotropic superconductors by introducing an anisotropic (tensorial) penetration depth.

## 1.2. HIGH-$T_C$ SUPERCONDUCTORS WITH LAYERED STRUCTURE

High-$T_c$ superconductors (HTSC) like $YBa_2Cu_3O_{7-\delta}$ (YBCO, with $T_c = 92.5\,\mathrm{K}$) and some Bi (BSCCO) and Tl cuprates with $T_c$ up to $125\,\mathrm{K}$, have very large $\kappa \approx 100$ and large anisotropy ratios $\Gamma = \lambda_c/\lambda_{ab}$, $\Gamma \geq 5$ for YBCO and $\Gamma > 100$ [19] for BSCCO. Here $\lambda_{ab}$ and $\lambda_c$ denote the penetration depths for currents flowing in the $ab$-plane or along the $c$-axis of these approximately uniaxial crystals, respectively. All these HTSC have a pronounced layered structure. This means the currents prefer to flow in the crystalline $ab$ plane, in layers formed by CuO. The flux lines want to align with these CuO planes in order to lower their magnetic energy, and the vortex cores prefer to run in the space between two CuO planes in order to lower their superconducting condensation energy [20]. Vortex lines which thread the CuO planes are formally composed of two-dimensional (2D) vortex disks ("pancake vortices") in the CuO planes [21-25] which are more or less coupled by Josephson currents flowing perpendicular to the CuO planes. If this coupling is weak a vortex line can decompose ("evaporate") into independent pancake vortices if their random displacements caused, *e.g.*, by pinning at oxygen vacancies or by thermal fluctuations are sufficiently large. Weak Josephson coupling between the CuO planes means large magnetic anisotropy. All these features of layered superconducors can be calculated from the powerful Lawrence-Doniach (LD) theory [26, 27], which generalizes the phenomenological GL and London theories to stacks of superconducting planes with zero thickness and with Josephson interaction between neighboring planes.

Numerous transitions of the vortex lattice in layered HTSC are predicted when $T$ or $B$ is changed. I summarize some of these ideas in short [28]:

1. A transition from 2D to 3D superconductivity is expected with increasing $T$ [26, 27] when the coherence length along the $c$-axis $\xi_c(T) \sim (1 - T/T_c)^{-1/2}$ equals the layer spacing $s$. For YBCO this should occur a few degrees below $T_c$, and for BSCCO very close to $T_c$.

2. Vortex-antivortex pairs nucleate spontaneously in a single superconducting layer or film in zero field at finite temperature. These pairs dissociate at a temperature $T_{BKT} = \Phi_0^2 s/(8\pi k_B \mu_0 \lambda_{ab}^2)$ (Berezinskiĭ-Kosterlitz-Thouless transition) since both their interaction and their entropy are proportional to $\ln(R/\xi_{ab})$ where $R$ is the specimen size [22, 23].

3. A vortex line along the $c$-axis "evaporates" into single pancake vortices at the same temperature $T_{BKT}$ where its thermally averaged radius diverges [21, 29, 30].

4. In analogy to the 2D BKT-transition, 3D vortices at $B = 0$ may nucleate spontaneously in the bulk since their fluctuations give an entropy contribution which reduces the free energy $F = U - TS$. The thermal fluctuation of the vortices influences the magnetization $M$ such that the curves $M(B_a, T)$ for all applied fields $B_a$ cross at the same temperature $T^*$ close to $T_c$ where $M = M^* = 4\pi k_B T^*/(\Phi_0 s)$ [31-32]. See also the recent papers [33].

5. A 2D vortex lattice of spacing $a$ in a film of thickness $d$ melts at $T_m^{2D} = a^2 d c_{66}/(4\pi k_B)$

due to the spontaneous nucleation of dislocations of energy $U = (da^2 c_{66}/4\pi)\ln(R/a)$ and entropy $S = k_B \ln(R/a)$ ($c_{66}$ is the shear modulus of the FLL) [34, 35]. In layered HTSC a similar transition is expected, with the film thickness $d$ replaced by the layer spacing $s$, $T_m^{2D} = a^2 s c_{66}/(4\pi k_B) = \Phi_0^2 s/(64\pi^2 k_B \mu_0 \lambda_{ab}^2)$. Thus, inserting the temperature dependence of $\lambda_{ab} \sim (1 - T^2/T_c^2)^{-1/2}$, one has $T_m^{2D}/T_{BKT} = \lambda_{ab}^2(T_m^{2D})/[8\pi\lambda_{ab}^2(T_{BKT})]$ [22, 23].

6. Thermal fluctuations of the vortex positions in the layers (at $z = z_n$, $n$ = integer) lead to the destruction of phase coherence between adjacent layers. The fluctuation of the phase differences $\delta_n$ of the order parameters $\psi_n$ in neighboring layers becomes large, $\langle \delta_n^2 \rangle \approx 1$ (average over $x$, $y$, $n$, and time) when $T$ reaches $T_0 \approx a\Phi_0^2/(4\pi\mu_0 k_B \lambda_{ab}\lambda_c) \propto B^{-1/2}$, which is typically smaller than the 2D melting temperature $T_m^{2D}$ [31, 36].

7. According to [37] these phase fluctuations renormalize the Josephson coupling between neighboring layers to a smaller value as $T$ increases. As consequences, the effective anisotropy ratio $\Gamma_{eff}$ and penetration depth $\lambda_{c,eff}$ increase and the maximum possible current density along the $c$-axis $J_{z,max} = J_0\langle \sin \delta_n(x, y)\rangle = (\Phi_0/2\pi\mu_0 s\lambda_{c,eff}^2) \propto J_0 \exp[-B/B_D(T)]$ decreases and vanishes at a "decoupling field" $B_D(T) = 8\Phi_0^3 /[\mu_0 k_B T s e\lambda_c^2(T)]$.

8. When $\mathbf{B}$ is nearly parallel to the $a,b$-plane a "lock-in transition" will switch $\mathbf{B}$ exactly into the $ab$-plane since the vortex lines gain energy by having their core in between the layers. This means the vortex kinks stretch and the pancake vortices run out from the specimen [24, 25, 38, 39].

9. Spontaneous thermal nucleation of vortex rings in the $ac$ or $bc$-planes and phase transitions for $B$ parallel to the layers are predicted in various papers [40].

10. Further transitions are expected when vortex pinning is considered. For theories of dislocation-mediated 2D collective pinning and depinning transitions see [22, 23].

## 1.3. FLUX FLOW AND PINNING OF FLUX LINES

An electric current density $\mathbf{J}$ through the superconductor exerts a Lorentz force density $\mathbf{B} \times \mathbf{J}$ on the flux-line lattice which causes the vortices to move with mean velocity $\mathbf{v}$. This vortex drift dissipates energy and thus generates an electric field $\mathbf{E} = \mathbf{B} \times \mathbf{v}$ where $\mathbf{B}$ is the flux density or magnetic induction in the sample. The dissipation is caused by two effects which give approximately equal contributions: (a) By dipolar currents which surround each moving flux line (eddy currents) and have to pass through the normal conducting vortex core. (b) By the retarded relaxation of the order parameter $\psi(\mathbf{r})$ when the vortex core moves. Since at low $B$ the dissipation of the vortices is additive and since at the upper critical field $B_{c2}(T)$ the flux-flow resistivity $\rho_{FF}$ has to reach the normal conductivity $\rho_n$, one approximately gets $\rho_{FF} \approx \rho_n B/B_{c2}(T)$. A more quantitative treatment of this flux dissipation uses time dependent Ginzburg-Landau theory. For reviews of flux motion see [41-42], and for extensions to layered and anisotropic superconductors [43-45].

In real superconductors at small current densities $J < J_c$ the flux lines are pinned by inhomogeneities in the material, e.g., by dislocations, vacancies, interstitials, grain boundaries, precipitates, irradiation defects, or by a rough surface, Fig. 1. Only when $J$ exceeds a critical value $J_c$ do the vortices move and dissipate energy [46]. Pinning of flux lines has two important consequences:

1. The current-voltage curve of a superconductor in a magnetic field is highly non-linear, with $E = 0$ for $J < J_c$ and $E = \rho_{FF}J$ for $J \gg J_c$. For $J$ slightly above $J_c$ various shapes of $E(J)$ are observed, depending on the type of pinning and on the sample geometry. Often a good approximation for $J \geq J_c$ is $E(J) = 2\rho_{FF}[1 - (J_c/J)^p]^{1/p}$ with, $e.g.$, $p = 1$ or 2.

2. The magnetization curve $M(B_a)$ exhibits a hysteresis, Fig. 2. When $B_a$ is increased or decreased the magnetic flux enters or exits until a *critical slope* is reached like in a pile of sand, namely, a maximum and nearly constant gradient of $B = |\mathbf{B}|$. More precisely, in this *critical state* the *current density* reaches a maximum value $J_c$; one has $\mathbf{J} = (\partial H/\partial B)\nabla \times \mathbf{B} \approx \mu_0^{-1}\nabla \times \mathbf{B}$, where $H(B) \approx B/\mu_0$ is the (reversible) magnetic field which would be in equilibrium with the induction $\mathbf{B}$. The critical state is often described by the "Bean model" [47], which assumes a $B$-independent $J_c$ and disregards demagnetizing effects; these become important in flat superconductors in perpendicular magnetic field (Sct. 2).

In general, the current density in type-II superconductors can have three different origins: (a) *Surface currents* (Meissner currents, shielding currents) flowing within the penetration depth $\lambda$, (b) a *gradient* of the flux-line density, or (c) a *curvature* of the flux lines or field lines. The latter two contributions are easily seen by writing $\nabla \times \mathbf{B} = \nabla B \times \hat{\mathbf{B}} + B(\nabla \times \hat{\mathbf{B}})$ where $\hat{\mathbf{B}} = \mathbf{B}/B$. In bulk samples typically the gradient term dominates, $J \approx \mu_0^{-1}\nabla B$, but in films the current is carried almost entirely by the *curvature* of the magnetic filed lines.

Note that in general the flux lines (vortex lines) need not be parallel to the field lines. In particular, in thin films the vortices can be nearly straight and perpendicular to the surface even when the field lines are strongly curved by an applied current [48, 49].

## 2. Critical Currents

### 2.1. LONGITUDINAL GEOMETRY

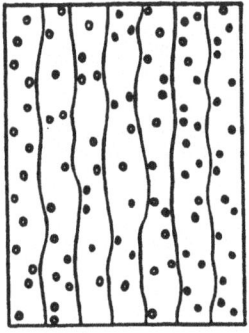 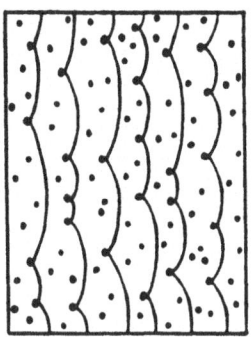 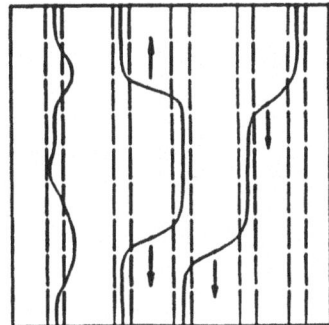

Figure 1: *Left:* Flux lines deformed by weak random pins, *e.g.*, oxygen vacancy clusters. *Middle:* Flux lines pinned by strong point pins. The Lorentz force acts to the right. *Right:* Flux lines pinned by columnar pins can move by the formation of kinks which nucleate by thermal fluctuation or by large current density as described in [78].

130

The critical state of the pinned FLL in long specimens in parallel field is described in detail, e.g., in the review [50]. In this longitudinal geometry demagnetizing effects are negligible, except near the ends of the specimen. For an infinite slab or cylinder the problem is one-dimensional, say $\mathbf{J} = \hat{z}J(x)$ for the slab or $\mathbf{J} = \hat{z}J(r)$ for the cylinder. When the applied field $H_a = B_a/\mu_0$ is increased or decreased the current density $J$ in this simple case only can take the values 0, $J_c$, or $-J_c$ within the Bean model. This means the flux-density gradient $dB(x)/dx \approx \mu_0 J$ is then either zero or constant, like in a sand pile. In particular, in an infinite slab of thickness $2a$ ($|x| < a$) in increasing field $H_a$ one has for the penetration depth $\delta = a - b$, current density $J(x) = H'(x)$, and internal field $H(x)$ for $H_a \leq J_c a$:

$$a - b = aH_a/H_c, \qquad H_c = J_c a \tag{3}$$

$$J(x) = \begin{cases} 0, & |x| < b \\ J_c\, x/|x|, & b < |x| < a \end{cases} \tag{4}$$

$$H(x) = \begin{cases} 0, & |x| < b \\ J_c\,(|x| - b), & b < |x| < a \\ H_a, & |x| > a \end{cases} \tag{5}$$

From (4) one gets the total negative magnetic moment per unit area of the slab,

$$M(H_a) = \int_{-a}^{a} x\, J(x)\, dx = J_c(a^2 - b^2) = J_c a^2 \left( \frac{2H_a}{H_c} - \frac{H_a^2}{H_c^2} \right). \tag{6}$$

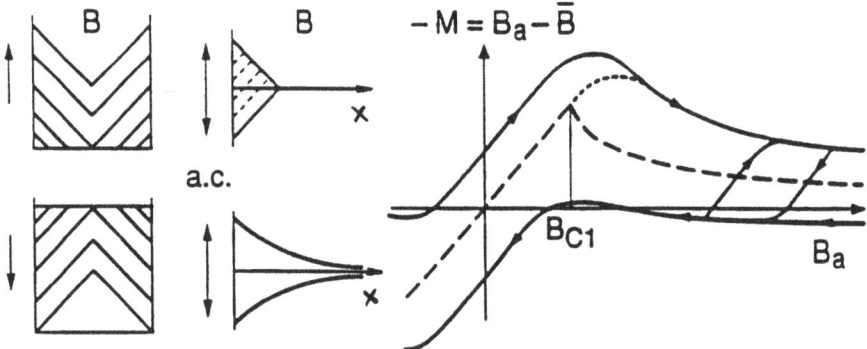

Figure 2: Field profiles (5) in a superconducting cylinder or slab with strong pinning in increasing (*left top*) and decreasing (*left bottom*) longitudinal applied field. In this Bean critical state model the field gradient is constant. Also shown is the field profile caused by an additional weak ac field near the surface in the Bean model (*middle, top*) and in the cases of elastic pinning, viscous drag, or thermally assisted flux flow (Sct. 5) (*middle, bottom*). Strong pinning leads to hysteretic magnetization curves (*right*) with the irreversible magnetization (solid lines) lying above and below the ideal reversible curve (dashed line).

Interestingly, the longitudinal currents and the transversal currents of the U-turn at the far-away ends of the slab give identical contributions to $M$, thus cancelling the factor $1/2$ in the general definition of $M$. The correctness of (6) is easily checked by considering the Meissner state ($H_a \ll H_c$, $b \to a$, ideal shielding) in which $M = M_{ideal} = 2aH_a$ equals the specimen width (volume per unit area) times the expelled field. The deviation of $M$ (6) from the ideal Meissner value is $M - 2aH_a \sim H_a^2$. For complete penetration ($|H_a| \geq H_c$, $b = 0$) one has $M = M_{max} = a^2 J_c$.

When $H_a$ is cycled with amplitude $H_0 \leq H_c$ the flux profile is as shown in Fig. 2. The magnetic moment for $H_a$ decreasing from $+H_0$ to $-H_0$ is then

$$M_{\downarrow}(H_a, H_0, J_c) = (H_a^2 + 4H_aH_c - 2H_aH_0 - H_0^2)/2J_c, \qquad (7)$$

cf. Eq. (23) below. The corresponding branch $M_{\uparrow}$ for $H_a$ increasing from $-H_0$ to $+H_0$ follows from symmetry, $M_{\uparrow}(H_a, H_0, J_c) = -M_{\downarrow}(-H_a, H_0, J_c)$. The dissipated power $P(H_0)$ is frequency $\nu$ times the area of the hysteresis loop, $P = \nu\mu_0 \oint M \, dH_a$. This yields for all amplitudes $H_0$, introducing a function $f(x)$ with $f(x \leq 1) = x^3/3$, $f(x \geq 1) = x - 2/3$,

$$P(H_0) = 4a\nu\mu_0 H_c^2 f(H_0/H_c). \qquad (8)$$

## 2.2. PERPENDICULAR GEOMETRY

Measurements on HTSC are often performed on films or flat monocrystalline patelets in *perpendicular* magnetic field in order to get a larger signal. In this perpendicular geometry, however, the original Bean model, Eq. (3-8), has to be modified drastically due to demagnetization effects even if the critical current density $J_c$ is assumed to be independent of the local flux density. I explain this well known but until recently little understood fact by considering the simple case of a long thin strip with thickness $d$ and width $2a$ ($|x| < a$, $|y| < d/2 \ll a$, $|z| < \infty$) in perpendicular field $H_a \| x$ [51].

When $H_a$ is small the strip is in the Meissner state and no flux can penetrate, *i.e.*, $H_y(x) = 0$ inside the strip and outside the strip close to its surface. The shielding current density is then $J(x, y) = J_s(x) \cosh(y/\lambda)/2\lambda \sinh(d/2\lambda)$ where $J_s(x) = \int_{-d/2}^{d/2} J(x, y) \, dy$ is the sheet current. All considerations of this section apply to both thin ($d < \lambda$) and thick ($d > \lambda$) films provided $d \ll a$. In general, the sheet current $J_s$ (flowing along $-z$) generates a parallel field at the specimen surfaces, $H_x(x, d/2) = -H_x(x, -d/2) = -\frac{1}{2}J_s(x)$ since the sheet current determines the jump of $H_x$. The perpendicular field component (flux density) $H_y$ in the plane $y = 0$ follows from Ampère's law, $H_y(x, d/2) = H_y(x, -d/2) \approx H_y(x, 0) \approx H(x)$ with

$$H(x) = \frac{1}{2\pi} \int_{-a}^{a} \frac{J_s(u) \, du}{x - u} + H_a. \qquad (9)$$

Eq. (9) may be inverted to give [52]

$$J_s(x) = \frac{2}{\pi} \int_{-a}^{a} \frac{H(u) - H_a}{x - u} \left( \frac{a^2 - u^2}{a^2 - x^2} \right)^{1/2} du. \qquad (10)$$

132

In the Meissner state or for small $H_a$ one has ideal screening of the applied field, *i.e.*, $H(x) \equiv 0$ for $-a < x < a$. This corresponding shielding current may be obtained from (10) or by conformal mapping of the strip cross section unto a circle. The result is [52]

$$J_{s\ ideal}(x) = 2xH_a/(a^2 - x^2)^{1/2}. \tag{11}$$

From (9) and (11) follows the field outside the strip,

$$H_{ideal}(|x| > a) = |x|H_a/(x^2 - a^2)^{1/2}. \tag{12}$$

The negative total magnetic moment per unit length $M = \int_{-a}^{a} xJ_s(x)\,dx$ [cf. Eq. (6)] from (11) is $M_{ideal} = \pi a^2 H_a$. Note that, as stated after Eq. (6), half of $M$ comes from the longitudinal currents flowing in the bulk and the other half from the transverse currents flowing near the ends of the strip where the current performs a U-turn, but due to $\text{div}\mathbf{J} = 0$, $M$ may be expressed in terms of the $z$-component of the sheet current only; this cancels the factor $1/2$ in the original definition of the magnetic moment $\frac{1}{2}\int \mathbf{J} \times \mathbf{r}\,d^3r$.

I consider now the model where the sheet current is limited to the critical value $J_c d$ which as in the Bean model is assumed to be independent of the flux density $\mu_0 H(x)$.

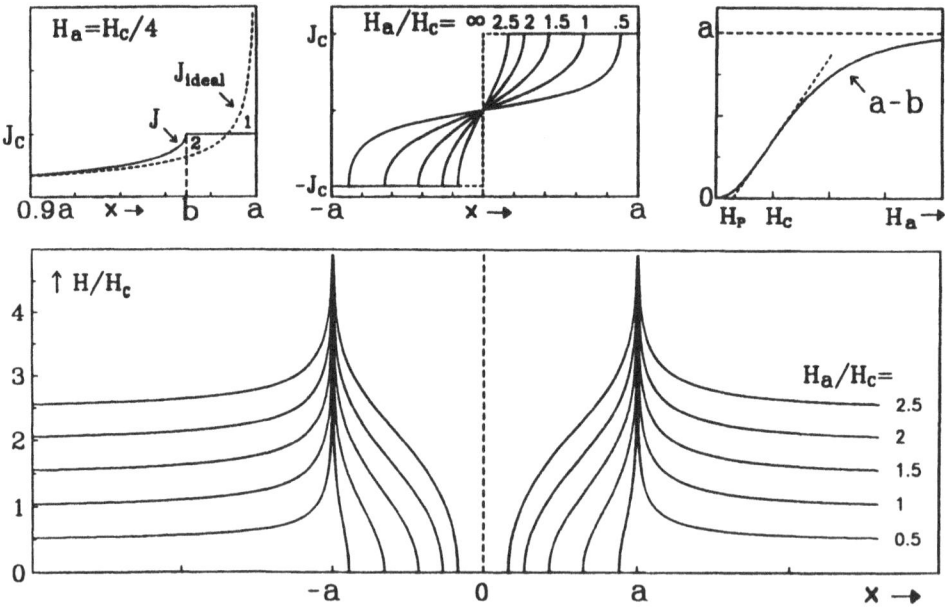

Figure 3: *Top left:* Saturation of the current density near the specimen edges at small applied field $H_a = H_c/4$. *Top middle:* Sheet current $J_s(x)$ in the strip for fields $H_a/H_c$ = 0.5, 1, 1.5, 2, 2.5. *Top right:* Penetration width $a - b = a - a/\cosh(H_a/H_c)$. *Bottom:* Perpendicular local magnetic field $H(x)$ for applied fields $H_a/H_c$ = 0.5, 1, 1.5, 2, 2.5.

Thus, when $H_a$ is increased the shielding current near the specimen edges saturates to $J_s(x > b) = J_c d$ and $J_s(x < -b) = -J_c d$ (Fig. 3) since flux starts to penetrate in form of flux lines such that $H(x) \neq 0$ for $|x| > b$ but still $H(x) \equiv 0$ for $-b < x < b < a$. The complete current distribution $J(x)$ of this two-dimensional problem with field independent $J_c$ can be obtained by conformal mapping as shown by Norris [53] for the similar problem of a strip with transport current. One first has to find the shielding current $J(x, x_0)$ in the region $|x| < b$ which compensates the magnetic field caused in $|x| < b$ by a pair of currents $I = J_c dx$ flowing at $x = x_0$ along $-z$ and at $x = -x_0$ along $z$. The final $J(x)$ in $|x| < b$ is then obtained by integrating the result

$$J(x, x_0) = x\, I\, (x_0^2 - b^2)^{1/2} / \pi (b^2 - x^2)^{1/2} (x_0^2 - x^2) \tag{13}$$

from $x_0 = b$ to $x_0 = a$ and adding a current $-2x H_a / (b^2 - x^2)^{1/2}$ [cf. (11)] which compensates $H_a$ in $|x| < b$. The condition that $J_s(x)$ be continuous, or $H(x)$ finite inside the specimen, then yields a relationship between $b$ and $H_a / J_c d$ which may be written as $(a^2 - b^2)^{1/2} = b \sinh(\pi H_a / J_c d)$ or $b = a / \cosh(H_a / H_c)$ or

$$(1 - b^2/a^2)^{1/2} \equiv c = \tanh(H_a / H_c) \tag{14}$$

where $H_c = J_c d / \pi$ is a critical field. The current and negative total magnetic moment in this critical virgin state (for $H_a$ increased from zero) are (Fig. 3)

$$J_s(x) = \begin{cases} (2 J_c d / \pi) \arctan[cy/(b^2 - x^2)^{1/2}], & |x| < b \\ J_c d\, x / |x|, & b < |x| < a \end{cases} \tag{15}$$

$$M(H_a) = J_c d a^2 c = J_c d a^2 \tanh(H_a / H_c), \tag{16}$$

with $c$ and $b$ from (14). From (9) and (16) the field $H$ and the total penetrated flux $\Phi$ are

$$H(x) = \begin{cases} 0, & |x| < b \\ H_c \operatorname{artanh}[(x^2 - b^2)^{1/2}/c\,|x|], & b < |x| < a \\ H_c \operatorname{artanh}[c\,|x|/(x^2 - b^2)^{1/2}], & |x| > a \end{cases} \tag{17}$$

$$\Phi(H_a) = 2\mu_0 H_c a \ln(a/b) = 2\mu_0 H_c a \ln\cosh(H_a / H_c). \tag{18}$$

The above integrations may be performed in the complex $z$-plane by substituting $z = cx/(x^2 - b^2)^{1/2}$. Note that in the limit $J_c \to \infty$ Eq. (14-18) reproduce the ideal-shielding results (11), (12), $M_{ideal} = \pi a^2 H_a$, and $\Phi_{ideal} = 0$. For *weak penetration* ($H_a \ll H_c$) one gets from (14-18) $b = a - a H_a^2 / 2 H_c^2$, $M = \pi a^2 H_a (1 - H_a^2 / 3 H_c^2)$, and flux $\Phi = \mu_0 a H_a^2 / H_c$. For *almost complete penetration* ($H_a \gg H_c$) one gets $b = 2a \exp(-H_a / H_c) \ll a$, $M = J_c d a^2 [1 - 2\exp(-2H_a / H_c)]$, $J_s(|x| < b) = (2 J_c d / \pi)\arcsin(x/b)$, $H(b < |x| < a/2) = H_c\operatorname{arcosh}|x/b|$, $H(2b < |x| < \infty) = H_a + H_c \ln|a^2/x^2 - 1|^{-1/2}$, and $\Phi = 2\mu_0 a (H_a - 0.69 H_c)$. Note the vertical slopes of $H(x)$ and $J(x)$ at $|x| = b$ and the logarithmic infinity of $H(x)$ at $|x| \to a$. The analytical solution $H(x)$ (17) thus proves a *vertical slope of the penetrating flux front* even for our model in which $J_c = $ const and the lower critical field $H_{c1} = 0$.

Comparing the results (14-17) for the strip with the results obtained very recently [54] for the critical state of a disk in perpendicular field one sees that the penetration width (14) and current distribution (15) exhibit the same functional dependences in both geometries if one replaces $x/a$ by $r/R$, $J$ by $2J/\pi$, and $J_c$ by $2J_c/\pi$, but the magnetic moment and field (obtained for the disk numerically in [54]) are different.

Let us now consider the critical state which is established when the applied field $H_a$ oscillates between the extremal values $\pm H_0$. It suffices to consider one half period, say, decreasing $H_a$. One can easily show that for $H_a$ decreasing from $+H_0$ to $-H_0$ the current distribution $J_\downarrow$, field $H_\downarrow$, flux $\Phi_\downarrow$, and magnetic moment $M_\downarrow$, follow from the "virgin" results (14) to (18) ($H_a$ increased from zero) by linear superpositions of the form (Fig. 4)

$$J_{s\downarrow}(x, H_a, J_c) = J_s(x, H_0, J_c) - J_s(x, H_0 - H_a, 2J_c) \tag{19}$$

$$H_\downarrow(x, H_a, J_c) = H(x, H_0, J_c) - H(x, H_0 - H_a, 2J_c) \tag{20}$$

$$\Phi_\downarrow(H_a, J_c) = \Phi(H_0, J_c) - \Phi(H_0 - H_a, 2J_c). \tag{21}$$

$$M_\downarrow(H_a, J_c) = M(H_0, J_c) - M(H_0 - H_a, 2J_c). \tag{22}$$

At $H_a = -H_0$ the original virgin state is reached again; $J$, $H$, $\Phi$ and $M$ have just changed sign. In the half period with increasing $H_a$ one has $J_{s\uparrow}(x, H_a, J_c) = -J_{s\downarrow}(x, -H_a, J_c)$, $H_\uparrow(x, H_a, J_c) = -H_\downarrow(x, -H_a, J_c)$, $\Phi_\uparrow(H_a, J_c) = -\Phi_\downarrow(-H_a, J_c)$, and $M_\uparrow(H_a, J_c) = -M_\downarrow(-H_a, J_c)$. The hysteresis loop for the magnetic moment is [cf. Eq. (7)]

$$M_{\downarrow\uparrow}(H_a) = \pm J_c a^2 \{\tanh(H_0/H_c) + 2\tanh[(H_a \mp H_0)/2H_c]\}. \tag{23}$$

At intermediate fields $|H_a| < H_0$ there is a new penetration width $b' = a/\cosh(|H_0 \pm H_a|/2H_c)$ inside which $H$ is frozen like in the Bean model though $J$ changes *everywhere*,

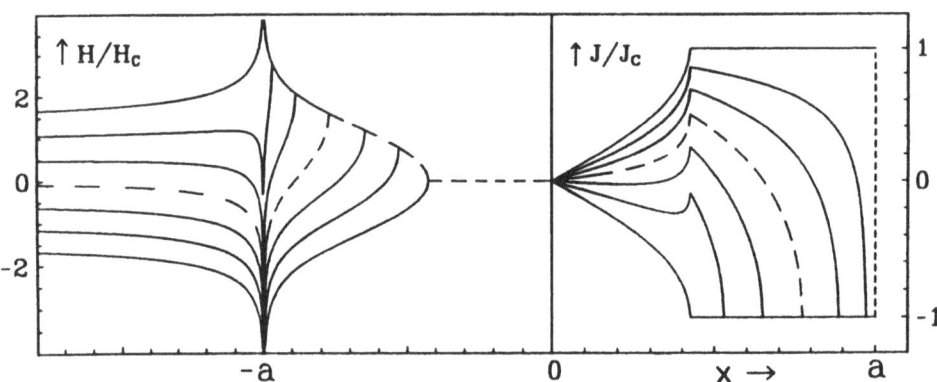

Figure 4: . Magnetic field $H(x)$ (*left*) and sheet current $J_s(x)$ (*right*) when the applied field $H_a/H_c$ is first increased to the value 1.5 (upper curve) and then monotonically decreased to the values 1, 0.5, 0 (trapped flux, dashed curve), -0.5, -1, -1.5 (lower curve). The corresponding curves for increasing $H_a$ are obtained by changing the sign of $H(x)$ and $J_s(x)$.

with the jump at $|x| = b = a/\cosh(H_0/H_c)$ persisting and new saturation occuring at $|x| > b'$.

For *general* magnetic history $H_a(t)$ ($t$ = time) the critical state of the strip is described by a similar superposition as (19–22) which now depends on $H_a(t)$ and on all previously reached maximum values of $|H_a(t)|$ such that each maximum wipes out the memory of all previously reached smaller maxima. Thus, the solution (14-18) *completely determines any possible critical state in the slab.*

Finally, I give the hysteretic losses of the strip when $H_a$ oscillates with frequency $\nu$ and amplitude $H_0$. From (16, 22) the energy dissipation $P = \nu\mu_0 \oint M(H_a)\,dH_a$ per unit length (frequency times area of the hysteresis loop) is

$$P(H_0) = 4\nu\mu_0 a^2 J_c H_0 \left(\frac{2}{h}\ln\cosh h - \tanh h\right) \tag{24}$$

with $h = H_0/H_c$. For $H_0 \ll H_c$ this gives $P = (2\pi\nu\mu_0 a^2/3H_c^2)H_0^4$ and for $H_0 \gg H_c$ $P = 4\nu\mu_0 a^2 J_c(H_0 - 1.386 H_c)$. The energy loss is thus initially very small, $P \sim H_0^4$.

The above results have important consequences. As shown for the first time experimentally and theoretically in [55], where the same problem was solved numerically, perpendicular flux penetrates into flat samples *delayed* as if there were a surface barrier, with penetration depth $a-b \sim H_a^2$. Extrapolation of $b(H_a)$ in Fig. 3 by a straight line through the inflection point defines a pseudo first critical field $H_p = H_c[\mathrm{arsinh}(1)+2^{1/2}-2] \approx 0.296 H_c$. If $J_c$ is high this $H_p$ can exceed the thermodynamic penetration field $\approx H_{c1}d/a$ by far. Thus, numerous measurements which for perpendicular $H_a$ show an up-turn of $H_{c1}(T)$ at low temperatures (where also $J_c$ greatly increases, for references see [56]) can be interpreted in terms of flux penetration controlled simply by constant volume pinning as calculated in the constant-$J_c$ model. Moreover, the above obtained cubic deviation of the magnetic moment from the ideal screening value $\delta M = M - \pi a^2 H_a \sim H_a^3$ explains, without the assumption of a surface barrier, why the "cubic fit" of $\delta M(H_a)$ proposed in [56] to get the expected BCS temperature dependence of $H_{c1}$ works so well. It appears that models of a surface barrier are not required in this case and all experimental data can be described in a natural way by the critical state model.

The slow initial penetration of the magnetic flux $\Phi \sim H_a^2$ can be explained by reconsidering the process of current saturation near the strip edges. The depth where the ideal screening current (11) exceeds $J_c d$ and the current saturates is $\delta_{est} \approx 2aH_a^2/J_c^2 d^2$ at $H_a \ll J_c d$. This estimate is smaller by a factor $\pi^2/4 \approx 2.5$ than the precise analytical value $a - b = aH_a^2/2H_c^2$ and *should not be used* in quantitative calculations. The physical reason for this deeper penetration is a redistribution of the currents after cutting off the area 1 in Fig. 3: In order to compensate for the decreased screening, more current (area 2 in Fig. 3) has to be added in the Meissner region. The magnetic moments of these two areas nearly compensate. Therefore, the simple cutoff gives a too large decrease of the magnetization with a wrong $H_a$ dependence $\delta M_{est} \sim H_a^2$ instead of the correct $\delta M \sim H_a^3$.

These analytical results for the strip show that the current and field profiles in thin type-II superconductors in perpendicular field $H_a$ qualitatively differ from the Bean model in many ways:

1. The flux penetration initially is *quadratic* (Bean: linear) in $H_a$, *i.e.*, it looks "delayed" as if there were a surface barrier.

2. The deviation of the magnetic moment from linearity and the remanent magnetic moment initially are *cubic* (Bean: quadratic) in $H_a$.

3. The ac losses initially grow with the *fourth power* (Bean: third) of the amplitude of $H_a$.

4. The penetrating flux front has *vertical* (Bean: constant finite) slope even in the model where a field-independent $J_c$ and vanishing lower critical field $H_{c1}$ are assumed.

5. When the flux has partly penetrated and a critical state with $J_s = J_c d$ is established near the edges of the specimen, the current flows over the entire width of the disk or strip in order to shield the central flux-free region (Bean: flux-free region is current free).

6. The screening current density is a continuous (Bean: piecewise constant) function but has a *vertical slope* at the flux front where it reaches the saturation value $J_c$.

7. As soon as the direction of the change of $H_a$ is reversed the sheet current $J_s$ falls below $J_c d$ *everywhere* (Bean: $J = J_c$); relaxation will therefore stop effectively, at least in thin films where the current density is uniform, $J = J_s/d$. In thick films ($d \gg \lambda$) the (integral) sheet current may be composed of current densities $\pm J_c$ or zero, flowing in various layers in which a usual current-induced Bean critical state occurs *across the specimen thickness*.

8. When $H_a$ is reversed there appears a spatial region where the driving force on the flux lines acts outwards *against* the pressure of the flux-density gradient (Bean: driving force = flux pressure).

Though some of these features were indicated in numerical calculations of the critical state in circular disks [57-59], most pecularities of perpendicular geometry were discovered and proven only by *analytical* calculations in the limit of zero thickness $d$. Such *exact* solutions are available for a current carrying strip [53] and for a disk [54] and strip [51] in perpendicuar field. A recent paper [60] compiles and completes these results and extends them to the simultaneous presence of an applied field $H_a$ and transport current $I$ in a strip or flat ring and to arbitrary magnetic history. The flux distribution at the surface of such flat superconductors can be observed by magneto-optics, *e.g.*, using the Faraday effect in thin layers of EuSe [59, 61].

# 3. Thermally Activated Depinning

## 3.1. PINNING FORCE AND PINNING ENERGY

At finite temperatures $T$ thermally activated depinning of flux lines may occur as predicted by Anderson [46]. In *conventional superconductors* this effect is observed only close to the transition temperature $T_c$ as *flux creep* [62]. Flux creep occurs in the critical state after the applied magnetic field is changed. The field gradient, and the persistent currents and magnetization, then slowly *decrease with a nearly logarithmic time law*. Formally, this flux creep is equivalent to a highly nonlinear, current dependent flux-flow resistivity, *e.g.*, $\rho \propto \exp(J/J_1)$ or $\rho \propto \exp(-J_2/J)$. Initially, the persistent currents in a ring feel a large $\rho$, but as the current decays, $\rho$ decreases rapidly and so does the decay rate $-\dot{J} \propto \rho(J)$.

In HTSC thermal depinning is observed in a large temperature interval below $T_c$. This "giant flux creep" [63-65] occurs mainly because (a) the superconducting coherence length $\xi$ ($\approx$ vortex core radius) is small, (b) the magnetic penetration depth $\lambda$ is large, and (c) these materials are strongly anisotropic; here $\xi = \xi_{ab}$, $\lambda = \lambda_{ab}$, and $\kappa = \lambda_{ab}/\xi_{ab}$ is used. All three properties decrease the pinning *energy* but tend to increase the pinning *force* as will be discussed now.

Small $\xi$ means that the elementary pinning *energy* $U_p$ of small pins, (*e.g.*, oxygen vacancies or clusters thereof) is small, of the order of $(B_c^2/\mu_0)\xi^3 = (\Phi_0^2/8\pi^2\mu_0\lambda^2)\xi$. The elementary pinning *force* $U_p/\xi$, however, is *independent* of $\xi$ in this estimate and is thus not necessarily small in HTSC.

Large $\lambda$ means that the stiffness of the FLL with respect to shear deformation and to short-wavelength tilt is small. Therefore, the flux lines can better adjust to the randomly positioned pins. This flexibility increases the average pinning force density. (The argument that a soft FLL or a flux-line liquid with vanishing shear stiffness cannot be pinned since it may flow around the pins, does not apply to the realistic situation where there are many more pins than flux lines.) The statistical summation of pinning forces at $T = 0$ [66] and $T > 0$ [67-69] requires the correct (non-local) elasticity theory of the FLL [13-18, 31, 70-74].

For $\kappa \gg 1$ and $B < 0.2B_{c2}$ the FLL with $B\|z\|c$-axis has a shear modulus $c_{66} = B\Phi_0/(16\pi\mu_0\lambda_{ab}^2)$ and a tilt modulus which depends on the wave vector $\mathbf{k} = (\mathbf{k}_\perp, k_z)$ of a periodic tilt deformation [31, 70, 71]

$$c_{44}(\mathbf{k}) = \frac{B^2}{\mu_0}\left[\frac{1}{1 + \lambda_c^2 k_\perp^2 + \lambda_{ab}^2 k_z^2} + \frac{f(k_z)}{\lambda_{ab}^2 k_{BZ}^2}\right] \tag{25}$$

$$f(k_z) = \frac{1}{2\Gamma^2}\ln\frac{\xi_c^{-2}}{\lambda_{ab}^{-2} + k_z^2 + \Gamma^2 k_0^2} + \frac{\ln[1 + k_z^2/(\lambda_{ab}^{-2} + k_0^2)]}{2k_z^2\lambda_{ab}^2} \tag{26}$$

where $k_0 \approx k_{BZ} = (4\pi B/\Phi_0)^{1/2}$ is a cutoff radius (radius of the Brioullin zone) and $\xi_c = \xi_{ab}/\Gamma$. Thus, for short tilt wavelengths $2\pi/k_z$ the tilt modulus in anisotropic HTSC may become very small, even smaller than the shear modulus,

$$c_{44}(k_z \gg \lambda_{ab}^{-1}) = (B\Phi_0/4\pi\mu_0\lambda_{ab}^2)\ln(1/k_z\xi_c) = [4\ln(\Gamma/k_z\xi_{ab})/\Gamma^2]\,c_{66}. \tag{27}$$

This means the correlation between the 2D vortex lattices in adjacent planes is easily destroyed be weak pinning or by thermal fluctuations, cf. Sct. 1.2 and 4.1.

Large material anisotropy effectively softens the FLL and *increases* the average pinning *force*, but it *decreases* the pinning *energy*. For example, long columnar pins generated by high-energy (500 MeV) heavy-ion irradiation perpendicular to the CuO planes in YBCO [75] or BSCCO [76] are most effective pins at low $T$ if the flux lines are parallel to the pins. However, at higher $T$, columnar pins can pin flux lines only in YBCO, while in the very anisotropic BSCCO the flux lines easily break into short segments or point vortices which then depin individually with very small activation energy [76-80], Fig. 1. As a consequence, BSCCO tapes are good superconductors only at $T = 4$ K. In principle, if $B$ could be kept strictly parallel to the layers, large $J_c$ and weak thermal depinning could be achieved even at $T = 77$ K, but this geometric condition is not easily realized.

## 3.2. THE KIM-ANDERSON THEORY

A novel feature of HTSC is that a linear (ohmic) resistivity $\rho$ is observed at small current densities $J \ll J_c$ in the region of thermally assisted flux flow (TAFF) [64, 65]. Both effects, flux creep at $J \approx J_c$ and TAFF at $J \ll J_c$, are limiting cases of Anderson's [46] general expression for the electric field $E(B,T,J)$ caused by thermally activated flux jumps out of pinning centers, which may be written as [81]

$$E(J) = 2\rho_c J_c \exp(-U/k_B T) \sinh(JU/J_c k_B T) \,. \tag{28}$$

In (28) $J_c(B)$ (the critical current density at $T = 0$), $\rho_c(B,T)$ (the resistivity at $J = J_c$), and $U(B,T)$ (the activation energy for flux jumps) are *phenomenological parameters*. The physical idea behind eqn. (28) is that the Lorentz force density $\mathbf{J} \times \mathbf{B}$ acting on the FLL *increases* the rate of thermally activated jumps of flux lines or flux-line bundels along the force, $\nu_0 \exp[-(U - W)/k_B T]$, and *reduces* the jump rate for backward jumps, $\nu_0 \exp[-(U + W)/k_B T]$. Here $U(B,T)$ is an activation energy, $W = JBVl$ the energy gain during a jump, $V$ the jumping volume, $l$ the jump width, and $\nu_0$ is an attempt frequency. All these quantities depend on the microscopic model, which is still controversial, but by defining a critical current density $J_c = JU/W = U/BVl$, only measurable quantities enter. Subtracting the two jump rates to give an effective rate $\nu$ and then writing the drift velocity $v = \nu l$ and the electric field $E = vB = \rho J$ one obtains (28).

For large currents $J \approx J_c$ one has $W \approx U \gg k_B T$ and thus $E \propto \exp(J/J_1)$ with $J_1 = J_c k_B T/U$. For small currents $J \ll J_1$ one may linearize the $\sinh(W/k_B T)$ in (28) and gets *ohmic* behavior with a thermally activated linear resistivity $\rho_{TAFF} \propto \exp(-U/k_B T)$. Combining (28) with the usual flux-flow resistivity $\rho_{FF}$ valid at $J \gg J_c$, or with the square-root result for a particle moving viscously across a one-dimensional sinusoidal potential (see appendix in [82]) one gets (Fig. 5)

$$\rho = (2\rho_c U/k_B T) \exp(-U/k_B T) = \rho_{TAFF} \quad \text{for } J \ll J_1 \quad \text{(TAFF)} \tag{29}$$

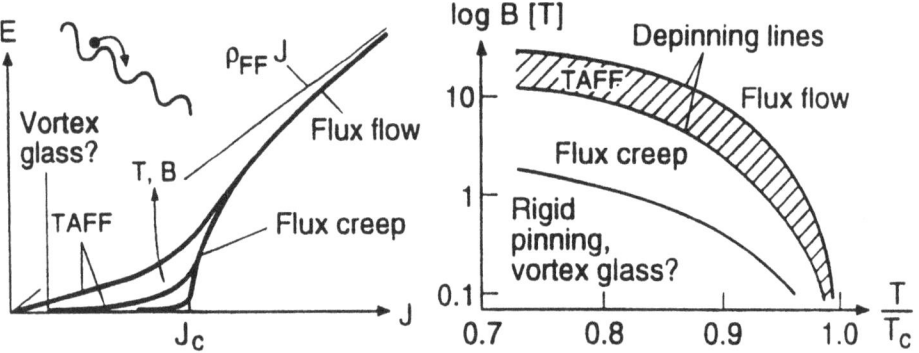

Figure 5: *Left:* Current–voltage curves. $E$ = electric field, $J$ = current density. $J \gg J_c$: flux flow, $E \approx J\rho_n$; $J \approx J_c$: flux creep, $E \propto \exp(J/J_1)$; $J \ll J_1 < J_c$: thermally assisted flux flow (TAFF), $E \approx J\rho_n \exp(-U/k_B T)$, or, at low $T$: vortex glass, $E \propto \exp[-(J_2/J)^\alpha]$. The insert shows a tilted periodic pinning potential with jumping flux-line bundle. *Right:* Depinning lines in the field–temperature plane separate the region of flux flow (with complete depinning) from the regions of flux creep (with slow logarithmic relaxation) and rigid pinning (with hysteretic behavior). Near the depinning lines (dashed area) thermally assisted flux flow (TAFF) occurs and the resistivity is linear.

$$\rho = \rho_c \exp[(J/J_c - 1)U/k_BT] \propto \exp(J/J_1) \qquad \text{for } J \approx J_c \quad \text{(flux creep)} \qquad (30)$$

$$\rho = \rho_{FF}(1 - J_c^2/J^2)^{1/2} \approx \rho_{FF} \approx \rho_n B/B_{c2}(T) \qquad \text{for } J \gg J_c \quad \text{(flux flow)}. \qquad (31)$$

The linear TAFF regime (29) is seen in experiments if $B$ and $T$ are sufficiently large [83, 84-88]. At lower $T$ and $B$ non-linear resistivity is observed [83, 85-92]. It appears that in the TAFF regime the FLL is in a "liquid" state [69], *i.e.*, it has no shear stiffness; therefore, elastic deformations of the FLL at different points are not correlated. This assumption leads to an activation energy $U$ which does not depend on the current density, or to an effective barrier $U_{eff} = (1 - J/J_c)U$, cf. (30).

## 3.3. COLLECTIVE CREEP AND VORTEX GLASS SCALING

Theories of collective pinning [67-69] going beyond Anderson's model (28) predict that the thermally jumping volume $V$ of the FLL depends on the current density $J$ and becomes infinitely large for $J \to 0$. As a consequence, also the activation energy diverges, *e.g.*, $U \propto V \propto 1/J^\alpha$ with $\alpha > 0$, thus the resistivity becomes truly zero as $J \to 0$. This result follows for weak random pinning if the FLL is treated as an elastic medium. A diverging activation energy is also obtained in theories of depinning of vortices from the space between the CuO layers by a kink mechanism [93, 94] and from columnar pins [77-80].

A similar result is arrived at by the "vortex glass" picture [95-96]. Its basic idea is that if there is a glass-transition temperature $T_G$ in the vortex-pin system similar to that in theories of spin glasses, then a characteristic length $\xi_G$ (the size of the jumping volume) in the FLL should diverge as $\xi_G \propto |T - T_G|^\nu$ ($\nu \approx 1$) when $T$ approaches $T_G$. The vortex-glass picture predicts scaling laws, *e.g.*, the electric field should scale as $E\xi_G^{z-1} = f_\pm(J\xi_G^{D-1})$ where $z \approx 4$, $D$ is the spatial dimension, and $f_\pm(x)$ are scaling functions for the regions above and below $T_G$. For $x \to 0$ one has $f_+(x) = $ constant and $f_-(x) \to \exp(-x^{-\mu})$. At $T_G$, a power-law current–voltage curve is expected, $E \propto J^{(z+1)/(D-1)}$, thus

$$\rho \propto J^{(z+1)/(D-1)-1} \qquad \text{for } T = T_G \qquad (32)$$

$$\rho \propto \exp[-(J_2/J)^\alpha] \qquad \text{for } T < T_G . \qquad (33)$$

In the theory of collective pinning [67, 97] there is no explicit glass temperature, but the picture is similar since collective creep occurs only below a "melting temperature" $T_m$ above which the FLL looses its elastic stiffness. Thus $T_m$ has a similar meaning as $T_G$. A vortex glass state should not occur in 2D flux-line lattices [22, 23]. More details may be found in [96-99] and in the review paper [74].

Experiments which measure the magnetization decay or the voltage drop with high sensitivity appear to confirm this scaling law in various HTSC in an appropriate range of $B$ and $T$. For example, by plotting $T$-dependent creep rates $\dot{M} = dM/dt$ in reduced form, $(1/J)|1 - T/T_G|^{-\nu(z-1)}\dot{M}$ versus $J|1 - T/T_G|^{-2\nu}$, van der Beek at al. [87] in BSCCO measured $T_G = 13.3\,\text{K}$, $z = 5.8 \pm 1$, and $\nu = 1.7 \pm 0.15$, Fig. 6. Very detailed curves $\rho(J)$ for three YBCO samples with different pinning (without and with irradiation with protons or Au ions) are presented by Worthington *et al.* [83] for different fields $B_a$ with $T$ as parameter. In the sample with intermediate pinning two transitions are seen in $\rho(J)$, a "melting transition" at $T_m$ (*e.g.*, $T_m \approx 91.5\,\text{K}$ at $B = 0.2\,\text{T}$, $J = 10^5\,\text{Am}^{-2}$) and a "glass transition" at $T_m \approx 90\,\text{K}$ (above case) or $T_m = 84.92\,\text{K}$ (strong pinning sample, very sharp transition at $B = 4\,\mathrm{T}$, $J \leq 4 \times 10^5\,\mathrm{Am^{-2}}$). The FLL phase in between these two transitions was named "vortex slush".

## 3.4. THE CREEP RATE

The decay of shielding currents, or of the magnetization, in principle is completely determined by the geometry and by the resistivity $\rho(T, B, J)$. In superconducting rings and hollow cylinders in axially oriented magnetic field one simply has $\dot{M} \propto \dot{J} \propto \rho(T, B, J)$. Within the Kim-Anderson model (28) the decay of persistent currents can be calculated analytically for all times [100, 101]. A large range of electric fields $E = 10^{-13}$ to $10^{-1}\,\mathrm{V/m}$ was measured in [91] by combining current–voltage curves with highly sensitive measurements of decaying currents in rings of YBCO films (3 mm diameter, 0.1 mm width, 200 nm thickness).

The current dependent activation energy $U(J)$ may be extracted from experiments by the method of Maley and Willis [92], Fig. 6, see e.g. [87]. The Kim-Anderson model (28) originally means an effective activation energy $\propto J_c - J$, Eq. (30), and actually corresponds to a zig-zag shaped pinning potential. As shown by Beasley et al. [62] a more realistic smooth potential yields $U \propto (J_c - J)^{3/2}$. Collective creep theory yields $U \propto 1/J^\alpha$ with $\alpha > 0$ depending on $B$ and the pinning strength. For single-vortex pinning one predicts $\alpha = 1/7$, for short jumps $l < a$ $\alpha \approx 7/9$ [67], and for long jumps $l > a$ $\alpha \approx 1/2$ [68]. Other experiments suggest a logarithmic dependence $U \propto \ln(J_2/J)$ (corresponding to the limit $\alpha \to 0$) for which the creep rate can be calculated analytically [102]. Combining the collective creep result $U(J) = U_c(J_c/J)^\alpha$ for $J \ll J_c$ with the Kim-Anderson formula $U(J) = U_c(1 - J/J_c)$ for $J \approx J_c$ one obtains for rings and cylinders the decaying current

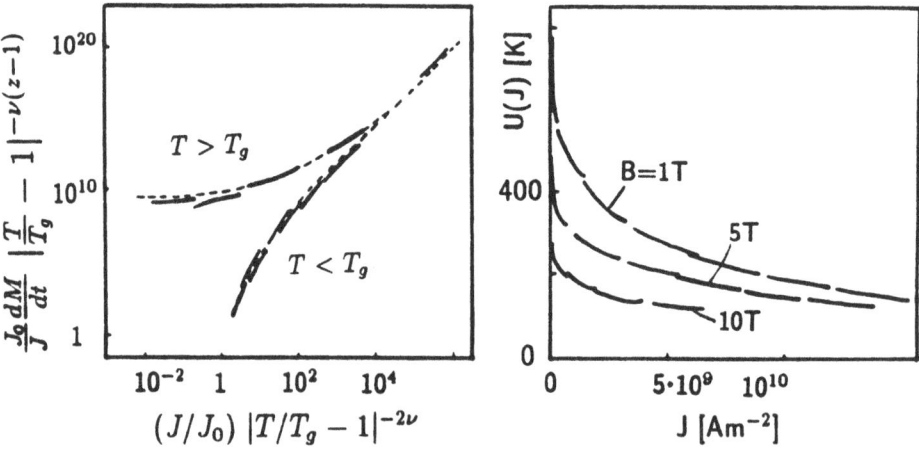

Figure 6: *Left:* Scaling plot (Sct. 3.3) of the decay rate (in units A/ms) versus current density (in units $J_0 = k_B T/\phi_0$), from [87]. *Right:* The (nearly temperature independent) activation energy $U(J)$ extracted from various experiments on the same sample at different temperatures $T$ for fields $B$ of 1, 5, and 10 Tesla. Schematic from [87].

density

$$J(t) \approx J_c \left[ 1 + \alpha \frac{k_B T}{U_c} \ln(1 + \frac{t}{t_0}) \right]^{-1/\alpha} \tag{34}$$

$$J(t) \approx J_c (t/t_0)^{-k_B T/U_c} \quad \text{for} \quad \alpha \ll 1 \tag{35}$$

where $t_0$ is an integration constant. From (35) one obtains a temperature dependent $J(T)$ and the normalized creep rate

$$J(T) \approx J_c \exp(-T/T_0), \quad T_0 \approx U_c/[k_B \ln(t/t_0)] \tag{36}$$

$$S = -\frac{d \ln J}{d \ln t} \approx \frac{k_B T}{U_c + \alpha k_B T \ln(1 + t/t_0)} \tag{37}$$

Experiments on HTSC yield $T_0 \approx 10 \, \text{K}$ [103] and $U_c/k_B \approx 100$ to $1000 \, \text{K}$. Note that the decay rate $S$ (37) decreases with increasing time and saturates at $k_B T > U_c/[\alpha \ln(t/t_0)]$ to a value $1/[\alpha \ln(t/t_0)]$. Numerical solutions to the relaxation rates for various dependencies $U(J)$ are given in [101]. Good fits to relaxation rates are also achieved by fitting a spectrum of activation energies [104-107]. The success of theories which assume only *one* activation energy $U$ suggests that at given values of $B$, $J$, and $T$ essentially *one* effective $U$ of an entire spectrum determines the physical process under consideration.

## 4. Further Topics Related to Depinning

### 4.1. THERMAL FLUCTUATION OF THE FLUX-LINE LATTICE

In HTSC the thermal fluctuations of the flux-line positions $\langle u^2 \rangle$ can become large, comparable to the flux-line spacing $a$. According to the Lindemann criterion one expects melting of a lattice if $\langle u^2 \rangle/a^2 \geq c_L^2$ where $c_L \approx 0.05 \ldots 0.2$. The fluctuations $\langle u^2 \rangle$ were first calculated from local elasticity in [108] and in the extreme nonlocal limit in [109]. The correct calculation [15, 31, 110] yields a $\langle u^2 \rangle$ which practically coincides with [109] but is typically much larger than the local result [108]. This enhancement is caused by the dispersion of the tilt modulus $c_{44}(\mathbf{k})$ (25). One finds

$$\langle u^2 \rangle \approx k_B T \int_0^{k_{BZ}} dk_\perp \, k_\perp \int_0^\infty \frac{dk_z/2\pi^2}{c_{66}k_\perp^2 + c_{44}(\mathbf{k})k_z^2} \tag{38}$$

$$\langle u^2 \rangle \approx \left( \frac{k_B^2 T^2 \mu_0}{4\pi B \Phi_0 c_{66}} \right)^{1/2} \times \left( \frac{B\kappa^2/2}{B_{c2} - B} \right)^{1/2} \times \frac{\lambda_c}{\lambda_{ab}}. \tag{39}$$

With $c_{66} = B\Phi_0/(16\pi\mu_0\lambda_{ab}^2)$ (for $B \ll B_{c2}$) and $k_{BZ}^2 = 4\pi B/\Phi_0$ inserted, (39) yields $\langle u^2 \rangle \approx k_B T \mu_0 \lambda_{ab} \lambda_c (4\pi/B\Phi_0^3)^{1/2}$. The fluctuation (39) is written in the form: Local result [108] times a typically large nonlocal correction factor $(B\kappa^2/2B_{c2})^{1/2} \approx (B \ln \kappa/4B_{c1})^{1/2} \gg 1$ times the anisotropy ratio $\lambda_c/\lambda_{ab} = \Gamma \gg 1$.

At present it is not quite clear whether a (hypothetical) pin-free 3D FLL has a sharp melting transition and what the properties of this line-liquid would be. The problem is even more complex when pinning is included. The Lindemann criterion $\langle u^2 \rangle^{1/2}/a = c_L \approx 0.05 \ldots 0.2$ and similar conditions for melting [110, 111] yield a rather low 3D "melting temperature" $T_M^{3D}$. The Lindemann criterion does not work in the ideal 2D FLL of thin films since $\langle u^2 \rangle$ diverges there. Finite pinning removes this divergence. Formally one may write $\langle u^2 \rangle_{film} = (k_B T/4\pi^2) \int_{BZ} d^2 k_\perp /(c_{66} k_\perp^2 + \alpha_L)$ where $d$ is the film thickness and $\alpha_L$ is the elastic pinning restoring force density (Labusch parameter [50], cf. Sct. 5.1) which in general has to be determined self-consistently by a theory of collective pinning. Computer simulations of the 3D FLL indicate a first order melting transition [112-114].

Recent resistivity data [83, 89, 115] as a function of $T$, $B$, and $J$ appear to indicate a melting transition in weak-pinning YBCO. However, as will be discussed in Sct. 6, several experiments which were interpreted as evidence for a phase transition can be explained by the rather abrupt onset of thermally activated depinning when $T$ or $B$ are increased, in particular some ac experiments and experiments on vibrating superconductors in a magnetic field [116-118], see also the review [119] and Sct. 6.3. However, other such experiments on very weak-pinning untwinned YBCO crystals show a very sharp damping peak which has been identified with FLL melting [120].

### 4.2. TUNNELING OF VORTICES OUT OF PINS AT T=0

In numerous experiments the creep rate as a function of $T$ tends to a finite value at zero temperature. Flux creep observed at low temperatures [121] in principle may be explained by usual thermal activation from smooth shallow pinning wells [122] but it may also indicate "quantum tunneling" of vortices out of the pins [97, 123-126]. The thermal depinning rate $\propto \exp(-U/k_B T)$ is now replaced by the tunneling rate $\propto \exp(-S_E/\hbar)$ where $S_E$ is the Euklidean action of the considered tunneling process. Tunneling of vortices differs from tunneling of particles by the smallness of the inertial mass of the vortex; this means the vortex motion is overdamped, there are no oscillations or resonances. This overdamped tunneling is treated by Ivlev *et al.* [124] and Blatter *et al.* [125]. Griessen *et al.* [126] show that the dissipative quantum tunneling theory of Caldeira and Legget [127] with the usual vortex viscosity $\eta$ inserted reproduces the main results of [124-125].

Blatter and Geshkenbein [97] give a very general theory of collective creep and (vortex-mass dominated) tunneling in anisotropic and layered superconductors, see also the review [74]. A further related topic is the idea that the FLL can melt not only by thermal fluctuations but also by quantum fluctuations, which are predicted to modify the theoretical "melting line" [128].

# 5. Linear AC Response of the Pinned Flux-Line Lattice

## 5.1. LONDON AND CAMPBELL PENETRATION DEPTHS

The FLL interacts with an applied dc or ac magnetic field and with induced or applied currents only at the specimen *surface*, in a layer of thickness $\approx \lambda$, the magnetic penetration depth. Any applied field generates a surface shielding current $J_s$ which exerts a *pressure* $J_s B$ on flux lines oriented parallel to the surface, and a *tangential force* $J_s \Phi_0$ on each flux-line end sticking out from the surface (magnetic monopole). A small ac magnetic field or current, induced or applied by contacts, generates *compressional waves* [129] in the FLL when the constant field $B_a$ (which generates this FLL) is parallel to the surface, and it generates *bending waves* when $B_a$ is perpendicular to the surface, Fig. 7. If the FLL is weakly pinned then such compressional or tilt waves penetrate with exponentially decreasing amplitude to a depth $\lambda_C = \lambda_{11} = (c_{11}/\alpha_L)^{1/2}$ or $\lambda_{44} = (c_{44}/\alpha_L)^{1/2}$, respectively, where $c_{11}$ and $c_{44}$ are the elastic moduli of the FLL for uniform (k = 0) compression and tilt and $\alpha_L$ (Labusch parameter) is the elastic restoring force of the pins per unit volume.

When the GL parameter $\kappa$ and induction $B$ are not too small, say $\kappa > 2$ and $B_a \geq B > 2B_{c1}$, the vortex fields overlap strongly and therefore one has $c_{44} = BB_a/\mu_0 \approx B^2/\mu_0 \approx c_{11}$. In this case $\lambda_{11}$ and $\lambda_{44}$ both equal the Campbell penetration depth $\lambda_C = (B^2/\mu_0\alpha_L)^{1/2}$ [50, 129]. The response of the FLL to small dc and ac fields is, therefore, *independent* of the orientation of $B_a$ with respect to the plane specimen surface if pinning and viscous drag are isotropic. For *anisotropic* superconductors the results derived below for a planar surface still apply if the appropriate components of the tensorial quantites $\lambda$, $\alpha_L$, and $\eta$ (viscosity of the FLL, see below) are taken.

## 5.2. LINEAR AC PENETRATION DEPTH

By distorting the FLL, an applied dc or ac field or current can penetrate much deeper into the superconductor than if the FLL were rigidly pinned. With ideally pinned flux lines a superconductor behaves as if it were in the Meissner state, and any small field change or current decreases exponentially over the penetration depth $\lambda$. For weak elastic pinning, however, the perturbation field $B_1$ can penetrate to the larger Campbell depth $\lambda_C \gg \lambda$.

A general treatment of the penetration problem has to account for the *image vortices*, which are introduced to satisfy the boundary condition that $J$ flows parallel to the surface. It has further to account for the *nonlocalities* of the elastic response of the FLL and of the applied force, which originate from the long range $\lambda$ of the interaction of the vortices with

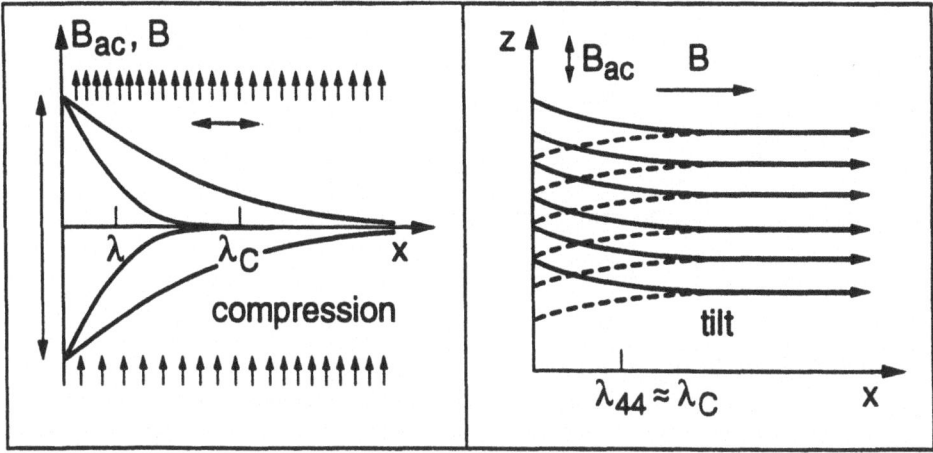

Figure 7: Compressional waves (*left*) and tilt waves (*right*) of the flux-line lattice (indicated by arrows) penetrate to a depth $\lambda_C \gg \lambda$ if pinning is weak. $\lambda$ is the penetration depth for Meissner shielding currents.

each other and with the surface currents, and possibly with a nonlocal collective pinning force, smeared over a correlation volume $V_c$ of the FLL. The penetration of a perturbation field $B_1(x)$ into a superconducting half space $x \geq 0$ was obtained by using an exponential ansatz [130, 131] or Fourier transforms [18, 132] for $B_1(x)$ and for the vortex displacement field $u(x)$. For local pinning the solution is simply

$$B_1(x \geq 0) = B_s \exp(-x/\lambda_{ac}) \tag{40}$$

$$\lambda_{ac}^2 = \lambda^2 + c_{11}/\alpha_L = \lambda^2 + \lambda_C^2 . \tag{41}$$

For nonlocal pinning, i.e., when the Labusch parameter $\alpha_L(\mathbf{k})$ is dispersive, one gets [132]

$$B_1(x \geq 0) = B_s \int_{-\infty}^{\infty} \frac{(-ik/\pi) \exp(ikx) \, dk}{k^2 + [\lambda^2 + B^2/\mu_0 \alpha_L(k)]^{-1}} . \tag{42}$$

Eq. (40-42) apply to arbitrary orientation of the dc field $B_a \approx B = (B_x, 0, B_z)$ w.r.t. the surface since $\lambda_{11} \approx \lambda_{44} = (B_x^2 + B_z^2)/\mu_0 \alpha$ for $B > 2B_{c2}$ if the ac field is $B_s \| z$. If $B_s \| y$, i.e., if the in-plane components of the dc and ac fields are perpendicular to each other, then only tilt waves are generated and the penetration depth is reduced to $\tilde{\lambda}_{ac}^2 = \lambda^2 + B_x^2/\mu_0 \alpha_L$. For general orientation of the ac field the penetrating ac field becomes $B_1 = [0, B_{sy} \exp(-x/\tilde{\lambda}_{ab}), B_{sz} \exp(-x/\lambda_{ab})]$.

## 5.3. ACCOUNTING FOR FLUX FLOW

If pinning is absent one has $\alpha_L = 0$ and thus $\lambda_{ac} = \lambda_C = \infty$. The flux lines then can move freely, feeling only a viscous drag force density $-\eta \mathbf{v}$ where $\mathbf{v} = \partial \mathbf{u}/\partial t$ is the vortex velocity with $\eta = B^2/\rho_{FF}$ the flux-flow viscosity, and $\rho_{FF} \approx \rho_n(T)B/B_{c2}(T)$ the flux-flow resistivity (31). The superconductor then behaves like a normal conductor with anisotropic linear resistivity $\rho = \rho_{FF}$ for currents $\perp B$, and $\rho = 0$ for currents $\| B$. Consequently, an ac field of frequency $\omega/2\pi$ will penetrate to a skin depth $\delta = (2\rho_{FF}/\omega\mu_0)^{1/2}$. One may say that the flux lines diffuse with diffusivity $D = \rho_{FF}/\mu_0$, thus $\delta = (2D/\omega)^{1/2}$. More precisely, using the induction laws $\dot{\mathbf{B}} = -\nabla \times \mathbf{E}$ ($\mathbf{E}$ = electric field) and $\mathbf{E} = \mathbf{B} \times \mathbf{v} = \mathbf{B} \times \mathbf{J} \times \mathbf{B}/\eta = (B^2/\eta)\mathbf{J}_\perp = \rho_{FF}\mathbf{J}_\perp$ one finds [81] that the internal magnetic field $\mathbf{B}(\mathbf{r}, t)$ follows the equation of motion

$$\dot{\mathbf{B}} = \mu_0^{-1} \nabla \times \rho_{FF} \hat{\mathbf{B}} \times \hat{\mathbf{B}} \times \nabla \times \mathbf{B} . \tag{43}$$

For $\mathbf{B}(\mathbf{r}, t) \approx \text{const} = \mathbf{B}_0$, (43) may be linearized to give

$$\dot{\mathbf{B}} = D\nabla^2 \mathbf{B} + \rho_{FF} \nabla \times \mathbf{J}_\| + O(|\mathbf{B} - \mathbf{B}_0|^2) \tag{44}$$

where $D = \rho_{FF}(B_0)/\mu_0$. Thus, in geometries where the current component $\mathbf{J}_\|$ parallel to $\mathbf{B}$ vanishes, flux flow is equivalent to a *linear diffusion of the flux lines*, $\dot{\mathbf{B}} = D\nabla^2 \mathbf{B}$. This was shown previously for the special case of parallel field lines, $B \| z$ [62, 65].

Flux flow is easily incorporated into the ac penetration problem by replacing $\alpha_L$ with $\alpha_L + i\omega\eta$ in (41) or (42). This adds to the elastic force density $-\alpha_L u$ the drag force $-\eta v = -\eta \partial u/\partial t = -\eta i \omega u$ where $\lambda_{ac}$, $u(x, t) = u(x) \exp(i\omega t)$, and $B_1(x, t) = B_1(x) \exp(i\omega t)$ are now complex quantities, but only the real parts of $u$ and $B_1$ (40) have physical meaning. One thus gets

$$\lambda_{ac}^2 = \lambda^2 + B^2/\mu_0(\alpha_L + i\omega\eta). \tag{45}$$

For $\eta = 0$ (45) reproduces (40, 41) and for $\alpha_L = 0$ it gives

$$B_1(x,t) = B_s \exp(-x/\delta) \cos(x/\delta - \omega t) \tag{46}$$

where $\delta$ is the skin depth from above.

## 5.4. ACCOUNTING FOR FLUX CREEP

In high-$T_c$ superconductors (HTSC) at sufficiently large temperatures $T$ *thermally activated depinning* is observed in form of flux creep, which causes the persistent currents or the irreversible (pinning-caused) magnetization to decay. At large current densities $J \gg J_c$ ($J_c$ = critical current density) this decay is caused by the usual flux flow (31) and is described by (44), yielding $J \propto \exp(-t/\tau)$ with $\tau = d^2/\pi^2 D$ for specimens of thickness $d$ at times $t \gg \tau$ [81, 65]. Near $J = J_c$ the current–voltage curve is highly nonlinear, $V \propto \exp(J/J_1)$ (30), and the decay is nearly logarithmic, $J \propto const. - \ln(t/t_0)$ (34).

I consider here the *linear* (ohmic) ac response, which occurs at $J \ll J_c kT/U$ in the region of thermally assisted flux flow (TAFF) (31). The TAFF resistivity is strongly temperature dependent, $\rho_{TAFF} \approx \rho_{FF} \exp(-U/kT)$ where $U(B, T)$ is an activation energy. In the TAFF regime, Eq. (43-46) apply with $\rho_{FF}$ replaced by $\rho_{TAFF}$, i.e., $\rho$, $D = \rho/\mu_0$, and $1/\eta = \rho/B^2$ are reduced by a factor $\exp(-U/kT)$ with respect to the usual (pin-free) flux-flow region.

Thermally activated depinning, or any other linear relaxation mechanism, can be incorporated into the complex penetration depth (45) by noting that after a sudden step-like displacement of the vortex lattice the elastic pinning force decreases in time, say exponentially with a relaxation time $\tau \gg \tau_0$, e.g., $\tau = \tau_0 \exp(U/kT)$, where $\tau_0 = \eta/\alpha_L$ is the flux-flow relaxation time of pinned flux lines. The elastic pinning force density $P(t)$ caused by a uniform FLL displacement $u(t)$ is then

$$P_{el}(t) = -\alpha_L \int_0^\infty \dot{u}(t - t') \exp(-t'/\tau) \, dt'. \tag{47}$$

In frequency space this means

$$\tilde{P}_{el}(\omega) = -\alpha_L \, \tilde{u}(\omega) \, i\omega \, /(i\omega + \tau^{-1}). \tag{48}$$

Therefore, flux flow and creep are considered by replacing $\alpha_L$ with $\alpha_L/(1 - i/\omega\tau) + i\omega\eta$. This gives our final result for the complex ac penetration depth

$$\lambda_{ac}^2 = \lambda^2 + \frac{B^2}{\mu_0} \left( \frac{\alpha_L}{1 - i/\omega\tau} + i\omega\eta \right)^{-1} \tag{49}$$

or, when $\tau = \tau_0 \exp(U/kT) \gg \tau_0 = \eta/\alpha_L$,

$$\lambda_{ac}^2 = \lambda^2 + \lambda_C^2 \frac{1 - i/\omega\tau}{1 + i\omega\tau_0}. \tag{50}$$

From (50) the complex surface impedance is obtained as $Z_s = i\omega\mu_0\lambda_{ac}$. Note that the result (50) does not specify the physical nature of the creep and thus applies also to quantum creep where $\tau = \tau_0 \exp(S_E/\hbar)$, Sct. 4.2.

## 5.5. LINEAR AC SUSCEPTIBILITY

Extending the above considerations to superconducting films ($|x| \leq d/2$) [18, 65, 81, 132-134] and cylinders ($r \leq R$) [135] in parallel ac field one obtains instead of (40)

$$B_1(|x| \leq d/2) = B_s \cosh(x/\lambda_{ac})/\cosh(d/2\lambda_{ac}) \tag{51}$$

$$B_1(r \leq R) = B_s I_0(r/\lambda_{ac})/I_0(R/\lambda_{ac}) \tag{52}$$

with $\lambda_{ac}$ from (50). $I_0(u)$ is a modified Bessel function. Averaging $B_1$ over the specimen one gets the complex ac susceptibility for slabs and cylinders

$$\mu(\omega) = \frac{\langle B_1(x) \rangle}{B_s} = \frac{\tanh(u)}{u} \quad \text{with} \quad u = \frac{d}{2\lambda_{ac}} \tag{53}$$

$$\mu(\omega) = \frac{\langle B_1(r) \rangle}{B_s} = \frac{2 I_0'(u)}{u \, I_0(u)} \quad \text{with} \quad u = \frac{R}{\lambda_{ac}} \tag{54}$$

where $I_0'(u) = dI_0/du$. The decomposition of the general results (51) to (54) into real and imaginary parts is straightforward. In the inductive case $1/\tau \ll \omega \ll 1/\tau_0$ the argument $u$ in (53, 54) is *real*; for the film one then has $\mu(\omega) \approx 1/u = 2(\lambda^2 + \lambda_C^2)^{1/2}/d$ for $d \ll \lambda$, and $\mu(\omega) \approx 1 - u^2/3 = 1 - d^2/12(\lambda^2 + \lambda_C^2)$ for $d \gg \lambda$. In the ohmic (diffusive) cases at low and large frequencies, the known skin effect behavior results, with a complex argument $u = (1 + i)(\omega d^2/8D)^{1/2}$ where the flux diffusivities are

$$D = \rho_{TAFF}/\mu_0 = \lambda_C^2/\tau \quad \text{for} \quad \omega \ll 1/\tau \tag{55}$$

$$D = \rho_{FF}/\mu_0 = \lambda_C^2/\tau_0 \quad \text{for} \quad 1/\tau_0 \ll \omega. \tag{56}$$

One then has explicitly for the slab [18, 65, 133, 134]

$$\mu(\omega) = \mu' - i\mu'' = \frac{(\sinh v + \sin v) - i(\sinh v - \sin v)}{v\,(\cosh v + \cos v)} \tag{57}$$

with $v = (\omega d^2/2D)^{1/2} = d/\delta$, Fig. 8a. One has $\mu(\omega) = 1$ for $\omega = 0$ and $\mu(\omega) \approx (1 - i)/v$ for $|v| \gg 1$. The dissipative part $\mu''(\omega)$ has a maximum $\mu''_{max} = 0.41723$ at $v = 2.2542$ corresponding to $\omega\tau_d = 0.97 \approx 1$ with $\tau_d = d^2/\pi^2 D$ or to $d/2\delta = 0.887 \approx 1$ with $\delta = (2D/\omega)^{1/2}$. For perpendicular geometry see Sct. 6.2.

## 5.6. COMPLEX RESISTIVITY

From (40) the Maxwell equations for the current density $J(x) = \mu_0^{-1} \partial B/\partial x = B_1(x)/\mu_0\lambda_{ac}$ and the electric field $E(x) = i\omega\lambda_{ac}B_1(x)$ (**J** and **E** are along $y$ when **B** is along $z$) yield the ac resistivity $\rho_{ac}(\omega) = E/J = i\omega\mu_0\lambda_{ac}^2$, which is independent of $x$,

$$\rho_{ac}(\omega) = i\omega\mu_0\lambda^2 + \rho_{TAFF} \cdot (1 + i\omega\tau)/(1 + i\omega\tau_0) \tag{58}$$

with $\rho_{TAFF} = \rho_{FF}\tau_0/\tau \approx \rho_{FF}\exp(-U/k_BT) \ll \rho_{FF}$, Fig. 8b. Thus,

$$\rho_{ac}(\omega) \approx \rho_{TAFF} \qquad \text{for } \omega \ll 1/\tau \tag{59}$$

$$\rho_{ac}(\omega) \approx i\omega\mu_0(\lambda^2 + \lambda_C^2) \quad \text{for } 1/\tau \ll \omega \ll 1/\tau_0 \tag{60}$$

$$\rho_{ac}(\omega) \approx \rho_{FF} \qquad \text{for } 1/\tau_0 \ll \omega. \tag{61}$$

For *low and high* frequencies, the resistivity is *real* and independent of $\omega$; the flux motion in this case is *diffusive* [81] with diffusivity $D = \rho_{ac}/\mu_0$ as in (55, 56). The ac penetration depth is then a *skin depth* $\delta = (2D/\omega)^{1/2}$ as in normal conducting metals. For *intermediate* frequencies, $\rho_{ac}$ is imaginary and $\propto \omega$; the surface currents are then nearly loss free due to strong elastic pinning; the shielding in this case is like in the Meissner state but with a larger penetration depth $(\lambda^2 + \lambda_C^2)^{1/2}$.

## 5.7. ALGEBRAIC MEMORY FUNCTION

In recent ac experiments on ceramic BSCCO [136, 137] in the range $2\,\text{Hz} \leq \omega/2\pi \leq 2\,\text{MHz}$ both the real and imaginary parts of the susceptibility $\mu(\omega)$ (54) below $T = T_l \approx 70\,\text{K}$ could be fitted very well by replacing the above exponential decay of the Labusch parameter, $\alpha_L(t) = \alpha_L\exp(-t/\tau)$ (47) yielding $\alpha_L(\omega) = \alpha_L i\omega\tau/(1+i\omega\tau)$ (48), by an algebraic decay of the form

$$\alpha_L(t) = \alpha_L/(1 + t/\tau)^\beta \tag{62}$$

yielding for $\omega\tau \ll 1$ instead of (49)

$$\alpha_L(\omega) = \alpha_L(i\omega\tau)^\beta\,\Gamma(1-\beta) \tag{63}$$

$$\lambda_{ac}^2 = \lambda_C^2/[(i\omega\tau)^\beta\,\Gamma(1-\beta)] \tag{64}$$

where $\Gamma(1-\beta)$ is Euler's gamma function. Interestingly, in these experiments the effect of temperature on the dynamics is entirely determined by that of the exponent $\beta(T) \approx$

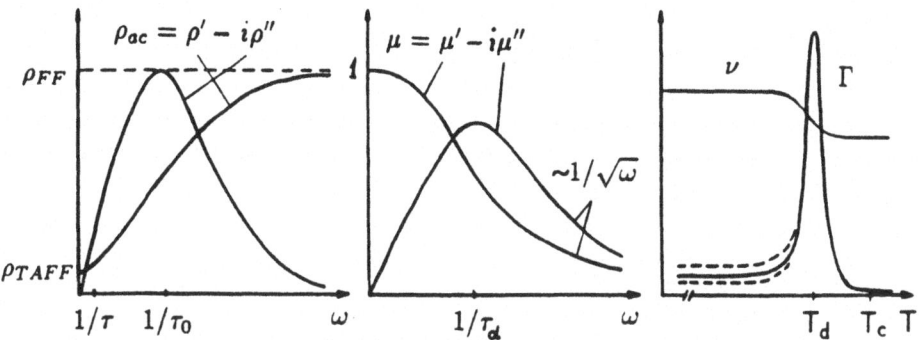

Figure 8: *Left:* Complex ac susceptibility $\mu(\omega) = \mu' - i\mu''$ (39) of slabs in the TAFF state $\omega \ll \tau^{-1}$, see text. *Middle:* Complex ac resistivity $\rho_{ac}(\omega) = \rho' - i\rho''$ (38). *Right:* Resonance frequency $\nu$ and attenuation $\Gamma$ of a HTSC performing tilt vibrations in constant magnetic field. The dashed lines indicate the amplitude dependence of the hysteretic losses.

$1/(1 + U/k_BT)$, *cf.* Eq. (35), while $\tau \approx 4 \cdot 10^{-12}$ s (for $B = 1$ T) is nearly independent of $T$. Moreover, $\beta(T)$ overlaps with exponents determined from the algebraic decay of the time dependent magnetization of this sample at the same field [138]. Behr, Kötzler et al. [136] find a depinning limit ($\beta = 1$) for $T_l \approx 70$ K $\leq T \leq T_c \approx 90$ K and a glass transition with $\beta \approx 0.07$ at $T_g \approx 24$ K.

## 5.8. NON-LINEAR RESPONSE

In ac experiments on type-II superconductors there is always a weak linear damping from (58) even in the shielding state (60) where $\mathrm{Re}\{\rho_{ac}\} = \rho_{TAFF} + \rho_{FF}\omega^2\tau_0^2 \ll \mathrm{Im}\{\rho_{ac}\}$. But often a much larger *nonlinear* (amplitude dependent) attenuation is observed, in particular in the very sensitive measurements on vibrating superconductors, Sct. 6.3. These *hysteretic* losses ($\sim \omega$) typically *exceed* the linear viscous drag losses ($\sim \omega^2$), in particular at low frequencies.

The hysteretic dissipation of the pinned FLL is given by the area of a hysteresis loop in the force–displacement curve $P(u)$ times the frequency $\nu = \omega/2\pi$ [139]. In contrast, the viscous (linear) dissipation density is $\eta v^2$, which means *constant attenuation* $\Gamma$ defined by the free relaxation $u(t) \propto \cos(\omega t)\exp(-\Gamma t)$, or by the resonance curve $u(\omega) \propto (\omega^2 + \Gamma^2)^{-1}$ at constant driving force. Hysteretic damping at *small* amplitudes $u$ exhibits $\Gamma \propto u^n$, $n > 1$, but at *large* $u$ it corresponds to a constant frictional force, thus $\Gamma \propto 1/u$.

The non-linear losses originate from the non-linearity of the current-voltage curve, *cf.*, *e.g.*, Eq. (28) and (30) and Fig. 5a. In the idealized case where the flux lines start to move only when $J = J_c$ is reached (Bean's assumption), the hysteretic losses in longitudinal and perpendicular geometry in the limit of small frequency are given by Eq. (8) and (24).

# 6. Depinning Lines

As can be seen in Fig. 5b, one can define an irreversibility line or depinning line $T_d(B)$ which in the field–temperature-plane separates the regions of irreversible and reversible magnetic behavior. Since this line is related to the rate of linear or non-linear relaxation of a current or flux distribution in the superconductor, it depends on the time and length scales probed by a given experiment. Larger lengths and frequencies shift the depinning line to higher $T$ and $B$ [18]. Numerous types of experiments define and measure such depinning lines, which are all slightly different and, therefore, in general do not reflect a pure material property like a melting transition of the FLL or other phase transitions.

The dependence of $T_d(B)$ on the specimen size (or grain size) and measuring rate (or frequency) is typically weak, *e.g.*, logarithmic when $T$ and $B$ appear in the exponent of a thermally activated creep or diffusion rate. This dependence may, therefore, be easily overlooked. An observed sharp damping peak or another distinct feature of a measured quantity may be interpreted as indicative of a phase transition only if its independence of the frequency and specimen size has been checked carefully, or if its dependence on pinning can be ruled out as is the case in the careful experiments by Farrell [120].

## 6.1. FROM MAGNETIZATION CURVES

Irreversibility or depinning lines are sometimes defined by the vanishing of the irreversibility in magnetization curves. This definition depends on the accuracy with which a tiny hysteresis can be detected and on the measuring rate. Clearly, if the magnetization curve could be measured more and more slowly, then the width of the hysteresis in principle would shrink to zero because of flux creep, Sct. 3. The same statement applies to very thin specimens, in which the critical state relaxes more rapidly than in bulk specimens. Anyway, the vanishing of hysteretic behavior at higher fields or temperatures is quite conspicuous and gave the first hint on a "glass-like" behavior of HTSC [140]. A related criterion is the vanishing of the difference between "field cooled" and "zero field cooled" magnetization. Typically these criteria give depinning lines which lie well below the lines defined by the more sensitive criteria described in the following sections.

## 6.2. FROM LINEAR AC RESPONSE

A less ambiguous definition of depinning lines is by the maximum in the dissiative part $\mu''(\omega, T, B)$ of the ac susceptibility, Sct. 5.5, at constant frequency $\omega/2\pi$, see, e.g., [105]. From Eq. (53, 54, 57) it is clear that such depinning lines depend on the specimen size $d$ or $R$. This criterion yields lines of constant $\omega d^2/D$, and thus of constant flux diffusivity $D(T, B) = \rho_{TAFF}/\mu_0$ if $\omega = $ const. When the flux diffusion is thermally activated one has $D(T, B) \sim \exp[-U(T, B)/k_B T]$ and the depinning line $T_d(B)$ depends logarithmically on $T$ and on the activation energy $U(T, B)$, and also logarithmically on $\omega$ and $d$.

A related criterion is the divergence of the ac penetration depth $\lambda_{ac}(\omega, T, B)$ (49) measured by the screening of an ac field by a superconducting film located between two coils [141]. A detailed theory of this experiment recently has been given by Clem [142], who finds that this *perpendicular geometry* (with specimen larger than the coils) can be described just by the complex and local ac resistivity (58) or penetration depth (49). This is so since the screening currents have to flow parallel to the film and, as a matter of principle, do not "see" the source of the ac magnetic field which induces these currents, and the ac field in this geometry does not see the edges of the film.

This situation changes when the specimen is smaller than the ac coil: In this case the effects of the specimen edge become important, and the theory of Sct. 2.2 has to be extended to time dependent fields and to flux diffusion and flux creep. This problem was solved very recently [143] for a superconducting or normalconducting strip or circular disk in perpendicular field. In the linear response region, the resulting complex permeability $\mu_\perp(\omega)$ looks very similar as the longitudinal $\mu_\parallel(\omega)$ (57), though the asymptotic behavior for large $\omega$ is different, $\mu_\perp(\omega) \sim 1/\omega$ rather than $\mu_\parallel(\omega) \sim 1/\sqrt{\omega}$, and the short-time behaviour is *not diffusive* as in longitudinal fields, Sct. 5.5.

When the applied perpendicular field is suddenly increased by $\Delta H_a$ at $t = 0$, then the shielding sheet current $J_s$ initially has the form (11) with the replacements $H_a \rightarrow \Delta H_a$ (strip) or $H_a \rightarrow (2/\pi)\Delta H_a$ and $x \rightarrow r$ (disk). Note that this shielding current *immediately appears in the entire flat specimen* and does not have to penetrate diffusively. The magnetic field initially (at $t = 0$) is completely expelled, but then penetrates from the edges. This penetration is faster than usual diffusion in the sense that at the center of the strip or disk $H$

appears much faster than given by the Gaussian spatial dependence of diffusive processes. At time $t \gg \tau_0$ both $J_s$ and $H - H_a$ decrease as $\exp(-t/\tau_0)$ where $\tau_0 = 0.2492ad/D$ for the strip ($D = \rho/\mu_0$). This relaxation exceeds the estimate $2ad/\pi^2 D$ of [134] by a factor of 1.23. Maximum ac dissipation in this perpendicular geometry occurs at $\omega\tau_0 = 1.11$ for the strip.

## 6.3. FROM VIBRATING SUPERCONDUCTORS

An elegant and very sensitive method to measure depinnig lines and detect possible phase transitions uses superconductors performimg tilt oscillations. Tilting a superconductor periodically in a constant magnetic field $B_a \approx B$ by an angle $\phi$ is equivalent to applying an ac magnetic field with amplitude $B_a \sin \phi$ *perpendicular* to $B_a$. Superconducting vibrating reeds [117-119, 139], and platelets glued on a high-quality silicon oscillator [116] or suspended on a wire [120, 144, 145], can therefore be understood by the same theory, Sct. 5, which explains ac experiments.

Vibrating superconductors exhibit a sharp peak in their attenuation (Fig. 8c) at a temperature $T_d(B)$ which was interpreted as a FLL melting temperature in [116] but usually indicates the onset of depinning. In *conventional* superconductors, this depinning-peak occurs near the upper critical field $B_{c2}(T)$, or temperature $T_c(B)$, where pinning has to go to zero. In HTSC, this depinning is thermally activated and reflects the maximum in the imaginary part of the ac susceptibility, *e.g.*, in Eq. (53) and (57) and Fig. 8a. Which lengths have to be inserted and how the appropriate flux diffusion modes look like is described in detail in [134].

As stated above, in many cases the dissipation is of the hysteretic type and thus *amplitude dependent*; but even then the peak should typically occur close to the linear damping peak, and a change of the vibration amplitude should hardly shift its position but may change its height. The attenuation peak occurs when the diffusion length of flux coincides with a typical length or thickness of the specimen. The peak is so sharp because the flux diffusivity $D(B,T)$ changes rapidly with $T$ and often also with $B$.

The resonance frequency $\nu$ of a vibrating superconductor typically *increases* when a field $B_a$ is applied, because $B_a$ "pulls" at the flux lines and these couple to the material by pinning, thus providing an additional restoring force. Above the damping peak the frequency enhancement $\delta\nu$ vanishes because of complete depinning. Thus, typically $\delta\nu \propto B_a^2$. However, for very short flux lines one has $\delta\nu \propto \alpha_L \propto B^{3/2}$ as shown in [118]. A restoring force of different type is the *anisotropy of the magnetization*, which in very weak-pinning specimens can even be *negative* and thus *reduces* $\nu$ when $T$ drops below $T_c$ or when $B$ is switched on [120]. The observed reduction of the resonance frequency in the superconducting state clearly indicates that this frequency change is not related to pinning, though the damping peak might well be: It is hard to imagine a mechanism by which *plastic shear deformations* of the FLL, which might give rise to internal friction even in the absence of pinning, can be excited by the surface shielding currents which couple the FLL to an applied ac field or periodic tilt of the specimen.

In recent experiments *two* damping peaks as a function of $T$ were observed with vibrating superconductors when the magnetic field was applied at an oblique angle [146, 147]. These peaks may belong to different diffusion modes of the FLL as predicted in [134] and confirmed in [147]. In *longitudinal* applied dc field, the tilt vibrations are equivalent to a transversal ac field leading to a flux-line tilt which penetrates (diffuses) from the ends of the reed along the reed length into the vibrating reed. In *transverse* applied dc field, the vortex tilt penetrates from the large surface and thus has to diffuse only along the small thickness of the reed. Obviously, the latter relaxation is faster and, therefore, in transverse dc field the damping peak occurs at lower values of the flux diffusivity, and thus at lower $T$ or $B$, than in longitudinal dc field. At intermediate angles of the applied field two damping peaks may occur simultaneously. This occurrence of two depinning lines in the same specimen at the same field shows that one has to be careful with the interpretation of such conspicuous novel observations and should understand macroscopic electrodynamics before a microscopic interpretation can be given. This statement applies also to critical currents in the perpendicular geometry of Sct. 2.2, and to the linear response to perpendicular ac magnetic fields [143].

## 6.4. FROM OTHER MEASUREMENTS

Various other methods may be conceived and have been used to obtain depinning lines. The conduction noise in HTSC films at constant current density and at a given frequency exhibits a sharp peak as a function of $B$ and $T$ [148]. This noise is caused by depinning processes: each "plucking" of a vortex releases elastic energy of the FLL, which then relaxes viscously with an exponential time law. Below $T_d(B)$ the noise is small since only few depinning processes occur; at $T = T_d(B)$ the noise is maximum; and above $T_d(B)$ it decreases again since the viscous motion of the thermally depinned vortices is smooth. The depinning line shifts to larger $T$ or $B$ when a higher frequency band is selected.

In all the above listed methods the FLL interacts with the applied field at its surface and then a deformation of the FLL penetrates into the specimen. A method which probes the FLL directly *inside* the specimen uses ultrasound. An ultrasonic attenuation peak and sound velocity enhancement have been used to define $T_d(B)$ [149]. Further methods that probe the FLL and its pinning and diffusion inside the superconductor are muon-spin rotation [150, 151] and nuclear magnetic resonance [152, 153]. Both methods in principle yield the probability to find a given magnetic field at the position of the implanted muon or the Thallium nucleus. Motion of the flux lines during the lifetime of the positive muon ($\tau_\mu = 2.2\,\mu s$) or during the precession of the nucleus may smear or narrow the observed spectrum or resonance line depending on the time scale of the local variation of the magnetic field compared to the width of the resonance line. A further useful tool to probe the flux-line arrangement are neutrons. Small-angle neutron-scattering very recently revealed the FLL structure in YBCO [154]. Neutron depolarization in classical [155] and HTSC [156] yields the correlation length of the flux distribution and the magnitude and direction of the average magnetization statically and dynamically. It is to be hoped that these methods will give valuable insight into the problem of structural transitions of the FLL caused by temperature and pinning.

# References

[1] A. A. Abrikosov, *Sov. Phys. - JETP* **5**, 1174 (1957).

[2] D. Cribier, B. Jacrot, L. M. Rao, B. Farnoux, *Phys. Lett.* **9**, 106 (1964).

[3] E. M. Forgan et al., *Physica C* **185-189**, 247 (1991); M. Yethiraj et al., *Phys. Rev. Lett.* **70**, 857 (1993).

[4] U. Essmann and H. Träuble, *Phys. Lett. A* **24**, 526 (1967); *Scientific American* **224**, 75 (March 1971).

[5] P. L. Gammel, D. J. Bishop, G. J. Dolan, J. R. Kwo, C. A. Murray, L. F. Schneemeyer, and J. V. Waszczak, *Phys. Rev. Lett.* **59**, 2592 (1987); P. L. Gammel, D. J. Bishop, J. P. Rice, and D. M. Ginsberg, *Phys. Rev. Lett.* **68**, 3343 (1992).

[6] L. Ya. Vinnikov, L. A. Gurevich, G. A. Yemel'chenko, and Yu. A. Ossipyan, *Solid State Comm.* **67**, 421 (1988).

[7] C. A. Bolle, P. L. Gammel, D. G. Grier, C. A. Murray, D. J. Bishop, D. B. Mitzi, and A. Kapitulnik, *Phys. Rev. Lett.* **66**, 112 (1991).

[8] P. G. De Gennes, *Superconductivity of Metals and Alloys* (Benjamin, New York, 1966).

[9] M. Tinkham, *Introduction to Superconductivity* (McGraw-Hill, New York, 1975).

[10] A. A. Abrikosov, *Fundamentals of the Theory of Metals* (North Holland, Amsterdam, 1988).

[11] J. R. Clem, *J. Low Temp. Phys.* **18**, 427 (1975); Z. Hao and J. R. Clem, *Phys. Rev. B* **43**, 7622 (1991).

[12] E. H. Brandt and U. Essmann, *phys. stat. sol. (b)* **144**, 13 (1987) (review on the FLL).

[13] E. H. Brandt, *J. Low Temp. Phys.* **26**, 709; 735 (1977); **28**, 263; 291 (1977).

[14] E. H. Brandt, *Phys. Rev. B* **34**, 6514 (1986).

[15] A. Houghton, R. A. Pelcovits, and A. Sudbø, *Phys. Rev. B* **40**, 6763 (1989).

[16] E. H. Brandt, *Physica B* **165 & 166**, 1129 (1990); *Physica B* **169**, 91 (1991).

[17] E. H. Brandt, *Int. J. Mod. Phys. B* **5**, 751 (1991) (review on the FLL in HTSC).

[18] E. H. Brandt, *Physica C* **195**, 1 (1992).

[19] J. C. Martinez, S. H. Brongersma, A. Koshelev, B. Ivlev, P. H. Kes, R. P. Griessen, D. G. de Groot, Z. Tarnavski, and A. A. Menovsky, *Phys. Rev. Lett.* **69**, 2276 (1992).

[20] M. Tachiki and S. Takahashi, *Solid State Comm.* **70**, 291 (1989); *Physica B* **169**, 121 (1991); L. Schimmele, H. Kronmüller, and H. Teichler, *phys. stat. sol. (b)* **147**, 361 (1988); S. Nieber and H. Kronmüller, *Physica C* **210**, 188 (1993).

[21] J. R. Clem, *Phys. Rev. B* **43**, 7837 (1991); S. N. Artemenko and A. N. Kruglov, *Phys. Lett. A* **143**, 485 (1990); K. H. Fischer, *Physica C* **178**, 161 (1991).

[22] M. V. Feigel'man, V. B. Geshkenbein, and A. I. Larkin, *Physica C* **167**, 177 (1990).

[23] V. M. Vinokur, P. H. Kes, and A. E. Koshelev, *Physica C* **168**, 29 (1990).

[24] L. N. Bulaevskii, M. Ledvij, and V. G. Kogan, *Phys. Rev. B* **46**, 366; 11807 (1992).

[25] D. Feinberg, *Physica C* **194**, 126 (1992).

[26] W. E. Lawrence and S. Doniach, Proc. 12th Internatl. Conf. on Low Temperature Physics LT12 (E. Kanda ed., Academic Press of Japan, Kyoto, 1971) p. 361.

[27] L. N. Bulaevskii, *Int. J. Mod. Phys. B* **4**, 1849 (1990) (review on LD theory).

[28] E. H. Brandt, *Statics and Dynamics of the Flux-Line Lattice in High-$T_c$ Superconductors*, in: NATO ASI, April 13-23, 1993, in Geilo, Norway: Phase Transitions and Relaxation in Systems with Competing Energy Scales. D. Sherrington and T. Riste, eds. (Kluwer, Dordrecht, The Netherlands, 1993) pp. 23-57.

[29] L. N. Bulaevskii, S. V. Meshkov, and D. Feinberg, *Phys. Rev. B* **43**, 3728 (1991).

[30] G. Hackenbroich and S. Scheidl, *Physica C*, **181**, 163 (1991); S. Scheidl and G. Hackenbroich, *Phys. Rev. B* **46**, 14010 (1992).

[31] L. I. Glazman and A. E. Koshelev, *Phys. Rev. B* **43**, 2835 (1991); D. S. Fisher, in: *Phenomenology and Applications of High-Temperature Superconductors*, K. S. Bedell et al. eds., (Addison-Wesley, New York, 1992) p. 287; S. Nieber and H. Kronmüller, *Physica C* **213**, 43 (1993).

[32] L. N. Bulaevskii, M. Ledvij, and V. G. Kogan, *Phys. Rev. Lett.* **68**, 3773 (1992); V. G. Kogan, M. Ledvij, A. Yu. Simonov, J. H. Cho, and D. C. Johnston, *Phys. Rev. Lett.* **70**, 1870) (1993); Z. Tešanović, L. Xing, L. N. Bulaevskii, Q. Li, and M. Suenaga, *Phys. Rev. Lett.* **69**, 3563 (1992).

[33] G. Blatter, B. Ivlev, and H. Nordborg, Vortex Fluctuations in Layered Superconductors and Thin Films (preprint); L. N. Bulaevskii and M. P. Maley, *Phys. Rev. Lett* **71**, 3541 (1993).

[34] B. A. Hubermann and S. Doniach, *Phys. Rev. Lett.* **43**, 950 (1979).

[35] D. S. Fisher, *Phys. Rev. B* **22**, 1190 (1980).

[36] L. I. Glazman and A. E. Koshelev, *Physica C* **173**, 180 (1991).

[37] L. L. Daemen, L. N. Bulaevskii, M. P. Maley, and J. Y. Coulter, *Phys. Rev. Lett.* **70**, 1167 (1993); *Phys. Rev. B* **47**, 11291 (1993); L. N. Bulaevskii, J. R.Clem, L. I. Glazman, and A. P. Malozemoff, *Phys. Rev. B* **45**, 2545 (1992).

[38] D. Feinberg and A. M. Ettouhami, *Int. J. Mod. Phys. B* **7**, 2085 (1993).

[39] A. E. Koshelev, *Phys. Rev. B* **48**, 1180 (1993).

[40] B. Horovitz, *Phys. Rev. Lett.* **67**, 378 (1991); *Phys. Rev. B* **45**, 12632 (1992); *Phys. Rev. B* **47**, 5947, 5964 (1993); G. Carneiro, *Phys. Rev. B* **45**, 2391; 2403 (1992); S. E. Korshunov, *Europhys. Lett.* **11**, 757 (1990).

[41] L. P. Gor'kov and N. B. Kopnin, *Sov. Phys.-Uspechi* **18**, 496 (1976).

[42] A. I. Larkin and Yu. N. Ovchinnikov, in: *Nonequilibrium Superconductivity*, D. N. Langenberg and A. I. Larkin, eds. (Elsevier, Amsterdam, 1986), p. 493.

[43] B. I. Ivlev and N. B. Kopnin, *Phys. Rev. B* **42**, 10052 (1990).

[44] J. R. Clem and W. M. Coffey, *Phys. Rev. B* **42**, 6209 (1990).

[45] Z. Hao and J. R. Clem, *IEEE Trans. Magn.* **27**, 1086 (1991).

[46] P. W. Anderson, *Phys. Rev. Lett.* **9**, 309 (1962); P. W. Anderson and Y. B. Kim, *Rev. Mod. Phys.* **36**, 39 (1964).

[47] C. P. Bean, *Rev. Mod. Phys.* **36**, 31 (1964); *J. Appl. Phys.* **41**, 2482 (1970).

[48] E. H. Brandt, Tilted and Curved Vortices in Anisotropic Superconducting Films, *Phys. Rev. B* **48** (1 Sept. 1993).

[49] E. H. Brandt and M. Indenbom, Vortex Tilt in Flat Superconductors with Transport Current, *Proc. of the XXth Int. Conf. on Low Temperature Physics, to appear in Physica B* (in print).

[50] A. M. Campbell and J. E. Evetts, *Adv. Phys.* **21**, 199 (1972) (review on flux pinning).

[51] E. H. Brandt, M. Indenbom, and A. Forkl, *Europhys. Lett.* **22**, 735 (1993).

154

[52] E. H. Brandt, *Phys. Rev. B* **46**, 8628 (1992).

[53] W. T. Norris, *J. Phys. D: Appl. Phys.* **3** 489 (1970).

[54] P. N. Mikheenko and Yu. E. Kuzovlev, *Physica C* **204** 229 (1993); se also: J. Zhu, J. Mester, J. Lockhardt, and J. Turneaure, *Physica C* **212**, 216 (1993).

[55] M. V. Indenbom, A. Forkl, H.-U. Habermeier, and Kronmüller H., *J. of Alloys and Compounds* **195**, 499 (1993).

[56] L. Burlachkov, Y. Yeshurun, M. Konczykowski, and F. Holtzberg, *Phys. Rev. B* **45** 8193 (1992).

[57] D. J. Frankel, *J. Appl. Phys.* **50** 5402 (1979); M. Daeumling and D. C. Larbalestier, *Phys. Rev. B* **40**, 9350 (1989); L. W. Connor and A. P. Malozemoff, *Phys. Rev. B* **43**, 402 (1991).

[58] S. Senoussi, *J. Phys. III (Paris)* **2** 1041 (1992) (review).

[59] H. Theuss, A. Forkl, and H. Kronmüller, *Physica C* **190** (1992) 345.

[60] E. H. Brandt and M. Indenbom, Type-II Superconductor Strip with Current in Perpendicular Magnetic Field, *Phys. Rev. B* **48**, 12893 (1993).

[61] Th. Schuster, M. R. Koblischka, B. Ludescher, N. Moser, and H. Kronmüller, *Cryogenics* **31** 811 (1991); L. A. Dorosinskii, M. V. Indenbom, V. I. Nikitenko, Yu. A. Ossip'yan, A. A. Polyanskii, and V. K. Vlasko-Vlasov, *Physica C* **203** (1992) 149.

[62] M. R. Beasley, R. Labusch, and W. W. Webb, *Phys. Rev.* **181**, 682 (1969); see also: C. Rossel et al., *Physica C* **165**, 233 (1990); P. Berghuis and P. H. Kes, *Physica B* **165 & 166**, 1169 (1990); P. Svedlindh et al., *Phys. Rev. B* **43**, 2735 (1991); M. Suenaga, A. K. Gosh, Y. Xu, and D. O. Welch, *Phys. Rev. Lett.* **66**, 177 (1991).

[63] Y. Yeshurun and A. P. Malozemoff, *Phys. Rev. Lett.* **60**, 2202 (1988).

[64] D. Dew-Hughes, *Cryogenics* **28**, 674 (1988).

[65] P. H. Kes, J. Aarts, J. van den Berg, C. J. van der Beek, and J. A. Mydosh, *Supercond. Sci. Technol.* **1**, 242 (1989).

[66] A. I. Larkin and Yu. N. Ovchinnikov, *J. Low Temp. Phys.* **43**, 109 (1979); E. H. Brandt, *J. Low Temp. Phys.* **64**, 375 (1986).

[67] M. V. Feigel'man, V. B. Geshkenbein, A. I. Larkin, and V. M. Vinokur, *Phys. Rev. Lett.* **63**, 2303 (1989).

[68] T. Nattermann, *Phys. Rev. Lett.* **64**, 2454 (1990); K. H. Fischer and T. Nattermann, *Phys. Rev. B* **43**, 10372 (1991).

[69] V. M. Vinokur, M. V. Feigel'man, V. B. Geshkenbein, and A. I. Larkin, *Phys. Rev. Lett.* **65**, 259 (1990); M. V. Feigel'man and V. M. Vinokur, *Phys. Rev. B* **41**, 8986 (1990); V. M. Vinokur, V. B. Geshkenbein, M. V. Feigel'man, and A. I. Larkin, *Zh. Eksp. Teor. Fiz.* **100**, 1104 (1991) [*Sov. Phys. JETP* **73**, 610 (1991)]; A. E. Koshelev, *Phys. Rev. B* **45**, 12936 (1992).

[70] A. Sudbø and E. H. Brandt, *Phys. Rev. B* **43**, 10482 (1991); *Phys. Rev. Lett.* **66**, 1781 (1991); *Phys. Rev. Lett.* **68**, 1758 (1992); *Phys. Rev. Lett.* **67**, 3176 (1991).

[71] E. H. Brandt and A. Sudbø, *Physica C* **180**, 426 (1991).

[72] E. Sardella, *Phys. Rev. B* **45**, 3141 (1992); *Phys. Rev. B* **44**, 5209 (1991).

[73] G. Blatter, V. B. Geshkenbein, and A. I. Larkin, *Phys. Rev. Lett.* **68**, 875 (1992).

[74] G. Blatter, M. V. Feigel'man, V. B. Geshkenbein, A. I. Larkin, and V. M. Vinokur, *Rev. Mod. Phys.* (in print).

[75] L. Civale, A. D. Marwick, T. K. Worthington, M. A. Kirk, J. R. Thompson, L. Krusin-Elbaum, Y. Sun, J. R. Clem, F. Holtzberg, *Phys. Rev. Lett.* **67**, 648 (1991); M. Konczykowski et al., *Phys. Rev. B* **47**, 5531 (1993); V. Hardy, D. Groult, J. Provost, and B. Raveau, *Physica C* **190**, 289 (1992); M. Konczykowski et al., *Phys. Rev. B* **47**, 5531 (1993).

[76] W. Gerhäuser, G. Ries, H. W. Neumüller, W. Schmidt, O. Eibl, G. Saemann-Ischenko, and S. Klaumünzer, *Phys. Rev. Lett.* **68**, 879 (1992); D. Prost et al., *Phys. Rev. B* **47**, 3457 (1993).

[77] D. R. Nelson and V. M. Vinokur, *Phys. Rev. Lett.* **68**, 2398 (1992).

[78] E. H. Brandt, *Phys. Rev. Lett.* **69**, 1105 (1992); *Europhys. Lett.* **18**, 635 (1992).

[79] I. F. Lyuksyutov, *Europhys. Lett.* **20**, 273 (1992).

[80] A. Kramer and M. L. Kulič, *Phys. Rev. B* **48** (1 Oct. 1993).

[81] E. H. Brandt, *Z. Physik B* **80**, 167 (1990).

[82] A. Schmid and W. Hauger, *J. Low Temp. Phys.* **11**, 667 (1973).

[83] T. K. Worthington, M. P. A. Fisher, D. A. Huse, J. Toner, A. D. Marwick, T. Zabel, C. A. Feild, and F. Holtzberg, *Phys. Rev. B* **46**, 11854 (1992).

[84] T. T. M. Palstra, B. Battlogg, R. B. van Dover, L. F. Schneemeyer, and J. V. Waszczak, *Phys. Rev. B* **41**, 6621 (1990).

[85] R. H. Koch et al. *Phys. Rev. Lett.* **63**, 1511 (1989); C. Dekker, W. Eidelloth, and R. H. Koch, *Phys. Rev. Lett.* **68**, 3347 (1992).

[86] Ph. Seng, R. Gross, U. Baier, M. Rupp, D. Koelle, R. P. Huebener, P. Schmitt, G. Saemann-Ischenko, and L. Schultz, *Physica C* **192**, 403 (1992).

[87] J. C. van der Beek, G. J. Nieuwenhuys, P. Kes, H. G. Schnack, and R. P. Griessen, *Physica C* **197**, 320 (1992).

[88] T. K. Worthington et al., *Phys. Rev. B* **43**, 10538 (1991).

[89] H. Safar, P. L. Gammel, D. A. Huse, D. J. Bishop, J. P. Rice, and D. M. Ginsberg, *Phys. Rev. Lett.* **69**, 824 (1992).

[90] P. L. Gammel, L. F. Schneemeyer, and D. J. Bishop, *Phys. Rev. Lett.* **66**, 953 (1991); N.-C. Yeh, D. S. Reed, W. Jiang, U. Kriplani, F. Holtzberg, A. Gupta, B. D. Hunt, R. P. Vasquez, M. C. Foote, and L. Bajuk, *Phys. Rev. B* **45**, 5654 (1992); H. Safar et al., *Phys. Rev. Lett.* **68**, 2672 (1992).

[91] E. Sandvold and C. Rossel, *Physica C* **190**, 309 (1992).

[92] M. P. Maley and J. O. Willis, *Phys. Rev. B* **42**, 2639 (1990).

[93] B. I. Ivlev and N. B. Kopnin, *J. Low Temp. Phys.* **80**, 161 (1990).

[94] S. Chakravarty, B. I. Ivlev, and Yu. N. Ovchinnikov, *Phys. Rev. Lett.* **64**, 3187 (1990); *Phys. Rev. B* **42**, 2143 (1990).

[95] M. P. A. Fisher, *Phys. Rev. Lett.* **62**, 1415 (1989).

[96] D. S. Fisher, M. P. A. Fisher, and D. A. Huse, *Phys. Rev. B* **43**, 130 (1991).

[97] G. Blatter and V. B. Geshkenbein, *Phys. Rev. B* **47**, 2725 (1993).

[98] A. P. Malozemoff and M. P. A. Fisher, *Phys. Rev. B* **42**, 6784 (1990).

[99] M. V. Feigel'man, V. B. Geshkenbein, A. I. Larkin, and V. M. Vinokur, *Phys. Rev. B* **43**, 6263 (1991).

[100] A. A. Zhukov, *Sol. St. Comm.* **82**, 983 (1992).

[101] H. G. Schnack, R. Griessen, J. G. Lensink, C. J. van der Beek, and P. H. Kes, *Physica C* **197**, 337 (1992).

156

[102] V. M. Vinokur, M. V. Feigel'man, and V. B. Geshkenbein, *Phys. Rev. Lett.* **67**, 915 (1991).

[103] S. Senoussi, M. Ousséna, G. Collin, and I. A. Campbell, *Phys. Rev. B* **37**, 9792 (1988).

[104] C. W. Hagen and R. Griessen, *Phys. Rev. Lett.* **62**, 2857 (1989); R. Griessen, *Phys. Rev. Lett.* **64**, 1674 (1990).

[105] A. Gurevich, *Phys. Rev. B* **42**, 4857 (1990).

[106] L. Niel and J. Evetts, *Europhys. Lett.* **15**, 453 (1991).

[107] H. Theuss, T. Reininger, and H. Kronmüller, *J. Appl. Phys.* **72**, 1936 (1992).

[108] D. R. Nelson and H. S. Seung, *Phys. Rev. B* **39**, 9174 (1989).

[109] M. A. Moore, *Phys. Rev.* **B 39**, 9174 (1989); **B 45**, 7336 (1992).

[110] E. H. Brandt, *Phys. Rev. Lett.* **63**, 1106 (1989).

[111] E. H. Brandt, *Physica C* **162–164**, 1167 (1989).

[112] Ying-Hong Li and S. Teitel, *Phys. Rev. Lett.* **66**, 3301 (1991); *Phys. Rev. B* **45**, 5718 (1992); *Phys. Rev. B* **47**, 359 (1993).

[113] R. E. Hetzel, A. Sudbø, and D. A. Huse, *Phys. Rev. Lett.* **69**, 518 (1992).

[114] R. G. Carneiro, R. Cavalcanti, and A. Gartner, *Phys. Rev. B* **47**, 5263 (1993).

[115] W. K. Kwok, S.Fleshler, U. Welp, V. M. Vinokur, J. Downey, and G. W. Crabtree, *Phys. Rev. Lett.* **69**, 3370 (1992).

[116] P. L. Gammel, L. F. Schneemeyer, J. V. Waszczak, and D. J. Bishop, *Phys. Rev. Lett.* **61**, 1666 (1988); comment: E. H. Brandt, P. Esquinazi, and G. Weiss, *Phys. Rev. Lett.* **62**, 2330 (1989); reply: R. N. Kleiman, P. L. Gammel, L. F. Schneemeyer, J. V. Waszczak, and D. J. Bishop, *Phys. Rev. Lett.* **62**, 2331 (1989).

[117] A. Gupta, P. Esquinazi, and H. F. Braun, *Phys. Rev. Lett.* **63**, 1869 (1989); *Physica B* **165 & 166**, 1151 (1990).

[118] J. Kober, A. Gupta, P. Esquinazi, H. F. Braun, and E. H. Brandt, *Phys. Rev. Lett.* **66**, 2507 (1991).

[119] P. Esquinazi, *J. Low Temp. Phys.* **85**, 139 (1991) (review on vibrating superconductors); *Sol. St. Comm.* **74**, 75 (1990).

[120] D. E. Farrell, J. P. Rice, and D. M. Ginsberg, *Phys. Rev. Lett.* **67**, 1165 (1991); R. G. Beck, D. E. Farrell, J. P. Rice, D. M. Ginsberg, and V. G. Kogan, *Phys. Rev. Lett.* **68**, 1594 (1992).

[121] J. G. Lensink, C. F. J. Flipse, J. Roobeek, R. Griessen, and B. Dam, *Physica C* **162-164**, 663 (1989); A. C. Mota, G. Juri, P. Visani, and A. Pollini, *Physica C* **162-164**, 1152 (1989); R. Griessen et al., *Cryogenics* **30**, 536 (1990); A. Fruchter et al., *Phys. Rev. B* **43**, 8709 (1991); M. Lairson et al., *Phys. Rev. B* **43**, 10405 (1991); A. C. Mota et al., *Physica C* **185-189**, 343 (1991).

[122] R. Griessen, *Physica C* **172**, 441 (1991).

[123] A. V. Mitin, *Zh. Eksp. Teor. Fiz.* **93**, 590 (1987) [*Sov. Phys. JETP* **66**, 335 (1987)].

[124] B. I. Ivlev, Yu. N. Ovchinnikov, and R. S. Thompson, *Phys. Rev. B* **44**, 7023 (1991).

[125] G. Blatter, V. B. Geshkenbein, and V. M. Vinokur, *Phys. Rev. Lett.* **66**, 3297 (1991).

[126] R. Griessen, J. G. Lensink, and H. G. Schnack, *Physica C* **185-189**, 337 (1991).

[127] A. O. Caldeira and A. J. Leggett, *Phys. Rev. Lett.* **46**, 211 (1981).

[128] G. Blatter and B. I. Ivlev, *Phys. Rev. Lett.* **70**, 2621 (1993).

[129] A. M. Campbell, *J. Phys. C* **4**, 3186 (1971).

157

[130] M. W. Coffey and J. R. Clem, *Phys. Rev. Lett.* **67**, 386 (1991); *Phys. Rev. B* **44**, 6903 (1991); *Phys. Rev. B* **45**, 9872 (1992).

[131] M. W. Coffey and J. R. Clem, *IEEE Trans. Magn.* **27**, 2136 (1991) and erratum **27**, 4396 (1991).

[132] E. H. Brandt, *Phys. Rev. Lett.* **67**, 2219 (1991); *Physica C* **185-189**, 270 (1991).

[133] A. E. Koshelev and V. M. Vinokur, *Physica C* **173**, 465 (1991).

[134] E. H. Brandt, *Phys. Rev. Lett.* **68**, 3769 (1992).

[135] J. R. Clem, H. R. Kerchner, and T. S. Sekula, *Phys. Rev. B* **14**, 1893 (1976).

[136] R. Behr, J. Kötzler, A. Spirgatis, and M. Ziese, *Physica A* **191**, 464 (1992).

[137] J. Kötzler, to be submitted.

[138] A. Spirgatis et al., *Cryogenics* **33**, 138 (1993).

[139] E. H. Brandt, P. Esquinazi, H. Neckel, and G. Weiss, *Phys. Rev. Lett.* **56**, 89 (1986); E. H. Brandt, P. Esquinazi, and H. Neckel, *J. Low Temp. Phys.* **63**, 187 (1986); E. H. Brandt, *J. de Physique, Colloque* **C8** (No. 27, Vol. 48), 31 (1987); E. H. Brandt, *Phys. Lett. A* **113**, 51 (1985).

[140] K. A. Müller, M. Takashige, and J. G. Bednorz, *Phys. Rev. Lett.* **58**, 1143 (1987);

[141] A. F. Hebard, P. L. Gammel, C. Rice, and A. Levi, *Phys. Rev. B* **40**, 5243 (1989).

[142] J. R. Clem and M. W. Coffey, *Phys. Rev. B* **46**, 14662 (1992).

[143] E. H. Brandt, Dynamics of Flat Superconductors in Perpendicular Magnetic Field, *Phys. Rev. Lett.* **71**, 2821 (1993).

[144] D. J. Baar and J. P. Harrison, *Physica C* **157**, 215 (1989); D. J. Baar, J. P. Franck, J. P. Harrison, Y. Lacroix, and M. K. Yu, *Physica C* **170**, 233 (1990).

[145] G. D'Anna, W. Benoit, J. Luzuriaga, and H. Berger, *Europhys. Lett.* **13**, 465 (1990); G. D'Anna et al., *Europhys. Lett.* **20**, 167 (1992); G. Canelli et al., *Phys. Rev. B* **38**, 7200 (1988); M. Weller et al., *Physica C* **162-164**, 953 (1989).

[146] C. Durán, J. Yazyi, F. de la Cruz, D. Bishop, D. B. Mitzi, and A. Kapitulnik, *Phys. Rev. B* **44**, 7737 (1991); J. Yazyi et al., *Physica C* **184**, 254 (1991).

[147] Y. Kopelevich, A. Gupta, P. Esquinazi, C. P. Heidmann, and H. Müller, *Physica C* **183**, 345 (1991); A. Gupta et al., *Phys. Rev. B* **48**, 6359 (1993).

[148] A. Maeda et al., *Physica B* **165 & 166**, 1363 (1990); E. S. Otabe, T. Matsushita, and K. Yamafuji, *IEEE Trans. Magn.* **27**, 1033 (1991).

[149] J. Pankert, *Physica C* **168**, 335 (1990); J. Pankert et al., *Phys. Rev. Lett.* **56**, 3052 (1990); P. Lemmens et al., *Physica C* **174**, 289 (1991); J. Pankert et al., *Physica C* **182**, 291 (1991); P. Lemmens et al., *Physica C* **185-189**, 2271 (1991).

[150] B. Pümpin, H. Keller et al., *Z. Physik B* **72**, 175 (1988); V. G. Grebinnik et al., *Hyperfine Interactions* **63-65**, 65 (1990); D. Zech et al., *Phys. Rev. B* **48**, 6533 (1993).

[151] E. H. Brandt, *Phys. Rev. B* **37**, 2349 (1988); *J. Low Temp. Phys.* **73**, 355 (1989); *Physica C* **162-164**, 257 (1989); *Phys. Rev. Lett.* **66**, 3213 (1991); M. Inui and D. A. Harshman, *Phys. Rev. B* **47**, 12205 (1993).

[152] F. Hentsch et al., *Physica C* **158**, 137 (1989).

[153] Y.-Q. Song et al., *Phys. Rev. Lett.* **70**, 3127 (1993).

[154] M. Yethirai et al., *Phys. Rev. Lett.* **70**, 857 (1993).

[155] H. W. Weber, *J. Low Temp. Phys.* **17**, 49 (1974).

[156] W. Roest and M. Th. Rekveldt, *Phys. Rev. B* **48**, 6420 (1993).

# PINNING AND DYNAMICS OF MAGNETIC VORTICES

P.H. KES

*Kamerlingh Onnes Laboratory, Leiden University*
*P.O. Box 9506, 2300 RA Leiden, The Netherlands*

## 1 Introduction

To relate the defect structure (materials disorder) to the maximum loss-free current density $j_c$ in the mixed state of a type II superconductor, a complicated theoretical problem has to be solved which may be devided into three parts: i) Starting from the equilibrium properties of the vortex lattice one has to find out what the interaction is between the individual defects and the vortex lattice (VL). This interaction yields the elementary pinning force $f(\vec{r})$ with its maximum value $f_p$. The results greatly depend on the kind of defects and the materials under investigation. ii) Next, one has to know how the VL depends to the pinning forces. This requires the determination of the elastic matrix and the elastic moduli of the VL for displacement fields $\vec{u}(\vec{r})$ of arbitrary wavelengths. iii) Finally, one has to sum over the effects of many pinning centers generally on random positions and with nonuniform $f_p$'s. Much progress has been made by considering the summation problem in terms of the theory for elastic media in a random potential. An extensive theoretical overview which deals with the above questions in great detail will appear soon [1].

In the first part of these lecture notes, Sections 2, 3 and 4, I will discuss a few model systems and show some experimental results to illustrate the physical picture. The second part, Sections 5 and 6, is about thermal fluctuations which lead to effects like thermal depinning and VL melting.

## 2 The theory of collective pinning

In recent papers some general considerations about the elementary pinning forces of several kinds of defects were presented [2]. Here, it suffices to consider point defects which couple to the variation of the superconducting order parameter $|\psi|^2$. Defects act as point pinning centers when their size is smaller than the length scale over which $|\psi|^2$ varies. According to the exact solutions of the Ginzburg-Landau (GL) equations [3], one should distinguish between the isolated vortex solution which is valid roughly up to fields $B \approx 0.2 B_{c2}$, and the Abrikosov solution valid above $0.2\,B_{c2}$. In the first case $|\psi|^2$ changes rapidly at the vortex core over the coherence length $\xi$, in the second case $|\psi|^2 \propto 1 + \cos\left(2\pi x/a_0\right)$ along the close-packed direction, with

*N. Bontemps et al. (eds.), The Vortex State, 159–174.*
© *1994 Kluwer Academic Publishers.*

$a_0^2 = \{(2/\sqrt{3})\phi_0/B\}^{\frac{1}{2}}$ the VL parameter. That the pinning interaction couples to $|\psi|^2$ implies that about $B \approx 0.2B_{c2}$ the interaction range of $f(r)$ is changing with increasing $B$ from $r_p \approx \xi$ to $r_p \approx a_0/2$ [4]; $r_p$ is the typical length scale over which $f(r)$ changes from $-f_p$ to $+f_p$.

When one considers a stiff and very large volume of the VL under the influence of random distributions of pinning centers, each distribution will result in a net collective force $F_c$ on the volume $V$. The average of the distribution of $F_c$ will be zero, and its fluctuation $\delta F_c = (n_p V < f^2 >)^{\frac{1}{2}}$, where $n_p$ is the concentration of pinning centers and $< f^2 >$ the mean square of $f(r)$ over a primitive vortex lattice cell with area $a_0^2$. We now can define the pinning strength $W \equiv n_p < f^2 > \approx \frac{1}{2}n_p f_p^2$ and express the volume pinning force $F_p(\equiv j_c B) = \delta F_c/V$ as

$$F_p = (W/V)^{\frac{1}{2}} \qquad (1)$$

If we estimate a typical $F_p$ by taking [2] $n_p = 10^{23} \mathrm{m}^{-3}$, $f_p = 10^{-14}\mathrm{N}$ and $V = 0.1$ mm$^3$, we obtain $F_p \simeq 10\mathrm{Nm}^{-3}$ and at 1T a very small $j_c$ of $10^{-5}$ Amm$^{-2}$. This estimate clearly contrasts experimental observations. Apparently, an important effect has not been taken into account, namely, the fact that the VL deforms in presence of pinning centers.

Deformations of the VL are controlled by the elastic moduli for compression, shear and tilt: $C_{11}$, $C_{66}$ and $C_{44}$ [5].

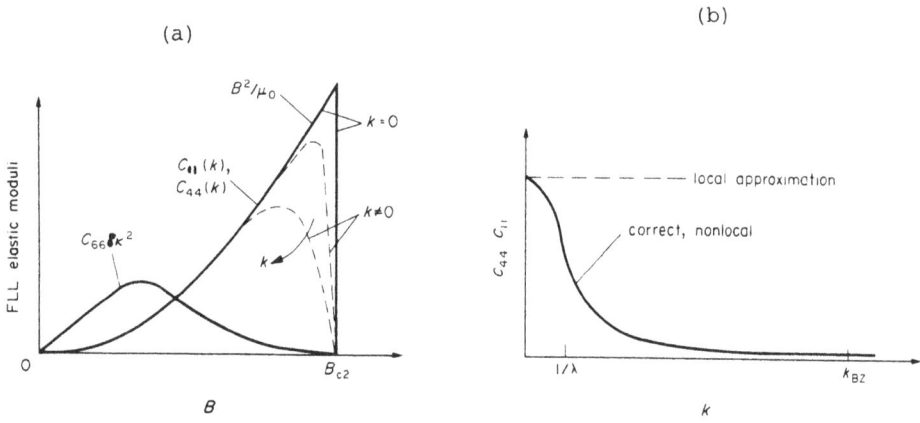

Figure 1: (a) The dispersive compression and tilt moduli $C_{11}$ and $C_{44}$ and the shear modulus $C_{66}$ vs flux density $B$ for various strain wave vectors $\vec{k}$ ($\kappa$ is the GL parameter). (b) Nonlocal $C_{44}$ and $C_{11}$ versus $\vec{k}$ (after Brandt and Evetts [5]).

Fig. 1a shows their typical field dependences and Fig. 1b their dependence on the wavevector of the deformation field, e.g. $u_x \exp(ik_y y)$ would be a shear wave of wavelength $2\pi k_y^{-1}$ and $u_x \exp(ik_z z)$ a tilt wave of wavelength $2\pi k_z^{-1}$ (the field is supposed to be along $z$). Fig. 1a shows that $C_{66} \ll C_{11}(0) \approx C_{44}(0) = B^2/\mu_0$ which means that longitudinal waves cost much more energy than transverse (shear) waves. Fig. 1b shows that short wavelength tilt deformations will also be less energetic than uniform deformations. The reason is that the position of the vortices and thus the deformation field is determined by the positions of the core, while the interaction range is given by the field dependent penetration depth $\lambda_h = \lambda/\sqrt{1-b}$, with $b \equiv B/B_{c2}$. Positional fluctuations are therefore largely decoupled from field and current fluctuations which decay over a distance $\lambda_h$ (this effect is called nonlocal elasticity [5]). Useful approximate expressions are derived by Brandt [5]: $C_{66} = (B_c^2/4\mu_0)b(1 - 0.58b + 0.29b^2)(1 - b)^2$ and $C_{44}(k_\perp, k_z) = C_{44}(0)/(1 + (k_\perp^2 + k_z^2)\lambda_h^2)$. Now suppose we could prepare a perfect VL and then gradually increase the strength of the randomly distributed pins. At first the VL would respond elastically. That is, everywhere the strain $\ll 1$. Over larger distances, however, the displacements may accumulate as in a random walk. This is expressed by the monotonous growth of the displacement correlation function $g(\vec{r}) \equiv < [u(\vec{r}) - u(0)]^2 >$ with distance $|\vec{r}|$ [6]. The dimensionality $D$ of the disorder determines the exponent in $g(r) \propto r^{4-D}$, see below. When $g$ reaches a value $\sim a_0^2$ the positional long range order is destroyed. For pinning it is more relevant to compare $g$ with $r_p^2$, since it tells us whether we can predict with any certainty which force a pinning center will exert when we exactly know the position of the vortices a distance $|r|$ away from this pinning center. The pinning becomes totally uncorrelated when

$$g(R_c, 0) = g(0, L_c) = r_p^2 \qquad (2)$$

Eq.(2) defines the transverse and longitudinal pinning correlation lengths (Larkin lengths) $R_c$ and $L_c$. Together, they determine the size of the Larkin domain (correlated region) in which the pinning is still correlated [7]

$$V_c \approx L_c R_c^2 \qquad (3)$$

This volume should be substituted in Eq.(1) to obtain the $F_p$ of the collective pinning theory [8].

The Larkin lengths may be estimated by equating elastic deformation and pinning energy densities:

$$\frac{1}{2}C_{44}(k_\perp) \approx \frac{1}{L_c}\frac{r_p^2}{L_c^2} = \frac{1}{2}C_{66}\frac{r_p^2}{R_c^2} = \left(\frac{W}{R_c^2 L_c}\right)^{\frac{1}{2}} r_p \qquad (4)$$

Note that $R_c$ and $L_c$ are the most relevant wavelengths of the deformation field to appear in the nonlocal expression of $C_{44}$. This leads to [9],

$$L_c = (C_{44}(0)/C_{66})^{\frac{1}{2}} R_c, \qquad L_c \gg R_c \gg \lambda_h \qquad (5a)$$

$$L_c = (C_{44}(0)/C_{66})^{\frac{1}{2}} R_c^2/\lambda_h, \qquad R_c \ll L_c \ll \lambda_h \qquad (5b)$$

These expressions together with (3) and (1) can be used to compute $R_c$ and $L_c$ from experimental $F_p$ data when the pinning strength is known. Especially, when the pinning is weak, $R_c \gg a_0 \approx 50\text{nm}$ and we estimate $L_c > 5\mu\text{m}$. If therefore, pinning is studied in thin-film amorphous superconductors with thickness $d$ in perpendicular field, it is expected that $L_c \gg d$. Under these conditions the VL remains ordered along the field direction and can be considered to be straight. This is the case of 2D collective pinning (2DCP) for which the expression for $R_c$ becomes [10]

$$R_c = 2r_p C_{66} \left\{ \frac{2\pi d}{W ln(w/R_c)} \right\}^{\frac{1}{2}} \tag{6}$$

Here $w$ is the macroscopic size of the film. Apart from the logarithmic correction, (6) follows from (4) by putting $C_{44} = 0$ and $L_c = \text{d}$.

If we now go on with our "gedanken" experiment and increase the pinning further relative to the VL interaction (note, that in practice this is achieved by either raising the field to $H_{c2}$ or lowering it to $H_{c1}$), the strains in the VL gradually increase and the Larkin lengths decrease. Restricting ourselves to 2DCP we can imagine that at a certain pinning strength the local flow stress at some points in the VL is exceeded. The flow stress $\tau_f$ of the triangular VL is not exactly known, $\tau_f$ is certainly less than $C_{66}/2\pi$. (Brandt [11] computes $\tau_f = 0.044 C_{66}$, a value we will also find below). The elastic instabilities arising at those sites may, for instance, create vacancy-interstitial pairs, when the transport current exerts a uniform force on the vortices. From the vacancies and interstitials edge-dislocation pairs may develop by the driving force [12]. Dislocations in 2D cause a softening of the shear modulus so that the vortices may more effectively adapt to the random potential caused by the pinning centers. Thus $F_p$ becomes larger than predicted by the 2DCP theory where $R_c$ was determined by purely elastic deformations. CP for a VL with topological disorder modeled by a square array of edge-dislocations, was worked out in [13] and indeed yields a steeper increase of $F_p$ vs $B$. For the 3D case topological disorder probably exists of edge- and screw- dislocation loops. It should be noted that the CP theory neglects the free energy decrease related to the entropy of topological disorder. It actually is a $T = 0$ theory.

When the pinning strength is increased further the dislocation density grows to such an extent that a description in terms of dislocations becomes meaningless, i.e. when the amorphous limit is reached. This limit is characterized by the criterion that the pin energy in (4) is larger than the shear energy, so that it can be expressed by putting $R_c = a_0$ and $C_{66} = 0$. The independently pinned objects now are single flux line segments of length $L_c$ given by

$$L_c \simeq \left\{ \frac{C_{44}(0)a_0^3 r_p}{\lambda_h^2 W^{\frac{1}{2}}} \right\}^{\frac{2}{3}} \tag{7}$$

In strong pinning materials this situation of "single vortex pinning" occurs very often. It gives rise to dome shaped $F_p$ vs $B$ curves which can be well described by a scaling relation $F_p \propto B_{c2}^n(T)b(1-b)$ with $n \approx 2.5$.

Before we shift to an overview of experiments on 2DCP, we should briefly mention the situation in the high- temperature superconductors (HTS). For an extensive review see [14]. The HTS are layered materials with weak superconducting coupling between the $CuO_2$ (single, double or triple) layers. Especially for Bi:2212, the situation at low temperatures is reasonably well understood [15,16]. For anisotropic superconductors with anisotropy $\Gamma = (m_z/m_{xy})$, the tilt modulus of the VL for short wavelength deformations in case the field is perperdicular to the layers, is reduced by a factor $\Gamma$ [17]. This leads to a considerable decrease of $L_c$. To understand large critical current densities $j_c \approx 5 \times 10^{10} Am^{-2}$, as observed in Bi:2212 at low $T$, one can invoke point defects in the $CuO_2$ layers as pinning centers, e.g. oxygen vacancies, and estimate $L_c$ from (7), (3) and (1) assuming $R_c = a_0$ and eliminating $W$ in favor of $j_cB$. One finds $L_c \ll s$, the periodicity of the layered structure, $s \approx 1.5$ nm. This means that the pinned object is as small as one single pancake vortex (see lecture Prof. Clem) and $V_c = a_0^2 s$ is the size of the Larkin domain. We estimate $W/B = 75 \times 10^{-4} N^2 m^{-3} T^{-1}$ and obtain for the pin energy $U_p = (WV_c)^{\frac{1}{2}} r_p \approx 20K$, if we take $r_p = \xi \approx 2nm$. The conclusion is that because the pinned object is extremely small, the $j_c$ is very large (only a factor 20 below the depairing current) and at the same time the pin energy is very small. The latter gives rise to fast flux creep governed by the Boltzmann factor $\exp[U(j)/k_BT]$ and a barrier height for hopping given by the collective creep theory [18], $U(j) = U_p[(j_c/j)^\mu - 1]$, with $\mu \sim 1$ [19]. In practice, the creep slows down when the current density has decayed so much that $U(j)/k_BT \approx 30$. This peculiar behavior of the HTS is thus mainly caused by the large anisotropy factor.

# 3  Collective pinning experiments

Experiments on thin amorphous films were the first indication that CP for D=2 can really be observed [20]. The main reasons are: no extended defects in amorphous films, only weak pinning centers, most likely quasi-dislocation loops, demonstrated by $j_c \sim 10^{-2} Amm^{-2}$. By preparing a series of identical films of a-$Nb_x$Ge $(x \sim 3)$ with thickness $d$ ranging from 70 nm to 5 $\mu$m, it was confirmed that the pinning indeed is 2D from the fact that $F_p d$ at the same reduced fields and temperatures is constant. A typical set of $F_p(b)$ curves in 2DCP is shown in Fig. 2 [21]. The solid line represents the theory, i.e. Eqs.(1) and (6) with $L_c = d$ and $W = W_0 b(1-b)^2$ which is the expected field dependence for $\delta T_c$-pinning [2]. The temperature dependent parameter $W_0$ follows from the fit at $b = 0.4$. On approaching $B_{c2}$ the experimental data show an upward deviation from the theoretical 2DCP prediction. Before explaining the reason for this peak, it is very useful to study the field dependence of $R_c/a_0$ determined from the data, see Fig. 3.

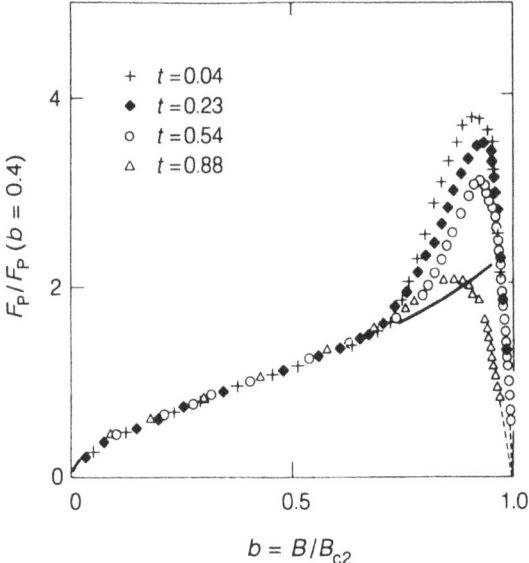

Figure 2: Normalized bulk pinning force versus $B/B_{c2}$ for a thin a-Nb$_3$Ge film (d = 1.24 $\mu$m, $T_c$ = 3.81 K) at various values of $t = T/T_c$.

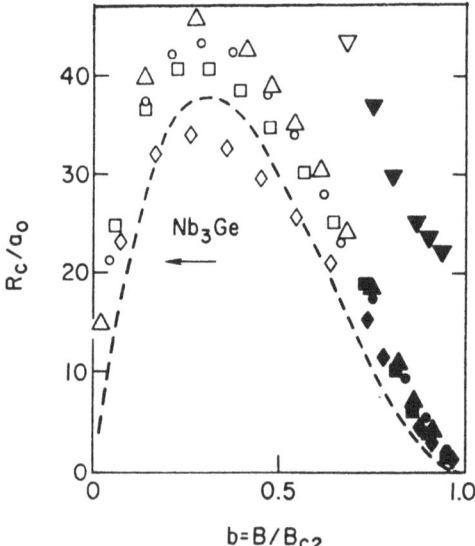

Figure 3: Normalized Larkin length plotted vs$b$ for an a-Nb$_3$Ge film (d = 0.46 $\mu$m, $T_c$ = 4.00 K) at $t = 0.79$ (circles), $t = 0.70$ (triangles), $t = 0.50$ (squares), and $t = 0.40$ (diamands). Open symbols refer to the region below the peak, filled symbols to the region of the peak. The dashed line shows the field dependence of $C_{66}$.

Typically, $R_c/a_0$ follows the field dependence of $C_{66}$, dashed line. It goes to zero both at small and large flux densities demonstrating the increase of disorder related to the softening of the vortex lattice for small $C_{66}$. At $b \approx 0.3$ a maximum is reached as high as 40, but values up to 200 have been determined for more homogeneous films, e.g. in 2H-NbSe$_2$ [22]. Larkin domains, thus, may contain as many as 1600 flux lines. To simulate collective behavior on the computer with a reasonable amount of precision would require a cell of at least 6400 vortices with at least one pinning center per flux line. This is far beyond the present capacity of computers and the conclusion that 2DCP does not follow from computer simulations is based on an overestimation of the validity of these simulations [23].

For all cases (different materials, film thicknesses, temperatures) in which 2DCP was observed at low fields, the onset of the peak effect coincided with the condition $R_c/a_0 \approx 20$.

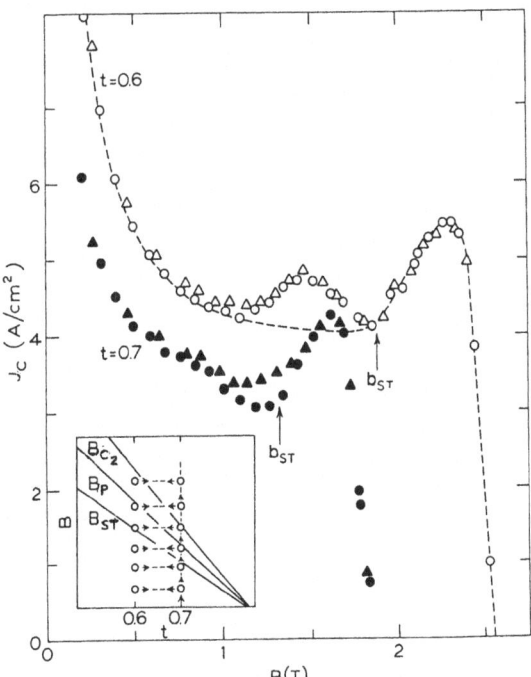

Figure 4: Isofield experiment for a thin a-Nb$_3$Ge film (1.24$\mu$m). The $j_c$ values are plotted versus $B$ for $t = 0.7$ (solid symbols) and $t = 0.6$ (open symbols) in increasing (circles) and decreasing (triangles) fields. The result of an isothermal measurement at $t = 0.6$ is given by the dashed line. The inset shows schematically the way of performing the isofield measurement.

It could be experimentally demonstrated [24] that for this amount of disorder the vortex lattice becomes elastically instable. The shear strength is locally exceeded giving rise to creation of edge-dislocations accompanied by history effects in the V,I curves around the onset field at low vortex velocities. The presence of edge-dislocations makes the shear modulus softer, so that the vortex conforms more adequately to the random pin potential causing an enhancement of $F_p$. The following picture arises: below the peak $R_c/a_0$ is determined by elastic deformations (elastic disorder) and Eq.(6) can be used. Above the onset field of the peak $R_c/a_0$ is predominantly determined by the density of edge dislocations (plastic disorder) and only Eq.(1) remains and can be used to compute $R_c/a_0$ (solid symbols in Fig. 3). Evidence for this model is provided by the results in Fig. 4. The open symbols display $j_c$ measurements at $t = 0.6$ after a fast cool down in constant field from $t = 0.7$ (solid symbols). Plastic disorder will be frozen in by this procedure resulting in a metastable state of higher disorder at the lower temperature. This produces an "image" peak below $b_{ST}$ (ST for structural transition) at $t = 0.6$ clearly discernable from the dashed line which represents the measured $j_c(B)$ for elastic disorder [24]. The image peak disappears gradually upon an increase of the vortex velocity comparable to an anneal treatment of the vortex lattice. To conclude this paragraph, it should be noted that most computer simulations [23] apply to the case of plastic disorder. If, in addition, a low density of pinning centers is assumed, the fluctuations of the random pinning potential are relatively large. The first vortex motion then develops along a path of weakest pinning and the pinning force is determined by the shear strength, see below. Since only part of the vortex lattice moves, the observed flow resistivity is a fraction of the resistivity corresponding to uniform flux flow $\rho_f$. Simulations in which a large density of pinning centers was assumed, showed that the onset of uniform flux flow is always triggered by partial vortex motion mediated by glide of edge-dislocations [25].

In thicker films or for stronger pinning a crossover from 2D disorder and 2DCP to 3D disorder and 3DCP has been detected [26]. At the crossover field that coincides with the condition $L_c(b) = d/2$, the pinning force shows a jumplike increase by an order of magnitude. The transition shows hysteresis. The 3D behavior cannot be described presuming elastic disorder and from that observation it is concluded that 3DCP, even for weak pinning, is dominated by plastic disorder of the vortex lattice, presumably a network of edge- and screw- dislocation loops. For further discussions we refer to [26].

# 4    Shear flow of 2D vortex lattice

Because of the smallness of $C_{66}$, plastic shear flow is a well-probable mechanism of vortex motion in materials with strong fluctuations of the random potential [27]. To study plastic flow Pruijmboom et al. [28] prepared a two-layer device consisting of a continuous weak-pinning a-Nb$_3$Ge layer and a discontinuous strong-pinning NbN layer with channels etched in it, see Fig. 5a. In the elastic continuum approximation the driving force on a vortex block of size $dlW$ in the channel is $jBdlW$.

This force is counteracted by the shear force exerted by the vortices fixed in the NbN layers on the sides of the block $2\tau_f dl$, where $\tau_f$ is the flow stress of the medium. Therefore, a voltage over the device develops when the current density exceeds a value $j_s$ given by

$$j_s = 2\tau_f/BW \tag{8}$$

Remaining questions for a real vortex lattice are the ratio $\tau_f/C_{66}$ and the effective width $W_{eff}$ of the channel through which the lattice planes move, see Fig. 5b. If we take $\tau_f = 0.05\ C_{66}$ and $W_{eff} = 90$ nm, which is the channel width as determined by SEM, we compute for $j_s$ the upper solid curve in Fig. 6. The experimental curves (see [28] for details) are not far below this line, but show some qualitatively different features. A striking oscillatory behavior is observed as a function of field and for field up the results deviate from the field- down data. As was explained in detail in [28], the oscillatory behavior is related to the commensurability between vortex lattice parameter and $W_{eff}$. The difference between field-up and field- down is linked to the two possible lattice configurations I and II, as sketched in Fig. 5b. Configuration II occurs for increasing field and as to the oxcillations in Fig. 6 has the larger period. The inset of Fig. 6 shows that upon reversing the field sweep direction an immediate change of vortex lattice configuration takes place from I to II and vice versa.

In [29] different channel patterns, i.e. an honeycomb and brickwall pattern [30], various geometries of pin mechanisms have been simulated. The honeycomb pattern is supposed to mimic the pinning by small grain boundaries in Nb3Sn, while the brickwall pattern represents the elongated defect structure of $\alpha$-Ti ribbons in drawn NbTi wires [31]. Similar behavior as shown above for the linear channels is observed, except that a geometry factor has to be introduced which describes the ratio between the actual and shortest path lengths of the vortices moving across the device. Other investigations concerning the transition from lattice to liquid are underway.

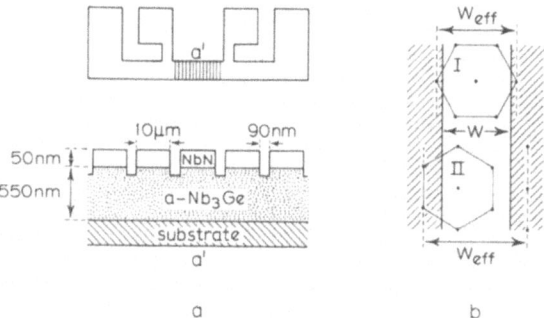

Figure 5: Sketch of the parallel channel array in a superconducting double layer (a) and of $W_{eff}$ for two different orientations (denoted I and II) of the FLL (b).

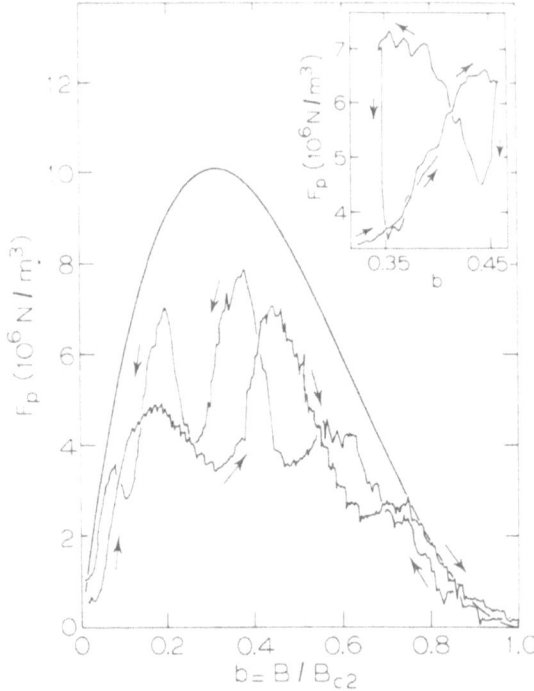

Figure 6: Recording traces of $F_p$ vs $b = B/B_{c2}$ from the dynamic measurements on the parallel channel array at 1.74 K. The solid line represents the continuum model. The inset displays the effect of reversing the field-sweep direction.

# 5 Melting

In the two final sections we will address the effect of thermal fluctuations on pinning and lattice properties. Raising the temperature gives rise to thermal vibration of the vortex lattice. The dynamics is overdamped, i.e. the viscous damping term in the kinetic equation is much larger than the mass term. When pinning is ignored the mean squared amplitude of the lattice vibrations is given by $< u^2 >_{th} \sim k_B T / a_0 C_{eff}$, where $C_{eff}$ is a combination of $C_{44}$ and $C_{66}$ as derived by Houghton et al. [17].

According to the Lindemann criterion the vortex lattice melts when $< u^2 >_{th}^{\frac{1}{2}} = c_L a_0$. The Lindemann constant $c_L$ has been estimated for $B \sim B_{c2}$ by Hikami et al. [32] to be 0.14. When both tilt and shear waves play a role melting is considered to be a 3D phenomenon, when only shear waves are important, e.g. in thin films, it is 2D. In the latter case one has [33]

$$< u^2 >_{th} = \frac{k_B T}{4\pi C_{66} d} ln \left( \frac{R_c}{a_0} \right)^2 \qquad (9)$$

However, the melting transition in 2D is not governed by a Lindemann criterion but rather follows from the condition for the unbinding of (edge) dislocation pairs as suggested by Berezinskii [34] and Kosterlitz-Thouless [35]. This theory was applied to 2D melting of a vortex lattice by Huberman and Doniach and Fisher [36]. According to this theory the melting line in the B,T plane is determined by

$$k_B T_m = A \frac{C_{66} a_0^2 d}{4\pi} \qquad (10)$$

where $A$ is a renormalization constant describing the softening of $C_{66}$ by anharmonicities and bound dislocation pairs. According to [39] $A = 0.64$. Evidence for the 2D melting transition of the vortex lattice has been recently found [37,38].

Berghuis [37] concentrated on dc and ac $IV$ measurements as a function of field in a-Nb$_3$Ge films of various thicknesses (note that Eq.(10) depends on $d$). From the 2DCP behavior below the melting line it follows that $R_c/a_0$ near melting is of disorder 10 to 20. This may not be large enough to justify the assumption that the influence of the random potential on melting may be neglected. From the experiments the field dependences of $F_p$, $\rho_f$ and $\rho_{ac}$ were determined, see Fig. 7.

For $F_p$ we used a $1\mu V/cm$ criterion as well as a criterion given by a linear extrapolation to $V = 0$ from the flux-flow regime. The flux-flow resistivity $\rho_f$ was determined from the slope of the $IV$ curves at a voltage corresponding to an average vortex velocity of 0.1m/s, and $\rho_{ac}$ followed from the slope at zero bias current. Fig. 7b shows how $B_{c2}$ is defined from the linear extrapolation of $\rho_f$ to $\rho_n$, as predicted for the dirty limit [40]. Well below this field the pinning force becomes zero at $B_0$ and at the same time we see that $\rho_{ac} = \rho_f$, i.e. the resistance has become ohmic above $B_0$. From Fig. 7c it follows that $\rho_{ac}$ increases exponentially crossing the $\rho_f$ curve at $B^*$. The fact that just below $B_0$ the $\rho_f$ data lie below the $\rho_f$ curve, supports the assumption that the exponential behavior is caused by a fast growing number of dislocation pairs moving by thermally activated hopping processes [37]. Finally, a third characteristic field that follows from the experiment, is $B_p$. At this field $F_p$ is maximum when determined from the $1\mu Vcm^{-1}$ criterion, while it marks the onset of the peak effect when the flux flow criterion is used.

In Fig. 8 we investigate which field behaves as the melting field by plotting $4\pi k_B T/U_m(B)$ vs $T$, where $U_m \equiv C_{66} a_0^2 d$. According to Eq.(10) we expect this quantity to be equal to $A = 0.64$, marked by the horizontal line. Both $B_0$ and $B^*$ behave in accord with (10), while $B_p$ clearly deviates. For practical reasons we defined $B^*$ as the melting field. Note, however, that the data for the thickest film lie significantly below the 0.64 line. We think that for this value of $d$ (=2.4 $\mu m$) tilt deformations are more important than shear distortions in determining the lattice vibrations. In fact, we are observing 3D melting and when these data are analyzed in terms of the Lindemann criterion a value $c_L = 0.161 \pm 0.005$ is obtained for thesee data points. For the thick films 3D melting may occur while the pinning is still 2D.

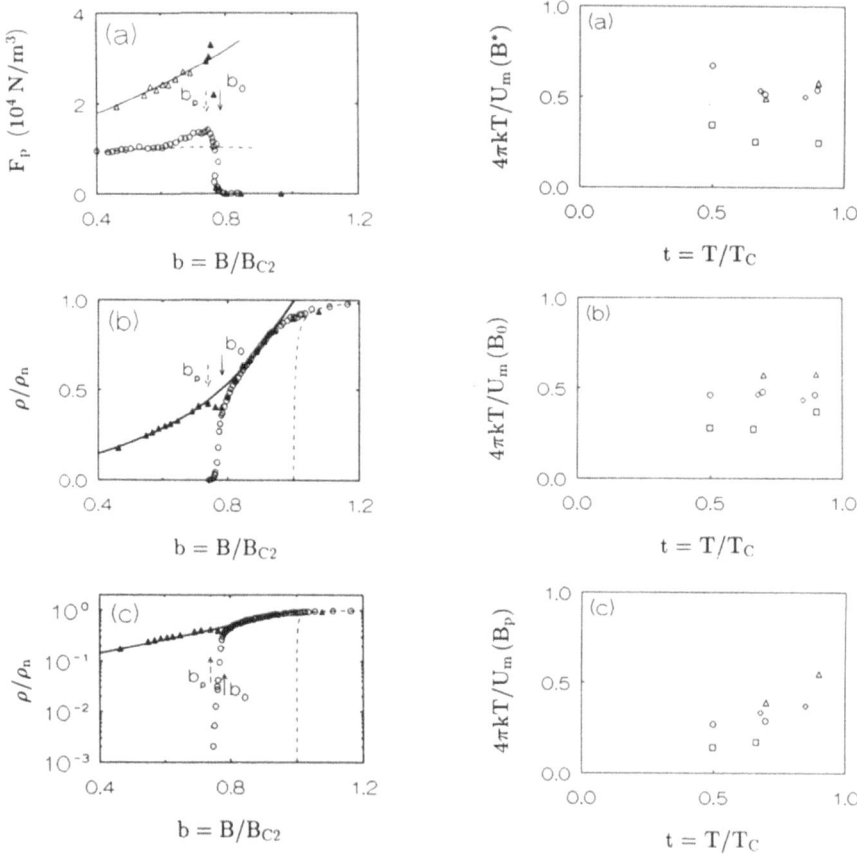

Figure 7: (left) (a) $F_p$ vs $b$ of a-Nb$_3$Ge film (d = 0.57$\mu$m, $T_c$ = 2.93 K) at $T$ = 2.49 K: 1$\mu$V/cm criterion (circles), linearly extrapolated $IV$ curves (triangles), 2DCP theory (solid and dashed lines). (b) $\rho/\rho_n$ vs $b$: flux flow data (triangles), zero bias $ac$ data (circles), expected flux flow behavior (solid line). (c) $\rho/\rho_n$ vs $b$ on semi logarithmic scale: symbols same as in (b).

Figure 8: (right) $(U_m/4\pi kT)^{-1}$ vs $T/T_c$ for various a-Nb$_3$Ge films: d = 93 nm, $T_c$ = 3.00 K (triangles), d = 205 nm, $T_c$ = 3.37 K (circles), d = 565 nm, $T_c$ = 2.93 K (diamants), d = 2.35$\mu$m, $T_c$ = 3.00 K (squares). The data are determined at $B^*$(a), $B_0$(b) and $B_p$(c).

# 6   Thermal depinning

Thermal vibrations also smear out the pinning potential. In fact, it is multiplied by a Debije-Waller factor [18,19]. Pinning becomes increasingly less effective when the temperature increases. The 3D case has been worked out in [18], the 2D case in [33]. In addition to Eqs.(6) and (9), a self-consistent scheme can be set up by introducing $r_p^2 = r_{p,0}^2 + <u^2>_{th}$, where $r_{p,0}$ denotes the value of $r_p$ at $T = 0$. As a result the critical current will sharply decrease at a temperature $T_p$ for which $<u^2>_{th} = r_{p,0}^2$:

$$k_B T_p = \frac{2\pi C_{66} d r_{p,0}^2}{ln(R_c/a_0)} \tag{11}$$

For large fields $r_{p,0} = a_0/2$ and a comparison of (11) and (10) yields $T_p > T_m$. The vortex lattice melts before depinning can take place. In the low-field limit, $b < 0.2$, $r_{p,0} = \xi$ and we obtain

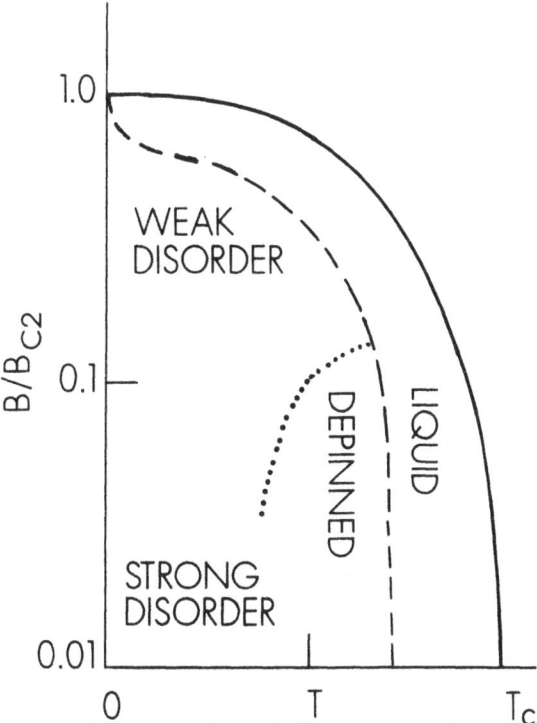

Figure 9: Phase diagram of a 2D vortex lattice. Denoted are the mean-field transition line $B_{c2}(T)$ (solid line), the 2D melting line given by Eq.(10) (dashed line) and the depinning line at low fields and for $R_c/a_0 > 10$ (dotted line) . For lower fields the disorder is larger and may be of influence on the melting transition.

$$\frac{T_p}{T_m} \approx \frac{8\pi b}{ln(R_c/a_0)}, \qquad b < 0.2 \tag{12}$$

From Fig. 3 it follows that $R_c/a_0 \approx 270 \, b(1- b)^2$ in this regime, so that $b \approx 0.13$ at $T_p = T_m$ and $R_c/a_0 \approx 27$.

We conclude that for small fields the depinning temperature may be reached before the melting transition which means that melting cannot be probed by means of transport measurements and if one tries deviations from (10) are to be expected.

The following phase diagram can now be constructed (see Fig. 9). In case pinning is weak, say $R_c/a_0 > 10$, so $b > 0.037$.

The melt line has been sketched according to Eq.(10) [36]. Below $b \simeq 0.13$ the depinning line marks the thermally assisted reduction of $j_c$. At $b \approx 0.037$ we have $T_p \approx 0.4 \, T_m$. Below this field we enter the strong pinning regime where the melting transition supposedly is overruled by 2D collective pinning. A more formal theoretical approach has been recently given by Matthew Fisher [41], and experimental evidence has been reported by Hebard and Palaanen [42].

# Acknowledgements

This work has been partly supported by the National Onderzoek Project Hoge-$T_c$ supergeleiding and by the Stichting voor Fundamenteel Onderzoek der Materie (FOM).

# References

[1] G. Blatter, M.V. Feigelman, V.B. Geshkenbein, A.I. Larkin and V.M. Vinokur, submitted to Rev. Modern Phys. (1993).

[2] P.H. Kes, in Concise Encyclopedia of Magnetic & Superconducting Materials, ed. J.E. Evetts (Pergamon, Oxford, 1992) p.163.

[3] E.H. Brandt, Phys. Stat. Solidi **51**, 345 (1972).

[4] E.H. Brandt, Phys. Rev. Lett. **57**, 1347 (1986).

[5] E.H. Brandt, Phys. Rev. **B34**. 6514 (1986); E.H. Brandt and J.E. Evetts, in Concise Encyclopedia of Magnetic & Superconducting Materials, ed. J.E. Evetts (Pergamon, Oxford, 1992) p.150.

[6] A.I. Larkin, Sov. Phys. - JETP **31**, 784 (1970).

[7] Since $a_0 > r_p$ , the positional correlation lengths exceed the Larkin lengths. This is clearly observed in decoration experiments on BiSCCO where the VL looks perfect while nevertheless the $j_c$ is large ($\sim 10^{10} \text{Am}^{-2}$) at low T.

[8] A.I. Larkin and Yu.N. Ovchinnikov, Sov. Phys. - JETP **38**, 854 (1974); A.I. Larkin and Yu.N. Ovchinnikov, J. Low Temp. Phys. **34**, 409 (1979).

[9] In the second case, (5b), the lengths themselves follow from a more extensive derivation, see [10].

[10] P.H. Kes and R. Wördenweber, J. Low Temp. Phys. **67**, 1 (1987); E.H. Brandt, J. Low Temp. Phys. **64**, 375 (1986).

[11] E.H. Brandt, J. Low Temp. Phys. **24**, 427 (1976).

[12] R. Seshadri and R.M. Westervelt, Phys. Rev. **B46**, 5150 (1992). In this paper nice examples of dislocation creation, interaction and annihilation are given.

[13] Mullock and J.E. Evetts, J. Appl. Phys. **57**, 2588 (1985).

[14] P.H. Kes, in "Phenomenology and Applications of High Temperature Superconductors", ed. K. Bedell et al. (Addison Wesley, New York, 1992) p.390.

[15] C.J. van der Beek and P.H. Kes, Phys. Rev. **B43**, 13032 (1991).

[16] C.J. van der Beek, M.P. Maley, M.J.V. Menken, A.A. Menovsky and P.H. Kes, Physica **C195**, 307 (1992).

[17] A. Houghton, R.A. Pelcovits and A. Sudbø, Phys. Rev. **B40**, 6763 (1989).

[18] M.V. Feigelman and V.M. Vinokur, Phys. Rev. **B41**, 8986 (1990).

[19] M.V. Feigelman, V.B. Geshkenbein, A.I. Larkin and V.M. Vinokur, Phys. Rev. Lett. **63**, 2303 (1989).

[20] P.H. Kes and C.C. Tsuei, Phys. Rev. Lett. **47**, 1930 (1981), and Phys. Rev. **B28**, 5126 (1983).

[21] I show the figure as originally published. More recent studies revealed that the field at which $F_p$ goes to zero, is not $B_{c2}$ but rather the field at which the vortex lattice melts. In fact $B_{c2}$ is about 10 percent larger (Section 5). It turns out, however, that the interpretation of the results is not really affected.

[22] P. Koorevaar, J. Aarts, P. Berghuis and P.H. Kes, Phys. Rev. **B42**, 1004 (1990). The title of this paper should read: "Tilt-modulus **reduction** ....".

[23] H.J. Jensen, A. Brass and A.J. Berlinsky, Phys. Rev. Lett. **60**, 1676 (1988); H.J. Jensen, A. Brass, Y. Brechet and A.J. Berlinsky, Phys. Rev. **B38**, 9235 (1988); A.-C. Shi and A.J. Berlinsky, Phys. Rev. Lett. **67**, 1926 (1991).

[24] R. Wördenweber, P.H. Kes and C.C. Tsuei, Phys. Rev. **B33**, 3172 (1986).

[25] A.E. Koshelev, Physica **C198**, 371 (1992).

[26] R. Wördenweber and P.H. Kes, Phys. Rev. **B34**, 494 (1986), and Cryogenics **29**, 321 (1989); see also Ref. [10]; P. Berghuis, R. Wördenweber and P.H. Kes,

Japan J. App. Phys. **26**, 1527 (1987).

[27] E.J. Kramer, J. Appl. Phys. **44**, 1360 (1973).

[28] A. Pruijmboom, E. van der Drift, S. Radelaar and P.H. Kes, Phys. Rev. Lett. **60**, 1430 (1988); Appl. Phys. Lett. **52**, 662 (1988); Cryogenics **29**, 232 (1989).

[29] A. Pruijmboom et al., Appl. Phys. Lett. **52**, 662 (1988).

[30] More recent studies (P. Berghuis et al., unpublished) on meander patterns where vortices are forced to partially move against the driving force, give similar results.

[31] P.J. Lee and D.C. Larbalestier, Acta Metall. **35**, 2523 (1987).

[32] S. Hikami, A. Fujita and A.I. Larkin, Phys. Rev. **B44**, 10400 (1991).

[33] V.M. Vinokur, P.H. Kes and A.E. Koshelev, Physics **C168**, 29 (1990).

[34] V.L. Berezinskii, Sov. Phys. - JETP **34**, 610 (1972).

[35] J.M. Kosterlitz and D.J. Thouless, in "Progress in Low Temperature Physics", ed. D.F. Brewer (North-Holland, Amsterdam, 1978) Vol. VII-B.

[36] B.A. Huberman and S. Doniach, Phys. Rev. Lett. **43**, 950 (1979); D.S. Fisher, Phys. Rev. **B22**, 1190 (1980).

[37] P. Berghuis, A. van der Slot and P.H. Kes, Phys. Rev. Lett. **65**, 2583 (190); P. Berghuis and P.H. Kes, Phys. Rev. **B47**, 262 (1993).

[38] A. Yazdani, W.R. White, M.R. Hahn, M. Gabay, M.R. Beasley and A. Kapitulnik, Phys. Rev. Lett. **70**, 505 (1993).

[39] S.W. de Leeuw and J.W. Perram, Physica **A113**, 546 (1982).

[40] A.I. Larkin and Yu.N. Ovchinnikov, in "Non Equilibrium Superconductivity", ed. P.N. Langenberg and A.I. Larkin (North-Holland, Amsterdam, 1986) p.493.

[41] M.P.A. Fisher, Phys. Rev. Lett. **65**, 923 (1990).

[42] A.F. Hebard and M.A. Paalanen, Phys. Rev. Lett. **65**, 927 (1990).

# PINNING AND CORRELATIONS IN THE VORTEX PHASE

*Analogies with the Spin-Glass problem*

J.P. Bouchaud

*Service de Physique de l'Etat Condensé, CEA - Saclay, and*

*TCM Group, Cavendish Laboratory, Cambridge.*

**Abstract**: We briefly review recent progress in the description of pinned vortex lattices (or other structures). The existence of at least *two* characteristic length scales is emphasized. Static (geometrical) correlation functions are discussed in connection with experiments. Some aspects of the dynamics of these pinned objects are also discussed: we mention in particular the possibility that non-stationary (aging) effects could occur, as in spin-glasses.

## I. INTRODUCTION

The pinning problem is of primary importance, both from a fundamental and technological point of view. If type II superconductors are to be used for their superconducting properties, vortices must be held fixed: their current-induced motion generate an electric field - and hence dissipation. The mechanical properties of metals and alloys strongly depend on how the dislocations are pinned by the precipitates: efficient pinning leads to hard (but brittle) materials. Both the amplitude and the frequency dependence of the susceptibility in commonly used magnets are partly determined by the way the Bloch walls are pinned by the impurities. In other cases, pinning should be avoided: the spreading of a liquid on a surface can be strongly inhibited by random chemical alterations. Among other examples, one should cite Charge Density Waves with impurities, magnetic bubble lattices on a rough substrate or froth in porous media; we shall call these various objects 'extended structures', which may classified according to their internal dimension $D$. $D = 1$ corresponds to lines (single vortex, polymer, dislocation...), $D = 2$ to Bloch walls or a two dimensional bubble lattice, and $D = 3$ to the Abrikosov lattice, for example.

*N. Bontemps et al. (eds.), The Vortex State, 175–192.*
© 1994 *Kluwer Academic Publishers.*

Although the above examples were all illustration of pinning by 'impurities' in a broad sense, one should keep in mind that pinning can also arise from the periodic nature of the underlying lattice. Dislocations in Ge or Si primarily sit at the bottom of the relatively steep (periodic) Peierls potential. Below a certain temperature, the interface between solid and liquid (say $^4$He) is 'facetted', i.e. locked to the lattice periodicity. In this case, the mobility of the interface, which determines the rate at which the crystal grows, is strongly reduced. A Peierls potential which pins the vortices also appears in layered superconductors: this is discussed in detail by D. Feinberg in this volume.

The two important questions one would like to address in this context are the following:

a) What is the conformation adopted by the extended structure in the presence of pinning? More precisely, one would like to compute the various geometrical correlations functions describing its equilibrium state.

b) How does the pinned structure respond to an external force? How much of the information obtained on the static problem can be used to infer the shape of the curve velocity versus applied force?

Many ideas developed in the context of disordered systems have been put forward to understand these issues - for a recent comprehensive review, see [1]. Although some open questions remain, point (a) appears to be reasonably well understood - the first part of these lectures will be devoted to a short review of the recent progress on this matter. Point (b) is however much more controversial - at least in the case of pinning by impurities - and a satisfactory theory is still to be developed. In the second part of these lectures, we shall discuss why such a theory might more subtle than anticipated, because of the importance of rare events which can lead to rather unexpected results. These arguments will be illustrated by a simple model and by experimental results on the non-stationary dynamics in spin-glasses (aging).

## II. MODELS AND METHODS FOR THE STATIC PROBLEM

We would thus like to describe the equilibrium configurations of a D-dimensional *elastic* structure in the presence of *quenched* random impurities. The two words in italics represent probably immediately some limitations to the whole approach. It is indeed possible that dislocations play an important rôle at large wavelengths even in the static problem - not to speak about the dynamical problem, where plastic flow is unavoidable [2,3]. The assumption of quenched, immobile impurities might be justified in some cases but not in others: in the case of vortex lattices, the position of the columnar defects in irradiated samples is given once and for all. Even point like defects like oxygen vacancies which are quite mobile at room temperature should be essentially unable to move at 100 K. In the opposite case, the disordered potential would be *time dependent* and lead to a cumulative effect. The impurities would diffuse towards the vortices and reinforce the pinning sites initially present, giving rise to interesting dynamical effects [4].

Notwithstanding these complications, one starts with a Hamiltonian of the form:

$$H(\{\vec{u}(\vec{x})\}) = H_{\text{elastic}}(\{\vec{u}(\vec{x})\}) + \sum_{\vec{x}} V(\vec{x} + \vec{u}(\vec{x})) \tag{1}$$

where $\vec{x}$ labels the unperturbed position of -say- the vortices and $\vec{u}(\vec{x})$ is an $N$-dimensional vector describing the displacement induced by the impurities or by thermal fluctuations. $\vec{x} + \vec{u}(\vec{x})$ is thus the actual position of the vortex. $V$ is the (random) potential created by the impurities, its order of magnitude will be taken to be $U_p$. The spatial structure of its correlation function should reflect the nature of the impurities: point-like defects create a short-range correlated potential, while columnar defects or grain boundaries introduce long-range correlations. Finally, the elastic part $H_{\text{elastic}}$ is a quadratic form of the displacement $\{\vec{u}(\vec{x})\}$ which, in the case of vortex lattices, is both non-local and anisotropic: there are three elastic moduli (tilt, compression and shear) which may depend quite strongly on the wavevector $q$ if $q^{-1}$ is less than the penetration depth (see the contribution of E. H. Brandt, and [1]).

The minimisation of Eq. (1) (relevant at low enough temperatures) is an optimisation problem: the structure wants to distort to make the most of the potential energy $V$ - this distortion is however limited by the elastic energy which prevents short-wave deformations. Relatively simple arguments allow to obtain a first qualitative picture of a typical configuration. At sufficiently small length scales, the distortion $u$ is expected to be small. Hence one may expand locally the potential energy to find: $V(\vec{x}+\vec{u}(\vec{x})) \simeq V(\vec{x}) - \vec{f}(\vec{x}) \cdot \vec{u}(\vec{x}) + ....$, with $f \simeq \frac{U_p}{\Delta}$ and for $u(\vec{x}) \ll \Delta$, $\Delta$ being the correlation length of the potential [$\Delta$ is of the order of the coherence length in the case of pinning by vacancies: see the discussion of P. Kes]. Assuming that $u$ only varies on the scale of $x$, and since $\vec{f}$ is random, the potential energy in a region of linear size $x$ is of order $\frac{U_p}{\Delta} \times u \times \sqrt{x^D}$, while the elastic energy is $Cx^D \times (\frac{u}{x})^2$, where $C$ is the elastic modulus (more care is needed in anisotropic or non-local cases). Balancing these two terms, one finds the original Larkin result, i.e.: $u \propto \frac{U_p}{\Delta C} x^{\frac{4-D}{2}}$, which suggests that the structure is strongly distorted by the impurities for all internal dimensions $D < 4$. This argument is however not valid if $u(x)$ exceeds $\Delta$, leading to a first crossover scale $\xi_\Delta \propto (\frac{\Delta^2 C}{U_p})^{\frac{2}{4-D}}$. The behaviour of $u(x)$ above this scale is much more subtle, and no satisfactory heuristic argument is available. The difficulty lies in estimating correctly the potential energy $V$. A very naive estimate would be $V \propto U_p \sqrt{x^D}$. This is certainly incorrect because $u$ is much larger than $\Delta$, and hence extremely small relative fluctuations of $u$ totally change the order of magnitude of $V$. Feigelman' et al. [5] have argued this effect should in a sense preaverage the random potential. Since the number of independent 'pinning sites' probed by each element of the structure is $\propto (\frac{u}{\Delta})^N$, it is reasonable to assume that the variance of the 'effective' potential is reduced by a factor $\frac{1}{\sqrt{(\frac{u}{\Delta})^N}}$ [if $V$ has short range correlations]. Balancing this modified potential energy with the elastic energy, one now finds $u \propto \left(\frac{\Delta^N U_p^2}{C^2}\right)^{\frac{1}{(4+N)}} x^{\nu_F}$, with $\nu_F = \frac{4-D}{4+N}$ [6]. Although the value $\nu_F$ is not expected to be exact in general, we shall see later that this value is precisely the one that one obtains with the replica variational appraoch introduced by Mézard and Parisi [7] (see also [8]). We may now define a second crossover scale $\xi$ such that $u(\xi) \simeq a_0$, where $a_0$ is the lattice spacing of the unperturbed structure. One finds : $\xi \propto \xi_\Delta (\frac{a_0}{\Delta})^{\frac{1}{\nu_F}}$. For $x < \xi$, the typical displacement $u$ is much smaller than the distance between vortices: hence the random potential probed by one vortex can be thought as independent from the one felt by its neighbour. This is no longer true when $x > \xi$, where new physical effects - in particular in the dynamics, see [9] - could occur (see below).

Although the above 'Flory' or 'Imry-Ma' arguments are powerful and instructive, one would like to be able to make more quantitative predictions. A particular motivation comes from the fact that a direct observation of vortex lattices is possible and that detailed information is available through image analysis [10]. This is also the case for magnetic bubble lattices [11]. Analytical calculations can be done along three different lines: perturbation theory, functional renormalisation group (FRG) and variational approaches. Perturbation theory (which amounts to expanding the random potential as a function of $u$) is well known to be unreliable in these cases: one can show for example that the value of the exponent $\nu$ is equal to the Larkin value $\frac{4-D}{2}$ to all orders of the perturbation theory, which violates bounds, rigourous results (in the context of the random field Ising model), and physical sense. The failure of this approach can be traced to the inability of perturbation theory to describe the existence of many different solutions (metastable states) to the minimisation of Eq. (1). FRG leads to interesting results to first order in $\epsilon = 4 - D$, but it is not clear how the approach can be extended to higher order. The difficulty may also lie in the proliferation of metastable states when $D$ significantly departs from 4. A more versatile approach (although technically demanding) is the replica variational theory, which we shall now briefly outline. We shall mainly focus here on the physical meaning of the calculation. More details can be found in refs. [7,12]. One starts as usual by representing the average partition function as:

$$< \log Z > \equiv \lim_{n \longrightarrow 0} \frac{< Z^n > -1}{n} \qquad (2)$$

The calculation of $< Z^n >$ amounts to computing the partition function of $n$ 'replicas' of the original vortex lattice but in which the randomness has disappeared and has been replaced by a fictitious 'interaction' between the replicas. This interaction term depends on the correlator of the potential and is in general quite messy. The Gaussian variational method consists in approximating this complicated interaction term by the 'best' quadratic form, in the sense that the free-energy calculated with the trial Hamiltonian is the closest possible to the true free-energy. More precisely, one looks for a trial Hamiltonian $H_0$ of the form:

$$H_0(\{u\}) = \int d^D x \, d^D x' \sum_{a,b=1,n} \sum_{\alpha,\beta=1,N} u_a^\alpha(\vec{x}) K_{a,b}^{\alpha,\beta}(\vec{x} - \vec{x}') u_b^\beta(\vec{x}') \qquad (3)$$

where $a, b$ label the replicas and $\alpha, \beta$ the $N$ different components of the displacement. The matrix $K$ (or more precisely its inverse) encodes the interesting correlation functions. For example,

$$\tilde{B}(\vec{x} - \vec{x}') \equiv < |\vec{u}(\vec{x}) - \vec{u}(\vec{x}')|^2 > = \lim_{n \longrightarrow 0} \sum_\alpha [K^{-1}]_{a,a}^{\alpha,\alpha} \qquad (4)$$

The whole difficulty of course lies in the curious $n = 0$ limit. In particular, $K_{a,b}$ becomes a $0 \times 0$ matrix. The simplest guess for the form of this matrix is the so-called 'replica symmetric' form:

$$K_{a,a} \equiv \tilde{K} \qquad K_{a,b \neq a} \equiv K \qquad (5)$$

from which one obtains a correlation function $\tilde{B}(\vec{x})$increasing as $x^{4-D}$ - which is again the perturbative result. Unfortunately, this simple ansatz leads to silly physical results. For example, the length scale $\xi$ is found to vanish at zero temperature - the lattice would be destroyed on all length scales by an arbitrary small amount of disorder! Furthermore, one can show [7] that this solution is in fact an unstable saddle point.

Hence, the rather natural form of $K$ given by (5) is inadequate. Exactly the same scenario occurs in the theory of spin-glasses, where an ansatz similar to Eq.(5) leads to an entropy catastrophy (negative entropy) at low temperatures. The solution to this paradox was proposed by Parisi in 1979. He showed that a certain hierarchical parametrization of the matrix $K$ could be given a meaning even in the limit of $0 \times 0$ matrices, and that a satisfactory solution could be found within this 'replica symmetry breaking' scheme. Furthermore, one can show that this rather fancy algebra in fact encodes in a relatively compact way many of the expected features of disordered systems: existence of many metastable states, strong fluctuations of certain quantities even in the thermodynamic limit, etc.. [13,7]. The difference between an energy landscape well described by a replica-symmetric solution and one which is not is schematized in figure 1: within the replica symmetric solution, the Boltzmann weight is pictured as a single Gaussian, whereas the replica symmetry breaking schemes allows to represent this Boltzmann weight as the superposition of many 'little Gaussians' of tunable position, height and width, parametrized by a function of the interval $[0,1]$ [7].

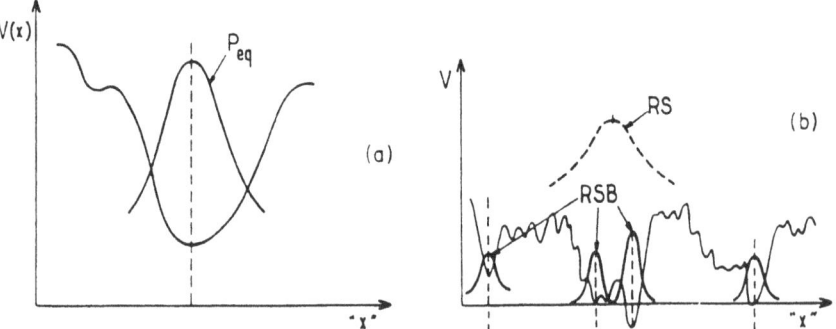

FIG. 1 Two different types of energy landscapes: (a) single-well potential, for which a Gaussian approximation is reasonable, and (b) the situation where many degenerate minima exist, and the description of the Boltzmann weight by a unique Gaussian (the center of which is optimized) is clearly very bad (replica-symmetric solution).

As for spin-glasses, the replica symmetry breaking scheme is found to give reasonable answers for the problem of pinned structures. One can in fact show that the variational approach becomes exact in the $N \longrightarrow \infty$ limit (with corrections $\simeq e^{-N}$, see [14]). From the solution to our variational equations and its interpretation, we obtain the following results:

*a. Geometrical correlation functions*

From high quality Bitter patterns, one can record the position $\vec{x} + \vec{u}(\vec{x})$ of every vortex within the frame of the picture (roughly 10 000 vortices). Hence one may compute from the picture the translational correlation function:

$$g_{\vec{k}}(\vec{x}) = < e^{i\vec{k} \cdot (\vec{x} + \vec{u}(\vec{x}) - \vec{u}(0))} > \qquad (6)$$

where the average here is over all the vortices taken as the 'origin' $\vec{x} = 0$. The lattice spacing is known accurately since $\sqrt{3}/2 H a_0^2 = \phi_0$ for a triangular lattice, where $H$ is the applied field and $\phi_0$ the flux quantum. $\vec{k}$ can thus be chosen to be a reciprocal lattice vector, and hence $\vec{k} \cdot \vec{x} = 2m\pi$ does not contribute in the above exponential. One thus has directly access to the characteristic function of the distribution of $\vec{u}(\vec{x}) - \vec{u}(0)$ for a given $\vec{x}$, which in our scheme is found to be Gaussian.

We find that $g_{\vec{k}}(\vec{x})$ should decay as a *stretched exponential* $\sim \exp -x^{2\nu}$, where the value of $\nu$ depends on the relative position of $x$ and the two crossover scales $\xi_\Delta, \xi$ defined above. We recover the Larkin regime $\nu = \frac{4-D}{2}$ for $x < \xi_\Delta$, and find that $\nu = \nu_F = \frac{4-D}{4+N}$ in the intermediate regime $\xi_\Delta < x < \xi$ [we have implicitly assumed that the pinning sites are point defects, and hence that no particular long-range correlations are present in $V(\vec{x} + \vec{u}(\vec{x}))$]. More precisely, in the case of three dimensional vortex lattices in layered supraconductors, we obtain

$$g_k(x) = \exp -(\frac{\gamma x}{\xi})^{\frac{1}{3}} \qquad (7)$$

where an angular average over the direction of $\vec{x}$ has been performed to compare directly with the experimental results. In Eq. (7), $\xi \equiv \frac{C_{66}^{3/2} C_{44}^{1/2} a_0^7 \Delta_c}{\Delta^2 U_p^2}$ ($\Delta_c$ is the interlayer spacing, and $C_{44}, C_{66}$ are, respectively, the tilt and shear moduli), and $\gamma$ is a pure number which is found to be rather large $\gamma \sim 9. \ 10^3$ (for $k = \frac{4\pi}{3}$). The point of these formulae is to show that rather precise predictions can be obtained, which allow to discuss quantitatively, for example, the assumption that pinning is due to oxygen vacancies.

Comparison with experiment should however be made with care [15]. First of all, one must be in a regime where the rôle of dislocations (absent in our model) can be neglected. This is true when the field is not too weak, and large regions where dislocations are totally absent can be obtained, both on vortex lattices and on magnetic bubble lattices [10,11]. Second, the samples must be thick enough so that the problem is indeed three dimensional, rather than being two dimensional. Specifically, the thickness $L_z$ must be large compared to $L^* = \sqrt{\frac{C_{44}}{C_{66}}}L_\perp$, where $L_\perp$ is the length scale probed in the plane. Numerically, one finds that $L^* \simeq \frac{L_\perp}{a_0} \times$ $1\mu m$. Surface pinning, long-range interaction between the magnetic charges [16] and dispersion of the elastic moduli (probably important up to a few microns in BSSCO) should also be discussed [17]. A simple-minded fit of the experimental data reported in [10] with Eq. (7) is however quite acceptable: one finds $\nu_{exp} \simeq 0.2-0.3$ instead of $\nu_F(N = 2, D = 3) = \frac{1}{6}$ or $\nu_F(N = 2, D = 2) = \frac{1}{3}$. In view of the rather large value of $L^*$ - between $1\mu m$ and $20\mu m$ - one might expect to be in the $D = 2 \longrightarrow D = 3$ crossover[17]. Note that $\nu_{exp}$ is in any case quite far from the Larkin result $\nu_L = \frac{1}{2}$, which would correspond to a simple exponential decay of $g_k(x)$ (see also [18]).

For the '69 Gauss' picture of ref. [10], the correlation length $\xi$ can be estimated to be $\simeq 5700\ a_0$, which is indeed compatible with the assumption of pinning by oxygen vacancies, with a reasonable density of $\sim 3\ 10^{17}\ m^{-2}$ per layer [19]. This value of $\xi$ also validates the assumption that the experimental data lies within the region $\xi_\Delta \ll x \ll \xi$. Indeed, $x_{max} \simeq 100\ a_0$ or $\frac{x_{max}}{\xi} \simeq 10^{-2}$; and $\xi_\Delta = (\frac{\Delta}{a_0})^6\xi$ is certainly very much smaller than $a_0 \simeq 1\mu m$ if $\Delta \leq 10nm$.

Let us add four remarks:

i) The field dependence of $\frac{\xi_\Delta}{a_0}$ and $\frac{\xi}{a_0}$ are very different: taking into account the field dependence of $C_{44}, C_{66}$, one finds $\frac{\xi_\Delta}{a_0} \propto H^2$ and $\frac{\xi}{a_0} \propto H^{-1}$. Note that the latter result is *not* borne out by the experiments, which rather suggest that $\frac{\xi}{a_0}$ *increases* with the field [10]. The data at low field might however be dominated by dislocations.

ii) The translational correlation function obtained on disordered magnetic bubble lattices [11] is again compatible with our prediction: in this case $\nu_{exp} \simeq 0.37 - 0.4$, which compares well with $\nu_F(N = 2, D = 2) = \frac{1}{3}$ or $\nu_{HH}(N = 2, D = 2) = \frac{2}{5}$ but is very far from $\nu_L = 1$.

iii) One can also study the decay of orientational order. Again in agreement with both the experiments on vortex lattices and on bubble lattices, we predict that the orientational order should not decay (at least in the absence of dislocations) as long as $\nu < 1$: we predict that the orientational correlation function should behave as $\simeq c + dx^{2\nu-2}$

iv) We have not yet discussed the asymptotic regime $x > \xi$. In our original papers [12], we claimed that the distortion field $u(x)$ would grow even faster in this regime, as $x^{\frac{4-D}{4}}$ with log corrections. The calculation leading to this result is however incorrect. Very recent papers [20,21], where a related Charge Density Wave (CDW) model is studied using the same replica approach, show that the growth of $u(x)$ is, for $x > \xi$, much weaker: $u(x) \propto \sqrt{\log x}$, in agreement with Nattermann's original analysis [22]. This would have the interesting consequence that the initial crystalline order is in fact *not destroyed at large length scales*: the growth of $u(x)$ is sufficiently slow to spare the Bragg peaks. It is however not obvious (at least to the author) that these two models describe the same physics, and preliminary numerical analysis of the $D = 1$ case reveals important differences. Furthermore, the elastic energy obtained with this CDW solution ($\propto Cx^{D-2}\log x$) is much smaller than the available potential energy ($x^{D/2}U_p u^{-N\zeta}$ - see [6]) which is somewhat unusual. This could signal a strong 'plastic' instability towards the formation of dislocations/defects at scales $x > \xi$, which is physically plausible since $u(x > \xi) \gg a_0$. If this is the case, the approach presented here must be quite radically modified to deal with this last regime.

*b. Energy scales and fluctuations.*

The replica symmetry broken solution provides quite a detailed picture of phase space structure. Some aspects in fact turn out to be similar to the predictions of the 'droplet model', which is based on phenomenological and scaling arguments. One finds that, for a system of linear size $\ell$, the typical (free-)energy difference between two metastable states is of order $\ell^\theta$ with $\theta = D - 2 + 2\nu$. More precisely, the probability to find a low lying state with free energy $-|f|$ is *exponential*:

$$P_\ell(f) = \frac{1}{\mathcal{F}(\ell)} \exp\left[-\frac{|f| - f_0(\ell)}{\mathcal{F}(\ell)}\right] \tag{8}$$

where $f_0(\ell)$ a reference free-energy and $\mathcal{F}(\ell) \propto C\xi^{-2\nu}\ell^\theta$: the typical free-energy difference is of order $C\ell^D(\frac{\delta u(\ell)}{\ell})^2$, where $\delta u(\ell)$ is the typical difference in displacement between the two states, which scales as $(\ell/\xi)^\nu$. The relation between $\theta$ and $\nu$ is probably exact in the regime $\ell < \xi$ but may be wrong in other situations, in particular when $\ell > \xi$ (see e.g. [23]). As will be illustrated below, Eq. (8) contains very useful information to discuss the dynamical response of these pinned structures. It also allows to understand why strong sample to sample fluctuations occur in the (static) response of these systems to an external force. The argument is as follows: if $\mathcal{N}_\ell \gg 1$ is the total number of metastable states at scale $\ell$, then the difference of free-energy between the lowest lying state and the first 'excited' state (or any two consecutive levels) is of order $\mathcal{F}(\ell)$ *independently* of $\mathcal{N}_\ell$. This is a specific property of the exponential distribution, Eq. (8). The probability that this difference is in fact of order $kT \ll \mathcal{F}(\ell)$ (near degeneracy) is of order $\frac{kT}{\mathcal{F}(\ell)}$. In these rare cases, a very small external perturbation might cause the system to 'jump' between these two states, thereby giving rise to an anomalously high susceptibility. This effect was discussed in [24,25,7], and along slightly different lines within the droplet model in [26,27].

*c. Extensions : Competition between lattice pinning and impurity pinning*

Another interesting situation which can be dealt with using the replica varia-
tional approach [28] is when the magnetic field is *parallel* to the layers (in the HTC materials or other layered superconductors - these layers may also be a periodic array of twin planes). The potential felt by the vortices now consists of two terms: the impurity pinning term $V(\vec{x} + \vec{u}(\vec{x}))$ already discussed above, and the periodic pinning term which is induced by the layered structure: vortices would rather sit *between* the superconducting planes. One should thus add to the Hamiltonian (1) a term of the form:

$$V_{\text{layers}} = V_0 \sum_{\vec{x}} \cos[2\pi \frac{x_1 + u_1(\vec{x})}{\Delta_c}] \qquad (9)$$

where the direction '1' is perpendicular to the layers, and $V_0$ the strength of the pinning to the layers. In the absence of impurities $U_p \equiv 0$, the physics of $H_{\text{elastic}} + V_{\text{layers}}$ is fairly well established. For $D < 2$, the structure is 'rough' for all temperatures, in the sense that the displacement $u(x)$ grows with distance (as $x^{\frac{2-D}{2}}$). At large length scales [29], the periodic potential (9) is efficiently averaged, and one has effectively $V_0 = 0$: there is no pinning. For $D > 2$, on the contrary, the structure is 'flat', in the sense that $u(x)$ is bounded, and that the structure is locked to the periodic potential. For $D = 2$, there exists a *roughening transition* $T_R = \frac{2}{\pi} C \Delta_c^2$ separating a flat low temperature phase from a rough high temperature phase [30].

For $D = 3$ structures such as the Abrikosov lattice [31], the phase diagram in the plane $(T, U_p)$ is represented in figure 2, and exhibits three different regions. Region I corresponds to effectively decoupled planes, where the effective number of degrees of freedom of the vortices is reduced to $N = 1$. As $U_p$ increases, one crosses a first-order transition line towards a phase (Region II) where the periodic potential $V_0$ disappears at large distances. Finally, at low temperatures and disorder, one finds a 'mixed phase' (Region III) where both aspects (periodic pinning and impurity pinning) are important. The transition between I and III is second-order. Increasing the external magnetic field allows to move 'North' in this diagram.

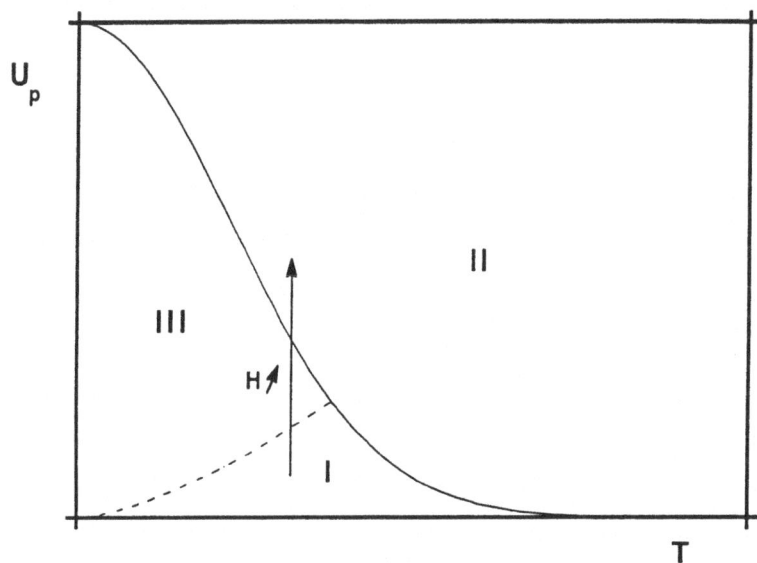

Fig. 2: Phase diagram in the plane $T, U_p$. See text for details

A very similar phase diagram holds for $D = 2$ structures, such as films cut perpendicularly to the layers, although the transition lines become crossovers, and the boundary between I and II crosses the $T$-axis at $T = T_R$.

Before turning to dynamics, let us quote another straightforward extension of the analysis to cases where disordered is strongly correlated - for example, the very important case of pinning by columnar defects (see D. Nelson's lecture, and

[32]). Another interesting problem is the competition between point disorder and columnar defects [33] .


## III. DYNAMICS: SIMPLE ARGUMENTS AND SURPRISES


Let us now discuss the truly important problem, which is the response of these pinned structures to an constant driving force $F$, and thus, in the case of dirty type II superconductors, the form of the $(I, V)$ curve. Apart from numerical calculations [3], three types of calculations are presently available:

• Perturbation theory (small pinning forces), a priori valid for high velocities or driving force [34,1].

• The 'correlation volume arguments' [35], which allow to predict the order of magnitude of the critical threshold $F_c$ (or critical current) at zero temperature. Recent FRG calculations predict that the velocity grows as $(F - F_c)^\phi$ for $F > F_c$, where $\phi$ is a critical exponent which is calculated to first order in $4 - D$ [36,37].

• Scaling arguments, which generalises to random systems the 'droplet nucleation' ideas in pure systems (see e.g. ref. [30]).

We shall restrict the present discussion to these scaling arguments, and show why, in certain cases, these arguments are inappropriate.

The basic idea is that an external force creates a 'slope' in the energy landscape, which progressively erases the energy barriers which pin the structure. A natural assumption is that the height of the energy barriers at scale $\ell$ is comparable to the depth of the metastable states, i.e $\mathcal{F}(\ell)$ (more general assumptions can be found in [38]). The external force might lower these barriers by an amount of order $\ell^D \times F \times u(\ell)$, which grows more rapidly than $\mathcal{F}(\ell)$ for large $\ell$. The highest barriers are thus those corresponding to $\mathcal{F}(\ell^*) = \ell^{*D} F u(\ell^*)$, or: $\ell^* \propto F^{-\frac{1}{D+\nu-\theta}}$. Assuming activated dynamics, one then writes that the velocity $\mathcal{V}$ is proportional to $\frac{u(\ell^*)}{\tau_0} \exp[-\frac{\mathcal{F}(\ell^*)}{kT}]$ (where $\tau_0$ is a microscopic timescale), or:

$$\log \mathcal{V} \propto F^{-\mu} \qquad \mu = \frac{\theta}{D + \nu - \theta} \qquad (9)$$

with an exponent $\mu$ depending on the relative position of $\ell^*(F)$, $\xi_\Delta$, $\xi$, etc.. (e.g. the length scales related with the dispersion of the elastic moduli or the finite thickness of the sample): we refer the reader to [1] for a thorough discussion. In the case of pinned vortex lattices, the force $F$ is proportional to the current $J$, and the velocity $\mathcal{V}$ to the voltage. The effective exponent $\mu(J)$ is expected to be a stepwise function, with plateaus at $\mu \sim 0.5, 0.8, 1.5, \ldots$ (for $D = 3$) and $\mu \sim 0., 0.5, 2, \ldots$ (for $D = 2$) as $J$ increases. Experimentally, one finds a rather smooth increasing function $\mu(J)$, which either starts around $\mu = 0.2$ at small $J$ [39], or decreases all the way to $\mu = 0$ [40,41].

Although physically compelling, the above arguments might lead to incorrect results. A first reason could be the presence (or the generation) of mobile defects in the structure at large length scales $\ell > \xi$. This is particularly the case in $D = 2$, where dislocations are point-like. Their activation energy $\mathcal{E}$ is thus finite and transport is ohmic with a resistivity $\propto e^{-\frac{\mathcal{E}}{kT}}$: see the numerical results of A. P. Young in this volume, and also the experiments of [9] in the small frequency (large length scales $\ell > \xi$) regime. So, transport might be $easier$ than suggested by the arguments leading to (9) in the presence of defects.

One the other hand, there are case where transport is $harder$ than anticipated, because of the existence of strong fluctuations. We shall illustrate this on the following toy model: a point particle (corresponding to $D = 0$) moving in a one dimensional ($N = 1$) random potential $V(x)$, according to:

$$\frac{dx}{dt} = -\frac{\partial V}{\partial x} + \eta(t) \tag{10}$$

where $\eta(t)$ is the Langevin noise representing the thermal bath. In order to mimic the situation discussed above where the barriers grow with distance, we shall assume that the $force$ $-\frac{\partial V}{\partial x}$ is a random uncorrelated variable, of variance $\sigma$ (Physical situations, for which this model is directly relevant, are reviewed in [42,43]). Hence the potential typically grows as $\sqrt{\sigma x}$. In the presence of an external force, the potential acquires a non-zero slope $-Fx$. Following the above arguments, one could say that the highest barrier to cross corresponds to a distance $x^*$ such that $Fx^* \simeq \sqrt{\sigma x^*}$ or $x^* \simeq \frac{\sigma}{F^2}$. The velocity is then found to be proportional to $\exp -\frac{\sqrt{\sigma x^*}}{kT}$, or $\log \mathcal{V} \propto F^{-1}$. In fact, as can be shown using various techniques [42], this result is wrong. What happens depends on the value of the parameter $\beta = \frac{FkT}{\sigma}$. For $\beta < 1$ (corresponding to small forces), the asymptotic velocity of the particle is $zero$. More precisely, the position of the particle grows as $t^\beta \ll t$. The $effective$ velocity for a system of size $L$ thus behaves as $L^{1-1/\beta}$, which gives, for small $F$: $\log \mathcal{V}(L) \propto \frac{\log L}{F}$. For $\beta > 1$, a finite velocity appears such that $\mathcal{V} \propto (\beta - 1)$.

The origin of this unusual sub-linear creep law is the fact that energy barriers are exponentially distributed: one finds that $P(\mathcal{E}) = \exp -[\frac{\mathcal{E}}{E_0}]$, with $E_0 = \frac{\sigma}{F}$. The distribution of local trapping times is thus, assuming $\tau = \tau_0 \exp[\frac{\mathcal{E}}{kT}]$, very broad:

$$\psi(\tau) \simeq_{\tau \gg \tau_0} \frac{\tau_0^\beta}{\tau^{1+\beta}} \tag{11}$$

When $\beta < 1$, the average trapping time is infinite - this is why naive arguments based on typical properties are not suitable in this case. The absence of a mean trapping time induces very strange properties, such as the explicit dependence of all observables on the experimental time scale $t_w$ -aging [44]- or on the size of sample $L$: this is because, for $\beta < 1$, a finite fraction of the total time $t_w$ is spent in the deepest trap visited between $t = 0$ and $t = t_w$ [44].

As discussed above, an exponential distribution of free energies is also found within the replica symmetry breaking scheme, and is thus expected to be a general feature of glassy systems (spin glasses, pinned vortices, etc...). Assuming again that barriers and free-energy share the same statistical properties, one may argue that relaxation and transport in these systems will exhibit aging and non-stationary effects.

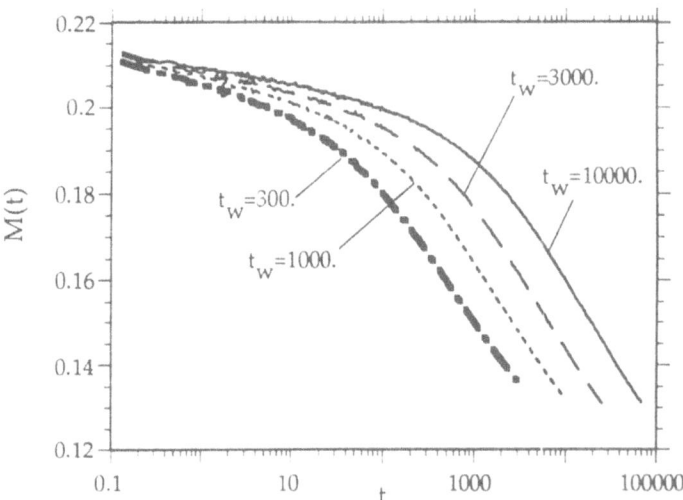

Fig. 3: Decay of the thermoremanent magnetisation in a spin-glass (after [45]),

for different waiting times $t_w$, defined as the time during which the magnetic field is left on. Note how the system 'stiffens' with $t_w$.

These effects have been studied in great detail in spin glasses: a very striking phenomenon is the time relaxation of the magnetisation after cooling down the system in the glassy phase and waiting for a time $t_w$ before cutting the field [45]: as shown in figure 3, the longer $t_w$, the slower the relaxation - in fact, the effective relaxation time $\tau(t_w)$ is nearly equal to $t_w$ itself (at least if $t_w$ is not larger than a very long 'ergodic time' $t_{\mathrm{erg}}$, see [44]). This can be accounted for if the lifetimes of the metastable states are distributed according to (11) with $\beta < 1$ [44]. The main question is to know whether such a broad trapping time distribution - present for $F = 0$ - can survive *in the presence of an external force* for truly extended structures $D \geq 1$. This might require going beyond the usual elastic description, as emphasized in [2,3], and including truly plastic effects (i.e. tearing). However, as illustrated by the above toy model, aging can only survive *if the external driving force is sufficiently weak*. It might also be that, as in pinned Charge Density Waves, aging is interrupted after a relatively short time [46]. We believe that it would be of great interest to discuss these issues experimentally: Do pinned vortex lattices exhibit aging (i.e. decay of the response function with time), which should be expected if they really enter a glassy state ? Are the $(I, V)$ curves and the critical current really independent of the time and size of the sample? We hope that the above discussion, emphasizing the importance and ubiquity of rare events in disordered systems, will motivate some systematical experiments - in line with spin-glasses and other glasses [47]- to investigate these matters.

*Acknowledgments.* I wish to thank Marc Mézard, Antoine Georges and Jonathan Yedidia for sharing with me their knowledge and skill. Needless to say, all the possible misjudgments appearing above are of my own bat. I had interesting exchanges with M. Feigel'mann, T. Giamarchi, J. Hammann, H. Jensen, P. Le Doussal and E. Vincent. I also wish to thank the organizers, lecturers and participants of the school for enlightning discussions on these matters, in particular D. Feinberg, A. Kapitulnik, P. Kes, D.S. Nelson, A.P. Young, L. Balents, K. A. Moler, S. Ryu and A. Yazdani.

# REFERENCES

1. G. Blatter, M. V. Feigel'mann, V. B. Geshkenbin, A. I. Larkin and V.M. Vinokur, "Vortices in High Temperature Superconductors", ETH preprint (June 1993).

2. S. Coppersmith, Phys. Rev. Lett. 65, 1044 (1990), Phys. Rev. B 44, 2887 (1991).

3. for a review, see H. J. Jensen, in Nato ASI Series, " Phase transitions and relaxation in systems with competing energy scales", Geilo 1993.

4. Known as dynamical strain aging in the context of dislocations.

5. M. V. Feigel'mann, V. B. Geshkenbin, A. I. Larkin and V.M. Vinokur, Phys. Rev. Lett, 63, 2023 (1989).

6. Feigelmann' et al. [5] have proposed a phenomenological form of the 'reduction factor': $(\frac{u}{\Delta})^{N\zeta}$. The choice $\zeta = 1/4$ reproduces the exact value $\nu = 2/3$ for $N = 1, D = 1$ and leads to the Halpin-Healy proposal for $\nu$: $\nu_{HH} = \frac{4-D}{4+N/2}$. [ T. Halpin-Healy, Phys. Rev. A 42, 711 (1990)]

7. M. Mézard, G. Parisi, J. Phys. (Paris) I 1, 809 (1991)

8. E. Shaknovich, A. Gutin, J. Phys A 22, 1647 (1989)

9. A. Yazdani, W. White, M. Gabay, M. R. Beasley, A. Kapitulnik, Phys. Rev. Lett. 70, 505 (1993)

10. D. G. Grier, C.A. Murray, C.A. Bolle, P.L. Gammel, D.J. Bishop, D. B. Mitzi and A. Kapitulnik, Phys. Rev. Lett 66, 2270 (1991)

11. R. Shesadri, R.M. Westervelt, Phys. Rev. B 46 5150 (1992)

12. J. P. Bouchaud, M. Mézard and J. Yedidia, Phys. Rev. Lett. 67, 3840 (1991) and Phys. Rev. B 46, 14686 (1992)

13. M. Mézard, G. Parisi, M.A. Virasoro, 'Spin Glasses and Beyond", World Scientific, Singapore (1987).

14. L. Balents, D. S. Fisher, preprint (1993)

15. We shall assume that the "effective temperature" (i.e the temperature at which the system falls out of equilibrium and is frozen in) is sufficiently small so that the above results, strictly speaking valid at $T = 0$, still apply.

16. D. A. Huse, Phys. Rev. B 46 8621 (1992)

17. In the regime where dispersion is important, one finds $\nu_F^{dis}(D = 3) = 1/3$: both the 2 D-3 D crossover and the dispersion of $C_{44}$ push $\nu$ towards $1/3$.

18. E. Chudnovski, Phys. Rev. B 43, 7831 (1991)

19. C. J. van der Beek, P. Kes, Phys. Rev. B 43, 13032 (1990)

20. T. Giamarchi, P. Le Doussal, preprint (1993)

21. S. E. Korshunov, Phys. Rev. B 48, 3969 (1993)

22. T. Nattermann, Phys. Rev. Lett. 64, 2454 (1990)

23. E. Medina, T. Hwa, M. Kardar, Y.C. Zhang, Phys. Rev. A 39, 3053 (1989)

24. M. Mézard, J. Physique (Paris) 51, 1831 (1990)

25. G. Parisi, J. Physique (Paris) 51, 1595 (1990)

26. D.S. Fisher, D.A. Huse, Phys. Rev. B, 43, 10728 (1991)

27. T. Hwa, D.S. Fisher, preprint (1993)

28. J.P. Bouchaud, A. Georges, Phys. Rev. Lett. 68, 3908 (1992)

29. Note that this length scale can be rather high: due to the large kink energy, dislocations are 'flat' over much more than 10 $\mu$ m in Si or Ge.

30. see, for example, P. Nozières, in "Solids Far From Equilibrium", C. Godrèche Editor, Cambridge University Press (1992).

31. The features reported here are valid in the limit where the distance between vortices is larger than the layer spacing $a_0 \ll \Delta_c$.

32. D. Nelson, V. Vinokur, Phys. Rev. Lett, 68, 2398 (1992), and preprint (1993)

33. L. Balents, M. Kardar, preprint (1993)

34. A. Schmid, W. Hauger, J. Low. Temp. Phys. 11, 667 (1973)

35. A. Larkin, Yu. N. Ovchinnikov, J. Low Temp. Phys. 34, 409 (1979)

36. T. Nattermann, H. Leschorn, L.H. Tang, S. Stepanow, J. Phys. II (France) 2, 1483 (1992)

37. O. Narayan, D. S. Fisher, Phys. Rev. B (to appear)

38. D. S. Fisher, M.P.A. Fisher, D. A. Huse, Phys Rev B 43, 130 (1991)

39. C. Dekker, P. Woltgens, R.H. Koch, B.W. Hussey and A. Gupta, Phys. Rev. Lett. 69, 2717 (1992)

40. C.J. van der Beek, P.H. Kes, M.P Maley, M.J.V. Menken and A.A. Menovsky, Physica C 195, 307 (1992)

41. In fact, the latter experimental data could be compatible with $\log \mathcal{V} \propto \text{cst} - J^{\alpha}$, with $\alpha \sim 0.5$, in the whole range of $J$.

42. see e.g. J.P. Bouchaud, A. Georges, Phys. Rep. 195, 127 (1990)

43. J. P. Bouchaud, A. Georges, Comments in Cond. Mat. Phys. 15, 125 (1991)

44. J. P. Bouchaud, J. Physique I 2, 1705 (1992)

45. for a review, see e.g. E. Vincent, J. Hammann, M. Ocio, p. 207 in "Recent Progress in Random Magnets", D.H. Ryan Editor, (World Scientific Pub. Co. Pte. Ltd, Singapore 1992)

46. K. Bijlakovic, J. C. Lasjaunias, P. Monceau, F. Lévy, Phys. Rev. Lett. 62, 1512 (1989) and 67, 1902 (1991)

47. L.C.E. Struick, "Physical Aging in Amorphous Polymers and Other Materials" (Elsevier, Houston, 1978)

# VORTEX DYNAMICS IN SUPERFLUIDS AND SUPERCONDUCTORS

E.B. SONIN and M. KRUSIUS
*Low Temperature Laboratory, Helsinki University of Technology,*
*02150 Espoo, Finland*

**ABSTRACT**    In recent years the structural and dynamic properties of quantized vortex lines have been increasingly investigated with an experimental resolution capable of distinguishing individual vortices. Measurements on the superflow of $^4$He and $^3$He superfluids or on vortices in high-$T_c$ superconductors have put our knowledge of quantized vorticity under detailed test. Here we emphasize the unifying features of vortices in both neutral and charged superfluids, while analyzing their shape and motion in a few example cases, where either one single vortex line or an ordered array of vortices is present. These examples illustrate the transition from single-vortex to collective vortex-array behavior and the consequences to the dynamic response from the vortices.

# 1   Introduction

Quantized vortices determine much of the properties of a number of ordered condensed media at low temperatures, such as the He superfluids [1], superconductors [2], and perhaps even the neutron superfluid [3] in a rapidly rotating compact star like the pulsar. These features reach technical importance in superconducting current conductors and other superconducting devices. In spite of these very different environments, where quantized vortices exist, they have general common properties:

- *Broken time invariance and topological stability:* A quantized vortex carries a topological charge which consists of an integer number of windings of a phase angle around the vortex axis. Thus a quantized vortex is topologically stable, it behaves like a trapped persistent current, which cannot be dissipated within the liquid. The topological stability distinguishes quantized vortices from vortices in a classical ideal fluid where vortices are stable only in the inviscid limit. The circulation determines the dynamic properties of vortices: their motion is not governed by Newton's second law, as is the case for time-invariant systems with inertia, but by an equation of motion which is dominated by the gyrotropic Magnus force proportional to the velocity of the vortex.

- *Elastic properties:* A vortex line behaves similar to an elastic string with tension along its axis. This line-tension force acts to reduce the vortex line to its minimum possible length. Moreover, an infinite assembly of collinear straight vortex lines, a vortex array, exhibits crystalline order, which often forms a triangular pattern in the transverse plane. Within this plane there exists rigidity with respect to shear deformation, similar to a crystalline lattice.

- *Long-range interaction between vortices:* The circulation of a quantized vortex gives rise to an interaction between vortices which is of long range, it decreases inversely with distance. In superconductors the range of the intervortex interaction extends only up to the London penetration depth, which, nevertheless, may exceed all other relevant length scales. In this case the intervortex interactions become of crucial importance even here. The long-range interaction produces a variety of collective effects, which are

*N. Bontemps et al. (eds.), The Vortex State,* 193–230.

expressed in the static and dynamic properties of the vortex array: in nonlocal effects and in the screening of externally applied perturbations.

- *Pinning:* Vortex lines can be pinned to irregularities at the boundary of the superfluid or at various types of pinning centers in the interior of the superconductor. In a vortex array elastic forces glue individual vortex lines together and reduce the effects from the random pinning forces. In the limit of weak pinning this leads to what is called collective pinning, where domains with correlated motion over the random pinning relief are formed.

The concept of a quantized vortex line was first introduced almost half a century ago. It explained some of the most basic features of both superfluids and superconductors. A fruitful exchange of ideas has existed between these two related fields of ordered condensed media. New understanding has sometimes emerged first in superfluids, sometimes in superconductors. We shall here try to follow a unified approach, which comprises the vortices of both systems, in the tradition first started by Fritz London [4] and later continued in other textbooks and reviews [5]. The study of quantized vortices is firmly rooted in classical hydrodynamics [6]: Most of the properties of vortices, which are known to hold for ideal classical liquids, may be directly transferred to quantum liquids. The important difference is quantization of circulation which transforms quantized vortex lines to stable objects, such that they can be individually identified and observed over the course of an experiment.

Historically large ensembles of vortex lines were the first to catch the attention, since such states were experimentally more accessible. Large ensembles of vortices fall between two extreme types of structure: 1) An ordered vortex array is created by steady rotation of a superfluid or by the application of a magnetic field to a superconductor in the mixed state, while 2) a chaotic vortex tangle is formed and sustained by supercritical flow in a state of superfluid turbulence. In both cases *collective effects* due to intervortex interaction are of importance. Collective effects were first searched for in rotating superfluids [7]-[10]. Investigations on vortices in rotating superfluids have played a special role since they represent ideally uniform media: Pinning is restricted to surfaces transverse to the rotation axis; there is no pinning in the bulk which would mask any of the intrinsic collective features, as is the case with a flux line lattice in a superconductor. Nevertheless, the results obtained for superfluids can be adapted to provide the starting point for more complicated cases.

Significant new developments have surfaced during recent years and much of this has been fuelled by the work on high-$T_c$ superconductors. Here a variety of different vortex structures, new kinds of pinning phenomena and vortex lattice arrangements have been discussed, including the melting of the crystalline vortex lattice to a fluid state at high temperatures. But also the superfluids have experienced new progress. Measurements on critical effects in the $^4$He superflow through a submicron-size orifice have clarified the role of nucleation and the onset of dissipation [11].

With the advent of the rotating $^3$He superfluids a new system and regime of superfluid hydrodynamics has become available for vortex studies. The tensorial order parameter of the various superfluid $^3$He phases consists of intercoupled orbital and magnetic parts [12]. This leads to a large variety of vortices with different topology and structure (7 different vortices have by now been experimentally identified) [13], but it also extends the experimental possibilities: NMR and the propagation of ultrasound are two powerful measuring techniques, which are not available for studies on superfluid $^4$He. The different superfluid phases of $^3$He and their many vortex structures provide a versatile environment for investigating the coexistence of different vortices, phase transitions between them and the ensuing structures of the vortex array. The superfluid coherence length is a factor of hundred larger for $^3$He than for $^4$He; as a result pinning in $^3$He is weaker. In $^3$He high critical velocities permit

Figure 1: Vortex line with a singular core and quantized circulation in an isotropic superfluid *(top)*. The radial distribution of the superfluid density $\rho_s$ is finite everywhere while that of the velocity $\vec{v}_s$ diverges, when $r \to 0$ *(bottom)*. In the superconductor $\vec{v}_s$ has an exponential cut-off at large radii $r$, when $r$ exceeds the Landau penetration depth $\lambda$.

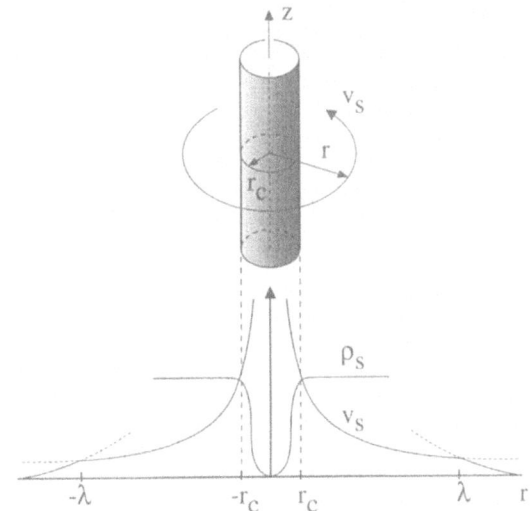

the study of metastable flow states with a fixed number of vortices and large vortex-free counterflow regions. Thereby the dynamic modes of the vortex array can be investigated in the absence of boundary interactions, in the presence of a fixed number of vortices.

Our goal here is to provide a brief introductory description of the dynamic properties of vortex lines and, in particular, to bridge the transition in behavior from a single vortex line to a dense ordered array of many vortex lines. First the properties of a solitary vortex line will be outlined. Then we shall illustrate how collective features arise as the vortex density is increased. The discussion is built around some recent experiments, in which the common denominator has been a measuring resolution down to individual vortex lines. We present a simple approach, which uses an intuitively transparent analogy with the mechanics of an elastic string. We believe that this picture provides good insight into the dynamics of vortices, in spite of the shortcoming that its quantitative preciseness would need to be studied separately in each case.

## 2 Quantized vortex line - basic concepts

### 2.1 Structure of quantized vortex line

The quantized vortex line was first introduced by Onsager [14] and Feynman [15] in the late forties - early fifties for superfluid $^4$He-II and by Abrikosov [16] in the mid fifties for the type-II superconductor in the mixed state. The simplest case is a straight vortex line with $2\pi$ circulation of the order parameter phase $\varphi$ around a singular core (Fig. 1). The superfluid velocity $\vec{v}_s = (\kappa/2\pi)\vec{\nabla}\varphi$ takes the form

$$\vec{v}_s = \frac{\vec{\kappa} \times \vec{r}}{2\pi r^2} \qquad (1)$$

and is divergent on approaching the vortex core. Here $\kappa = h/M$ is the circulation quantum, the circulation vector $\vec{\kappa} \parallel \hat{z}$ is directed along the vortex axis and the 2-dimensional position vector $\vec{r}$ is in the transverse plane, as shown in Fig. 1. In the kinetic energy the divergence in $v_s$ is compensated by the vanishing superfluid density $\rho_s$ within the core with a radius

$r_c$, which is on the order of the superfluid coherence length $\xi(T)$. The $\vec{v}_s$ field in Eq. (1) is curl-free everywhere except in the core, ie.

$$\vec{\nabla} \times \vec{v}_s = \vec{\kappa}\delta(\vec{r}) \tag{2}$$

where $\delta(\vec{r})$ is the 2-dimensional delta-function in the transverse plane.

In the superconductor the phase distribution around the vortex line is the same as in a neutral superfluid, but the gauge invariant expression for the superfluid velocity is

$$\vec{v}_s = \frac{\kappa}{2\pi}\left(\vec{\nabla}\varphi - \frac{2\pi\vec{A}}{\Phi_0}\right) . \tag{3}$$

Here $\vec{A}$ is the electromagnetic vector potential which determines the magnetic field $\vec{h} = \vec{\nabla}\times\vec{A}$, while $\Phi_0 = hc/(2e) = \kappa mc/e$ is the magnetic-flux quantum for Cooper pairs with mass $M = 2m$ and charge $2e$, where $m$ and $e$ are the mass and the charge of an electron.

By taking the curl of Eq. (3) one obtains instead of Eq. (2)

$$\vec{\nabla} \times \vec{v}_s = \vec{\kappa}\delta(\vec{r}) - \frac{\kappa\vec{h}}{\Phi_0}, \tag{4}$$

or in more customary notation

$$\vec{h} + \frac{4\pi\lambda^2}{c}\vec{\nabla} \times \vec{j}_s = \vec{\Phi}_0\delta(\vec{r}) , \tag{5}$$

where the magnetic-flux quantum vector $\vec{\Phi}_0 \parallel \hat{z} \parallel \vec{h}$. The electric supercurrent $\vec{j}_s = en_s\vec{v}_s$ is associated with the magnetic field by the Maxwell equation

$$\vec{j}_s = \frac{c}{4\pi}\vec{\nabla} \times \vec{h} . \tag{6}$$

Substituting Eq. (6) into Eq. (5) one obtains the London equation for the magnetic field around the vortex line:

$$\lambda^2\vec{\nabla} \times [\vec{\nabla} \times \vec{h}] + \vec{h} = \vec{\Phi}_0\delta(\vec{r}) . \tag{7}$$

Here $\lambda$ is the London penetration length given by

$$\lambda^{-2} = \frac{16\pi e^2 n_s}{mc^2} , \tag{8}$$

and $n_s$ is the superfluid electron density. At distances $r < \lambda$ the supercurrent is inversely proportional to $r$, as in a neutral superfluid, but at $r > \lambda$ the supercurrent and the magnetic field vanish exponentially (cf. Fig. 1). In the superconductor two length scales thus describe the structure of the vortex: $\lambda$ and $r_c$. In a type-II superconductor $\lambda > r_c$ and in a high-$T_c$ superconductor $\lambda \gg r_c$. Thus the properties of the latter are closer to those of a superfluid, which in turn represents the limit $\lambda \to \infty$.

The energy of the vortex line consists of the kinetic energy of the supercurrent, the condensate energy in the vortex core and the magnetic energy in the superconductor. The most important contribution is the kinetic energy of the supercurrent which is given per unit length by

$$\varepsilon = \int \frac{1}{2}\rho_s v_s(r)^2 \, d\vec{r} = \frac{\rho_s}{2}\int_{r_c}^{r_u} \frac{\kappa^2}{(2\pi r)^2} \, 2\pi r \, dr = \frac{\rho_s\kappa^2}{4\pi}\ln\frac{r_u}{r_c} . \tag{9}$$

The logarithmic divergence in the integral is cut by the core radius $r_c$ at small radii and by some upper cut-off radius $r_u$ which varies for different situations. Let us consider some examples:

- For a solitary straight vortex in a neutral superfluid $r_u$ is on the order of the container size (or the distance from the solid wall).

- In a superfluid with a dense vortex array $r_u$ is on the order of the intervortex distance.

- In a superconductor $r_u$ can additionally be equal to the penetration depth $\lambda$.

In all cases $r_u$ should be determined by the smallest among the space scales listed. For instance, in a dense vortex array in the mixed state of a superconductor the upper cut-off radius $r_u$ is determined by the intervortex distance $r_v \sim \sqrt{\Phi_0/h}$ which is much smaller than the penetration depth $\lambda$ at magnetic fields much larger than the lower critical field $H_{c1} = \Phi_0/(4\pi\lambda^2)\ln(\lambda/r_c)$. Then the dynamic properties of the vortex array are in many ways similar to those of a corresponding array in a neutral rotating superfluid.

The hydrodynamic theory, which is discussed here, is based on the assumption that it is legitimate to neglect in the vortex energy all terms except for the logarithmic term in Eq. (9). This applies even in the case of a curved vortex line, when a similar expression represents only the leading term proportional to the large logarithm $\ln(r_u/r_c)$. Therefore we do not include here other contributions to the energy. The energy per unit length in Eq. (9) can be understood as a line tension, similar to that of an elastic string. The line tension attempts to restore minimum length and to keep the vortex line straight. The concept of the elastic vortex line tension was introduced in the early days of superfluid hydrodynamics, starting with Hall [7] and Vinen [8], Andronikashvili [9], and Bekarevich and Khalatnikov [10], and will be used here to promote a simple intuitive picture of the vortex line.

## 2.2 Equation of motion of straight vortex line

The foundation for vortex dynamics was laid in the classical hydrodynamics of the 19th century [6]. Later it was applied to superfluid $^4$He [7,8] in the framework of the two-fluid model in which vortices belong to the superfluid fraction with the superfluid velocity $\vec{v}_s$ and density $\rho_s$, while normal excitations are described by the normal fraction with the velocity $\vec{v}_n$ and density $\rho_n$.

In classical hydrodynamics the equation of motion of a thin vortex line, around which the velocity is curl-free or irrotational, is derived from the momentum-conservation law for the fluid within a cylindrical volume coaxial to the vortex line [6]. Suppose an external force $\vec{f}$ acts on a unit length of the vortex line; the momentum balance is then expressed in the form

$$\vec{f} = \rho_s \vec{\kappa} \times (\vec{v}_s - \vec{v}_L) . \tag{10}$$

Here $\vec{v}_s$ is the external superfluid velocity field, in which the vortex line resides, ie. the velocity far from the vortex line which does not include the singular velocity field induced by the line itself. If the external force $\vec{f}$ vanishes, Eq. (10) yields Helmholtz's theorem: *the velocity $\vec{v}_L$ of the vortex line is equal to the fluid velocity at the point where the vortex line is located*, ie. $\vec{v}_L = \vec{v}_s$ and the vortex moves with the external flow.

The quantized vortex belongs to the superfluid component, but it also interacts with normal excitations when it moves relative to the normal fluid. This interaction gives rise to the mutual friction force which is inserted for the external force $\vec{f}$ in Eq. (10). Following the pioneering work of Hall and Vinen [17], the mutual friction force $\vec{f}$ is often expressed as a linear function of the counterflow velocity $\vec{v}_s - \vec{v}_n$:

$$\vec{f} = \frac{\rho_s\rho_n\kappa}{2\rho}B\left(\vec{v}_s - \vec{v}_n\right) + \frac{\rho_s\rho_n\kappa}{2\rho}B'\,\hat{z}\times\left(\vec{v}_s - \vec{v}_n\right) , \tag{11}$$

where $B$ and $B'$ are the mutual friction parameters parallel and perpendicular to the counterflow. Inserting this into Eq. (10) the vortex velocity is found to be given by

$$\vec{v}_L = \vec{v}_s - \frac{\rho_n}{2\rho} B' (\vec{v}_s - \vec{v}_n) - \frac{\rho_n}{2\rho} B \,\hat{z} \times (\vec{v}_s - \vec{v}_n) \ . \tag{12}$$

In many applications this expression is simplified if the normal fluid can be considered clamped, ie. $\vec{v}_n = 0$. In $^3$He the normal fluid is in all practical cases stationary with respect to the solid walls because of its large viscosity and is thus clamped to corotation with the rotating container. In superconductors the quasiparticle excitation gas is kept at rest by strong interactions with the crystalline lattice. Then the vortex velocity is

$$\vec{v}_L = \left(1 - \frac{\rho_n}{2\rho} B'\right) \vec{v}_s - \frac{\rho_n}{2\rho} B \,\hat{z} \times \vec{v}_s \ . \tag{13}$$

Instead of the counterflow velocity $\vec{v}_s - \vec{v}_n$ one can also write the mutual friction force as a function of $\vec{v}_L - \vec{v}_n$, or $\vec{v}_L$ if the normal fluid is at rest. In that case friction parameters different from $B$ and $B'$ are used: the connecting formulae are listed in [1,18].

For a superconductor one can rewrite Eq. (10) in another form by introducing the electric supercurrent $\vec{j}_s = en_s\vec{v}_s$ and the magnetic-flux quantum $\Phi_0 = \kappa mc/e$ instead of the superfluid velocity $\vec{v}_s$ and the quantum of circulation $\kappa$:

$$\frac{1}{c} \vec{\Phi}_0 \times \vec{j}_s = \frac{2en_s}{c} \vec{\Phi}_0 \times \vec{v}_L - \vec{f} \ . \tag{14}$$

In the theory of superconductivity the force on the left-hand side is called the Lorentz force, and the force transverse to the velocity $\vec{v}_L$ is called the Magnus force. In contrast, in the theory of superfluidity the term "Magnus force" is applied to the Galilean-invariant force on the right-hand side of Eq. (10). Very often for impure superconductors the forces on the right-hand side of Eq. (14) may be approximately reduced to only a friction force parallel to $\vec{v}_L$, ie.

$$\frac{1}{c} \vec{\Phi}_0 \times \vec{j}_s = -\eta \vec{v}_L \tag{15}$$

The relations connecting the friction coefficient $\eta$ to the Hall-Vinen parameters $B$ and $B'$ are thus

$$1 - \frac{\rho_n}{2\rho} B' = 0 \ , \qquad B = \frac{2\rho\rho_s\kappa}{\rho_n\eta} \ . \tag{16}$$

In the Bardeen-Stephen model [2] the friction coefficient $\eta$ is determined by the normal conductivity $\sigma_n$:

$$\eta = \frac{1}{c^2} \Phi_0 H_{c2} \sigma_n \ , \tag{17}$$

where $H_{c2}$ is the upper critical magnetic field.

## 2.3 Equation of motion of curved vortex line

The equation of motion, Eqs. (12) or (13), can be generalized to a curved vortex line. In this case the vortex line induces at some location $\vec{R}$ a superfluid velocity $\vec{v}_s(\vec{R})$ which is given by the Biot-Savart law

$$\vec{v}_s(\vec{R}) = \frac{\kappa}{4\pi} \int_L \frac{d\vec{R}_L \times (\vec{R} - \vec{R}_L)}{|\vec{R} - \vec{R}_L|^3} = \frac{1}{4\pi} \int_L dL \frac{\vec{\kappa}(\vec{R}_L) \times (\vec{R} - \vec{R}_L)}{|\vec{R} - \vec{R}_L|^3} \ . \tag{18}$$

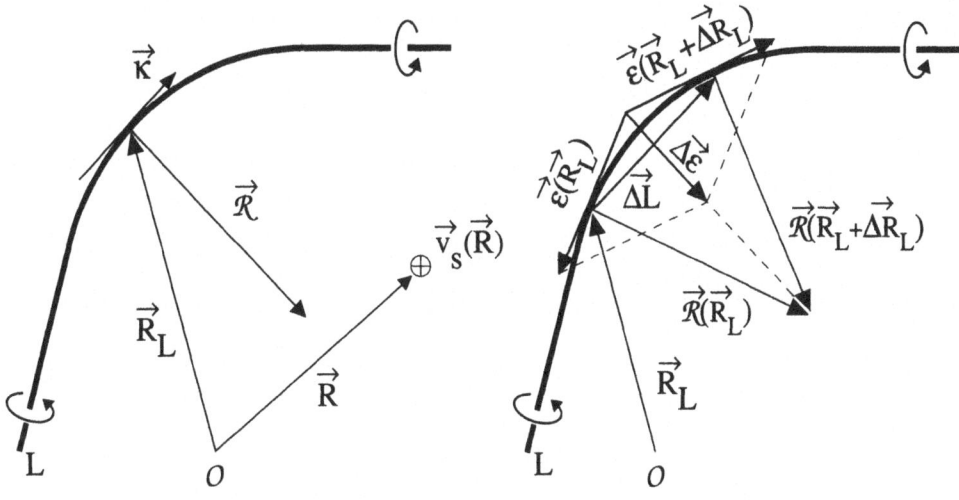

Figure 2: Curved vortex line. *On the left* the position vectors used in Eqs. (18) and (19) are defined. *On the right* it is outlined how the line-tension force in Eq. (28) is obtained due to the variation of the line-tension vector $\vec{\varepsilon}$ as a function of the arc length $L$. The magnitude of the line tension is considered to be constant, $\varepsilon(\vec{R}_L + \delta\vec{R}_L) = \varepsilon(\vec{R}_L)$, while the orientational variation results in a line-tension force $\vec{\Delta\varepsilon}$ which is oriented towards the center of curvature. Its magnitude is obtained from the congruent triangles: $\Delta\varepsilon/\Delta L = \varepsilon/\mathcal{R}$. Then the line-tension force per unit length, $\vec{\Delta\varepsilon}/\Delta L$, in the limit $\Delta L \to 0$ corresponds to the force on the right-hand side of Eq. (28).

Here $dL$ is a vortex line element and the integral is calculated along the vortex line, whose location is traced by the end point of the position vector $\vec{R}_L$. The vector $\vec{\kappa}(\vec{R}_L)) = \kappa\,d\vec{R}_L/dL$ is tangent to the vortex line at the end point of $\vec{R}_L$ and has the magnitude of the circulation quantum $\kappa$ (Fig. 2). Let us first neglect mutual friction. Then according to Helmholtz's theorem a point on the vortex line moves with the velocity $\vec{v}_L(\vec{R}_L) = d\vec{R}_L/dt$, which is determined by Eq. (18) in the limit $\vec{R} \to \vec{R}_L$ and $\vec{v}_L(\vec{R}_L) = \vec{v}_s(\vec{R}_L)$. In this limit a singularity of the form $1/|\vec{R} - \vec{R}_L|$ is present in the induced velocity field $\vec{v}_L(\vec{R})$. However, according to our analysis of the straight vortex line in Sec. 2.1, where this singularity is also present, it does not contribute to the velocity and should be ignored. But even this does not remove all singularities in the velocity field of the curved vortex: there remains a weaker logarithmic singularity $\sim \ln|\vec{R}_L - \vec{R}'_L|$ in the linear integral due to the curvature of the line. This is avoided by imposing a cut-off at distances of order $r_c$ from the point $\vec{R}_L$ for which the velocity $\vec{v}_L(\vec{R}_L)$ is determined. One is then left with a closed integral equation, which determines the motion of the vortex line,

$$\vec{v}_L(\vec{R}_L) = \frac{d\vec{R}_L}{dt} = \frac{\kappa}{4\pi}\int_{L'}\frac{d\vec{R}'_L \times (\vec{R}_L - \vec{R}'_L)}{|\vec{R}_L - \vec{R}'_L|^3} = \frac{1}{4\pi}\int_{L'}dL\frac{\vec{\kappa}(\vec{R}'_L) \times (\vec{R}_L - \vec{R}'_L)}{|\vec{R}_L - \vec{R}'_L|^3}, \qquad (19)$$

where the integral was regularized to avoid singularities as explained above. By analogy with electromagnetism, this velocity field corresponds to the vortex line energy

$$\mathcal{E} = \frac{\rho_s \kappa^2}{8\pi} \int_L \int_{L'} \frac{d\vec{R}_L \cdot d\vec{R}'_L}{|\vec{R}_L - \vec{R}'_L|} = \frac{\rho_s}{8\pi} \int_L dL \int_{L'} dL' \frac{\vec{\kappa}(\vec{R}_L) \cdot \vec{\kappa}(\vec{R}'_L)}{|\vec{R}_L - \vec{R}'_L|} \ . \tag{20}$$

Since the component of $\vec{v}_L$ along $\vec{\kappa}$ is of no physical meaning one can now rewrite Eq. (19) as

$$- \rho_s \vec{\kappa}(\vec{R}_L) \times \frac{d\vec{R}_L}{dt} = -\frac{\delta \mathcal{E}}{\delta \vec{R}_L} \ . \tag{21}$$

where the functional derivative $\delta\mathcal{E}/\delta\vec{R}_L$ is defined from the following variational identity:

$$\delta\mathcal{E} = \frac{\rho_s}{4\pi} \int_L dL \int_{L'} dL' \left[ -\delta\vec{R}_L \cdot (\vec{R}_L - \vec{R}'_L) \frac{\vec{\kappa}(\vec{R}_L) \cdot \vec{\kappa}(\vec{R}'_L)}{|\vec{R}_L - \vec{R}'_L|^3} + \frac{\delta\vec{\kappa}(\vec{R}_L) \cdot \vec{\kappa}(\vec{R}'_L)}{|\vec{R}_L - \vec{R}'_L|} \right] \ . \tag{22}$$

Using the definition of the vector $\vec{\kappa}(\vec{R}_L)) = \kappa \, d\vec{R}_L/dL$ and integrating by parts the second term in Eq. (22) we obtain that $\delta\mathcal{E} = \int_L dL \, \delta\vec{R}_L \cdot \delta\mathcal{E}/\delta\vec{R}_L$ with

$$\frac{\delta\mathcal{E}}{\delta\vec{R}_L} = \frac{\rho_s}{4\pi} \int_{L'} dL' \left[ -(\vec{R}_L - \vec{R}'_L) \frac{\vec{\kappa}(\vec{R}_L) \cdot \vec{\kappa}(\vec{R}'_L)}{|\vec{R}_L - \vec{R}'_L|^3} + \vec{\kappa}(\vec{R}'_L) \frac{\vec{\kappa}(\vec{R}_L) \cdot (\vec{R}_L - \vec{R}'_L)}{|\vec{R}_L - \vec{R}'_L|^3} \right]$$

$$= \frac{\rho_s}{4\pi} \int_{L'} dL' \frac{\vec{\kappa}(\vec{R}_L) \times [\vec{\kappa}(\vec{R}'_L) \times (\vec{R}_L - \vec{R}'_L)]}{|\vec{R}_L - \vec{R}'_L|^3} \ . \tag{23}$$

In mechanics the functional derivative of the energy with respect to displacement is a force by definition; here it is the force acting upon the unit length of the element $dL$, which arises from all other elements of the same curved vortex line. The net force is balanced by the gyrotropic Magnus force on the left-hand side of Eq. (21). Sometimes also an inertial force is added to the equation of vortex motion, but mostly it leads to insignificant corrections which we shall not discuss here.

This approach for determining the equation of motion of a curved vortex line, based on the Biot-Savart law, can be used for analytical calculations only in a restricted number of simple cases. One example is a vortex ring with radius $R$. In this case the integral in the expression Eq. (20) for the energy can be exactly calculated [19] if one specifies a model of the vortex core. For the case of a hollow core, whose size is determined by surface tension (see Ref. [1], Sec. 1.6), the energy per unit length of the line forming the vortex ring, $\varepsilon = \mathcal{E}/2\pi R$, is given by

$$\varepsilon = \frac{\rho_s \kappa^2}{4\pi} \left( \ln \frac{8R}{r_c} - 1 \right) = \frac{\rho_s \kappa^2}{4\pi} \ln \frac{8R}{er_c} \ . \tag{24}$$

This corresponds to an integration cut-off in Eq. (9) with $r_u = 8R/e$ (here $e = 2.7...$ is the basis of the natural logarithm and not the electron charge!). A vortex ring moves normal to its plane with the velocity $v_L$ which is determined from the radial component of Eq. (21), when it is written in cylindrical coordinates:

$$\rho_s \kappa v_L = \frac{\delta\mathcal{E}}{\delta\vec{R}_L}\bigg|_r = \frac{1}{2\pi R} \frac{d\mathcal{E}}{dR} = \left( \frac{\varepsilon}{R} + \frac{d\varepsilon}{dR} \right) \ . \tag{25}$$

Here we have used the relation between the radial component of the functional derivative and the force given by the derivative of the vortex ring energy with respect to the ring radius

$R$, ie. with respect to the expansion of the ring as a whole. The radial component of the functional derivative has the dimensions of force per unit length (or linear force density) and accounts for the local displacement of the vortex loop. Then according to Eq. (24)

$$v_L = \frac{\kappa}{4\pi R} \ln \frac{8R}{r_c} . \tag{26}$$

The energy $\varepsilon$ per unit length depends only weakly on the total length $2\pi R$ of the vortex ring. If we ignore this weak logarithmic dependence, ie. $d\varepsilon/dR \approx 0$, then the right-hand side of Eq. (25), $\varepsilon/R$, may be treated as a line-tension force similar to that of a deformed elastic string. This is a valid interpretation within the framework known as logarithmic accuracy, which means that a logarithm of the form $\ln(R/r_c)$ is large and corrections of relative order $\sim 1/\ln(R/r_c)$ can be neglected. Within this approximation a logarithm of the form $\ln(R/r_c)$ is treated as a bare number and

$$\varepsilon \approx \frac{\rho_s \kappa^2}{4\pi} \ln \frac{R}{r_c} , \qquad v_L \approx \frac{\kappa}{4\pi R} \ln \frac{R}{r_c} . \tag{27}$$

This approach may be carried over to the more general case of a vortex line of arbitrary shape. We simply replace the radius $R$ of the ring by the local radius of curvature $\mathcal{R}$ which varies continuously along the line, as opposed to the vortex ring with a constant $\mathcal{R} = R$ along the whole length of the line. Rewriting Eq. (25) in vector form one now obtains a differential equation

$$- \rho_s \vec{\kappa}(\vec{R}_L) \times \frac{d\vec{R}_L}{dt} = \frac{\vec{\mathcal{R}}}{\mathcal{R}^2} \varepsilon . \tag{28}$$

Here the position vector $\vec{R}_L$ traces the whole vortex line and $\vec{\mathcal{R}}$ is a vector which is directed towards the local center of curvature and is equal in magnitude to the local radius of curvature. It is determined by the variation of the tangent unit vector along the curved vortex line:

$$\frac{\vec{\mathcal{R}}}{\mathcal{R}^2} = \frac{1}{\kappa} \frac{d\vec{\kappa}}{dL} . \tag{29}$$

This approximation for the equation of motion of a vortex line is obtained from the general theory based on the Bio-Savart law by restricting the inner integral $\int dL'$ in the self-energy in Eq. (20) to some section of the vortex line around the point $\vec{R}_L$ with a length on the order of the local radius of curvature. This is justified by the fact that the contribution from this section is logarithmically large compared to the more distant sections of the vortex line, ie. $\mathcal{E} = \int dL\varepsilon$ and an upper cut-off $r_u \approx \mathcal{R}$ is inserted in the expression Eq. (9) for the vortex line tension $\varepsilon$. In this approximation the variation of $\varepsilon$ along the vortex line is neglected and the only contribution arises from the variation of the vortex line length, ie. $\delta \mathcal{E} = \int dL(\varepsilon/\kappa)\vec{\kappa} \cdot [d(\delta \vec{R}_L)/dL]$. Integrating by parts one obtains that $\delta \mathcal{E}/\delta \vec{R}_L = -(\varepsilon/\kappa)d\vec{\kappa}/dL$. This gives the line-tension force on the right-hand side of Eq. (28) which is balanced by the Magnus force on the left. In Fig. 2 it is shown how the line-tension force denoted by $\vec{\Delta\varepsilon}$ arises from the curvature of the vortex line.

The model, in which interactions between distant parts of the vortex line are neglected, is known already from classical hydrodynamics as the *localized induction approximation* [20]. It is based on the principle of logarithmic accuracy: the local radius of curvature is so substantial compared to the core radius that the logarithm of their ratio becomes a large number. The correction term to this model is on the relative order of the inverse of the large logarithm.

The equation of motion for the vortex line, either in the general form, Eq. (21), or in the localized induction approximation, Eq. (28), represents the condition for the mechanical balance of forces. But it should be emphasized, that in deriving these equations we did not use any *a priori* definitions of "forces" and thus the mechanical interpretation in terms of "forces" *a posteriori* is a matter of taste and semantics. In vortex dynamics the net force on an element of a vortex line unambiguously determines its velocity, similar to the correspondence between acceleration and force in time-invariant particle mechanics. This statement is true in a general meaning, also beyond the localized induction approximation, as expressed in Eq. (21). However, beyond the localized induction approximation the concept of a nonlocal generalized force in terms of the functional derivative in Eq. (23) is perhaps not so transparent and one may simply refer directly to Helmholtz's theorem by saying that a vortex moves with the velocity induced by the curved vortex line itself, which is given by

$$\vec{v}_L = -\frac{1}{\rho_s \kappa^2} \vec{\kappa}(\vec{R}_L) \times \frac{\delta \mathcal{E}}{\delta \vec{R}_L} \ . \tag{30}$$

To include mutual friction we may proceed as in Sec. 2.2 for the case of a straight vortex line. However, the velocity $\vec{v}_s$ in the expressions of Sec. 2.2 should now contain in addition to any external velocity field also the velocity induced by the curved vortex line itself. For their sum we use the local superfluid velocity $\vec{v}_{sl}$ determined from the Biot-Savart law, ie. given by the right-hand side of Eq. (30):

$$\vec{v}_{sl} = -\frac{1}{\rho_s \kappa^2} \vec{\kappa}(\vec{R}_L) \times \frac{\delta \mathcal{E}}{\delta \vec{R}_L} \ , \tag{31}$$

or in the localized induction approximation,

$$\vec{v}_{sl} = \frac{1}{\rho_s \kappa^2} \vec{\kappa}(\vec{R}_L) \times \frac{\vec{\mathcal{R}}}{\mathcal{R}^2} \varepsilon \ . \tag{32}$$

Note that $\vec{v}_L \neq \vec{v}_{sl}$, since Helmholtz's theorem does not hold in the presence of mutual friction. Equation (32) together with Eq. (13), which is now rephrased in terms of the local superfluid velocity $\vec{v}_{sl}$,

$$\vec{v}_L = \left(1 - \frac{\rho_n}{2\rho} B'\right) \vec{v}_{sl} - \frac{\rho_n}{2\rho} B \hat{z} \times \vec{v}_{sl} \ , \tag{33}$$

form a close set of equations, which describe the dynamics of a curved vortex line. This approach for taking into account the mutual friction force assumes that the radius of curvature exceeds the size of the volume around the vortex line where the mutual friction losses occur. Only then is Eq. (33) applicable, since it has been derived for a straight vortex line. Otherwise the mutual friction parameters depend on the radius of curvature. The volume where normal excitations scatter from the vortex, is not necessarily just the vortex core itself, but may in fact exceed it [18].

## 2.4 Boundary conditions for vortex line

The equation of motion of a vortex line has to be supplemented by boundary conditions when the line interacts with the boundary of the system. An evident hydrodynamic boundary condition for the flow of liquid is that the velocity at the boundary should not have a component normal to the wall. Consider a vortex near a parallel plane wall. Here the method of images becomes useful: An image vortex of equal configuration, but with opposite circulation, is placed on the other side of the wall and moves in unison with the real vortex

Figure 3: Vortex line pinned to a shallow protuberance on the boundary of the superfluid system. In the linear regime of pinning surface roughness is characterized by an asperity with a radius of curvature $\mathcal{R} = 1/b$ and a pinning force $\vec{f_p} = -b\vec{u}$, which depends on the displacement $\vec{u}(z)$ from the summit of the protuberance. The result is a harmonic pinning potential well as a function of displacement. In the limit of weak pinning (applicable to eg. $^3$He-B) the pinning strength $b$ is small, the radius of curvature of the pinning site is large and the bending of the vortex line is small.

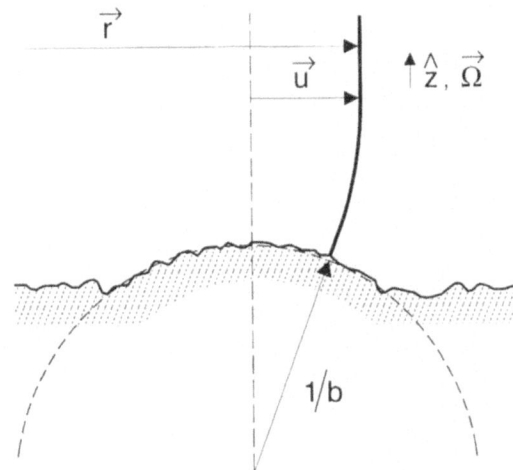

with respect to the wall [6]. The combination of the real vortex and its image allows one then to remove the wall from the problem. In the case of an intersecting wall the vortex line is continued into the wall in terms of its image (see Fig. 4 later). Here not only the line plus its image are continuous, but also their tangent is continuous everywhere. This means that the ideally flat boundary intersects a vortex line and its image continuation at right angles. If the $z$ axis is normal to the wall and the vortex displacement $\vec{u}(z)$ is measured in the $xy$ plane, then $d\vec{u}/dz = 0$ during the motion of the vortex line (in the absence of pinning).

In contrast, at a rough boundary a pinning force $\vec{f_p}$ appears which attempts to maintain the end of a vortex line fixed to an asperity on the surface (Fig. 3). For relatively weak pinning the boundary condition at the top (minus sign) and the bottom (plus sign) of the container may be expressed in the linearized form

$$\frac{d\vec{u}}{dz} = \mp b\vec{u}. \tag{34}$$

Here $b$ is the elastic pinning parameter, a phenomenological quantity, which has the geometrical interpretation that $1/b$ is equal to the radius of curvature of an axisymmetric protuberance [21]. The right-hand side in Eq. (34) represents a linear pinning force $\vec{f_p} = \mp b\vec{u}$, whose magnitude increases linearly with displacement from the summit of the protuberance, ie. the bottom of the potential well. This pinning force proportional to the displacement $\vec{u}$ is elastic, contrary to the surface friction force proportional to the velocity $\vec{v_L}$ in the boundary condition, which was originally suggested by Hall [7] and in a more generalized form by Bekarevich and Khalatnikov [10]. The frictional contributions add to the overall losses from bulk mutual friction and lead effectively to a renormalized mutual friction [21]. In the present context we shall neglect surface friction.

An interesting boundary problem of experimental importance we meet when quantized circulation, originally trapped around a thin wire, starts to peel off the wire. The vortex becomes disconnected from the wire, as shown in Fig. 4, such that below the termination point of the free vortex its continuation is the circulation trapped around the wire. Very close to the termination point at distances $r - r_w \ll r_w$, the surface of the wire appears like a flat wall and the vortex line is perpendicular to it. At distances $r \sim r_w$ from the wire, the situation becomes a complicated 3-dim boundary problem with vortex line curvature; here the image method does not provide the shape of the vortex line. At distances $r \gg r_w$ our

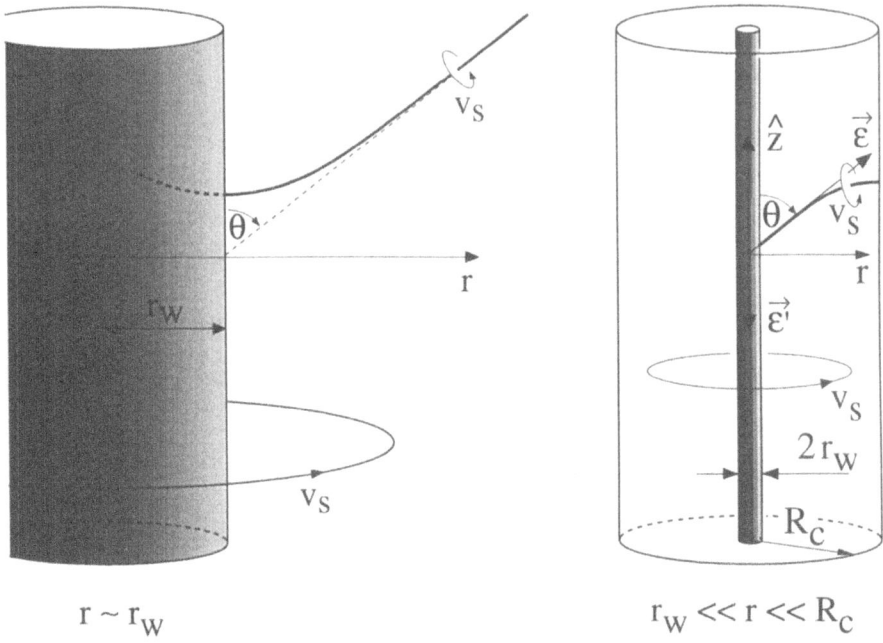

$$r \sim r_W \qquad\qquad r_W \ll r \ll R_C$$

Figure 4: Vortex line attached to a thin wire due to quantized circulation which is trapped around the lower section of the wire. This configuration has been observed in Vinen's vibrating wire cell, which consists of a long cylindrical container with a fine coaxial vibrating wire [8,22]. Initially a quantum of circulation has been trapped around the wire by rotating the equipment around the central axis. After trapping rotation is stopped and the circulation will persist as a trapped metastable state. Eventually at the upper end of the wire a free vortex segment will nucleate, which will become stretched between the wire and the cylindrical side wall of the container. The trapped circulation will now peel off the wire by "unwinding", ie. the free vortex both precesses around the wire and simultaneously moves slowly down along the whole length of the wire. Both the decay of the trapped persistent supercurrent and the period of the precessional motion of the free vortex segment can be measured from the vibrational properties of the wire, if it is not exactly centered (see Fig. 7). The length scales of the cylindrical geometry are specified by the radii of the container $R_c$ and of the wire $r_w$, with $R_c \gg r_w \gg r_c$. The profile of the vortex line can be examined on two different length scales: 1) On the scale $r \sim r_w$, the vortex line goes smoothly over into the image vortex (dashed continuation of vortex line), with the surface of the wire perpendicular at the connection point (ie. the wire appears like a flat wall). This scale is illustrated *on the left*. 2) At large distances $r \gg r_w$ the hydrodynamic approach becomes applicable, in which the wire is treated as a large vortex core. This situation is depicted *on the right* and is discussed in Sec. 3.4.

hydrodynamic analysis becomes feasible. One can then treat the wire with trapped circulation like a vortex with a core of radius $r_w$, which may be much larger than the core radius $r_c$ of the free vortex segment, but in any case much less than the container radius $R_c$. Then the line tensions of the free and trapped segments are different and are given, re-

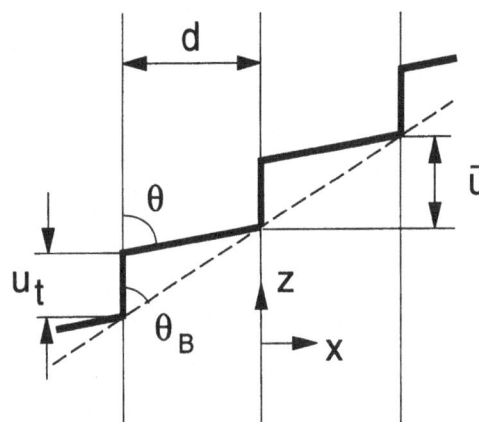

Figure 5: Vortex line partially trapped on parallel twin-boundary planes (vertical lines) in a single crystal of high-$T_c$ YBaCuO superconductor. The vortex line has on an average the orientation of the dashed line, but follows it in a tilted staircase pattern (thick solid line).

spectively, by

$$\varepsilon = \frac{\rho_s \kappa^2}{4\pi} \ln \frac{R_c}{r_c} \;, \qquad \varepsilon' = \frac{\rho_s \kappa^2}{4\pi} \ln \frac{R_c}{r_w} \;. \tag{35}$$

The line tension of the free vortex segment ($\varepsilon$) is larger than that of the trapped segment ($\varepsilon'$), which makes the partially trapped configuration possible. The inclination of the free vortex segment with respect to the wire is determined in the long distance limit ($r \gg r_w$) by the balance of the components of the line tension forces parallel to the wire at the point where the vortex disconnects from the wire:

$$\varepsilon \cos\theta = \varepsilon' \;. \tag{36}$$

Thus in general the free vortex segment is not normal to the wire. Only in the limit of very large wire radius $r_w$ compared to the core radius $r_c$ is the free vortex oriented perpendicular. The same boundary condition Eq. (36) holds for a vortex line trapped on a twin-boundary plane in a high-$T_c$ superconductor (Sec. 3.1).

The boundary condition Eq. (36) for the partially trapped vortex is here derived from the principle of mechanical balance for the line-tension force. But it can be also proven by minimization of the vortex line energy in the coordinate frame in which the vortex is at rest (simple examples of the derivation are given in Secs. 3.1 and 3.3). This is no surprise since the principle of mechanical balance itself is derived from the energy minimization. The accuracy of the boundary condition is the same as that of the localized induction approximation: the relative error is on the order of the inverse of the logarithm $\ln(R_c/r_w)$.

# 3 Static single-vortex problems

## 3.1 Vortex line trapping on twin boundary in YBaCuO single crystal

Let us first apply the hydrodynamic approach to static problems in which the vortex line is at rest with respect to the normal fluid (or the crystalline lattice in the superconductor). The simplest example is a vortex line tilted with respect to a twin boundary in a single crystal of high-$T_c$ YBaCuO superconductor. We assume that there are parallel twin boundaries with a spacing $d$ and normal to the $a-b$ plane (Fig. 5), the case which has been experimentally studied in Refs. [23]. Consider now pinning at a twin-boundary in a simplified model which neglects interactions between vortex lines [24,25].

When the vortex line is aligned along the twin boundary we denote its line tension with $\varepsilon'$ while in the bulk we use $\varepsilon$. The difference $\varepsilon - \varepsilon'$ is responsible for the pinning on the twin planes. As a result, the vortex line forms a tilted stair case pattern with periodically repeating straight segments. The segments of length $u_t$ run within the twin boundary planes. The segments of length $\sqrt{(\bar{u} - u_t)^2 + d^2}$ are stretched between two neighboring twin-boundaries (Fig. 5). Here $\bar{u}$ is the displacement of the vortex line along the twin-boundary per one period. The energy of the vortex line per one period is

$$E = \varepsilon' u_t + \varepsilon \sqrt{(\bar{u} - u_t)^2 + d^2} \ . \tag{37}$$

Finally the shape of the vortex line is determined by minimizing the energy with respect to $u_t$ at fixed average slope $\bar{u}/d$:

$$\frac{dE}{du_t} = \varepsilon' - \varepsilon \frac{\bar{u} - u_t}{\sqrt{(\bar{u} - u_t)^2 + d^2}} = 0 \ . \tag{38}$$

This is the boundary condition in Eq. (36), which was obtained in Sec. 2.4 from the mechanical balance of the line-tension forces and which we have now rederived by minimization of the energy. Here $\theta$ is the angle between the twin boundary plane and the free segment of vortex line, whereas $\theta_B$ is the average inclination angle of the vortex line (Fig. 5). The shape of the vortex line remains distorted by the twin-boundary pinning as long as $\theta_B < \theta$. When $\theta_B > \theta$ the energy of the vortex line is minimized with $u_t = 0$ and the twin boundary does not influence the shape of the vortex line. Thus $\theta$ is the critical angle for $\theta_B$ at which pinning on the twin boundaries starts. The boundary condition Eq. (36) is not restricted to the case of a single noninteracting vortex; it remains valid even for a dense vortex array in high magnetic fields (Sec. 5.4).

## 3.2 Energy barrier of vortex ring expansion

If there is superflow past a wall, vortex loops may spontaneously nucleate and dissipation becomes possible, via mutual friction, for instance. This process has a bottleneck: dissipation is possible if the vortex loop is able to expand without limitations crossing all the superflow stream-lines. However, the expansion of the vortex ring is resisted by an energy barrier. The top of the barrier corresponds to some vortex loop for which the energy has an extremum. This energy extremum is a point of unstable equilibrium, but to find the extremal loop configuration is a static problem which we are going to solve for uniform superflow.

In Sec. 2.3 we considered a vortex ring in the superfluid at rest. In the presence of superflow past a wall, the energy of the vortex loop in the laboratory frame, which is fixed with respect to the wall and where the normal liquid is clamped in the stationary state ($\vec{v}_n = 0$), is given by

$$\mathcal{E} = E + \vec{P} \cdot \vec{v}_s \ . \tag{39}$$

The energy $E = \varepsilon L$ arises from the line tension $\varepsilon$ in Eq. (27) and is proportional to the length $L$ of the line, whereas the momentum $P = \int \rho_s \vec{v}_s \, dV = (\rho_s \kappa / 2\pi) \int \vec{\nabla}\phi \, dV = \rho_s \kappa \vec{S}$ of the loop is proportional to the vector $\vec{S}$ normal to the plane of the loop and in magnitude equal to the area $S$ of the loop [1]. The energy Eq. (39) has an extremum for the particular configuration of a loop which encircles a fixed area with the shortest circumference. This is a vortex ring with $L = 2\pi R$, $S = \pi R^2$, and the radius $R$ given within the logarithmic accuracy by

$$R = \frac{\varepsilon}{\rho_s \kappa v_s} = \frac{\kappa}{4\pi v_s} \ln \frac{R}{r_c} \approx \frac{\kappa}{4\pi v_s} \ln \frac{\kappa}{v_s r_c} \ . \tag{40}$$

This corresponds to Eq. (27), when $v_L = v_s$. Bearing in mind that in the extremal configuration $\vec{P}$ is antiparallel to $\vec{v}_s$, the extremal vortex ring moves with the velocity $\vec{v}_L = -\vec{v}_s$ in a frame of the flowing superfluid, but is stationary in the laboratory frame. The energy $\mathcal{E}$ has its maximal value as a function of $R$ equal to

$$\mathcal{E}_b = \pi R\varepsilon = \frac{\rho_s\kappa^2}{4}R\ln\frac{R}{r_c} \approx \frac{\rho_s\kappa^3}{16\pi v_s}\ln\frac{\kappa}{v_s r_c}\;. \tag{41}$$

The energy in Eq. (41) represents the energy barrier which resists the growth of a vortex ring at a fixed value of $v_s$: at smaller $R$ the ring will shrink away, at larger $R$ it will start to expand. In a more general sense, this extremum is not a maximum, but a saddle point, since it is also a minimum with respect to deviations from circular shape.

Both processes, shrinking and expansion, should be calculated by including mutual friction. The expansion of the ring after nucleation in the rotating superfluid is discussed in Sec. 4.2. In the nucleation process the mechanism for crossing the energy barrier in Eq. (41), is controlled by thermal fluctuations or by quantum tunnelling [1,11]. In superfluid $^3$He the nucleation of a singular vortex core occurs practically only at the pair breaking velocity, ie. at the Landau limit for the break-down of superfluidity [13]. However, whatever the process which allows the system to overcome the nucleation barrier and to create an expanding vortex ring, a critical superflow velocity $v_s = v_c(T,p)$ has to be reached, which is a characteristic of the intrinsic nucleation event.

An important kinematic relation follows from Eq. (39) and the fact that a vortex ring corresponds to the energy extremum of a vortex loop in the laboratory frame. Since the total energy is invariant under small variations of the vortex ring radius, ie., $\delta\mathcal{E} = \delta E + \vec{v}_s \cdot \delta\vec{P} = 0$, and bearing in mind that $\vec{v}_s = -\vec{v}_L$, one obtains

$$\vec{v}_L = \frac{\partial E}{\partial \vec{P}}\;. \tag{42}$$

Thus the vortex ring may be treated as an excitation branch with the spectrum $E(P)$ and its velocity $v_L$ is the group velocity for such excitations.

## 3.3 Energy barrier of trapped vortex

The high nucleation barrier may be circumvented, when remnant vortices are present and only a lower energy barrier of trapping needs to be overcome. Remnant vortices may persist in suitable surface traps, if supercritical velocities have been applied previously. In $^4$He-II they may perhaps also exist as "primordial" vortices, precipitated in an inhomogeneous cooldown through the $\lambda$ transition. This is believed to be the reason, why the critical velocities for the superflow of $^4$He-II in tubes or containers of macroscopic size, have not been so high as expected [1]. When the superflow velocity is increased, the trapped vortices become mobile. This happens at a critical velocity, which is characteristic of the energy barrier of the trapped vortex configuration. Recently the successive detachment of solitary trapped vortex filaments has been observed in rotating $^3$He-B as single quantum events one by one. In this experiment the net result from emptying a trap is the addition of one more rectilinear vortex line to the cluster of vortices in the center of the container. The traps were formed by pore-like cavities on the cylindrical side wall of the rotating container. From here the trapped vortices came loose successively one after the other when the rotation velocity $\Omega$ was increased. The traps were reloaded when $\Omega$ was decreased and the vortices were pushed to the side wall for annihilation.

Here we shall consider a particular geometry for a trap, a sharp needle or a piece of wire, which sticks out of the side wall of the rotating container at some oblique angle $\alpha$ (Fig. 6): Around the wire there is trapped circulation up to some point, where it disconnects as a free

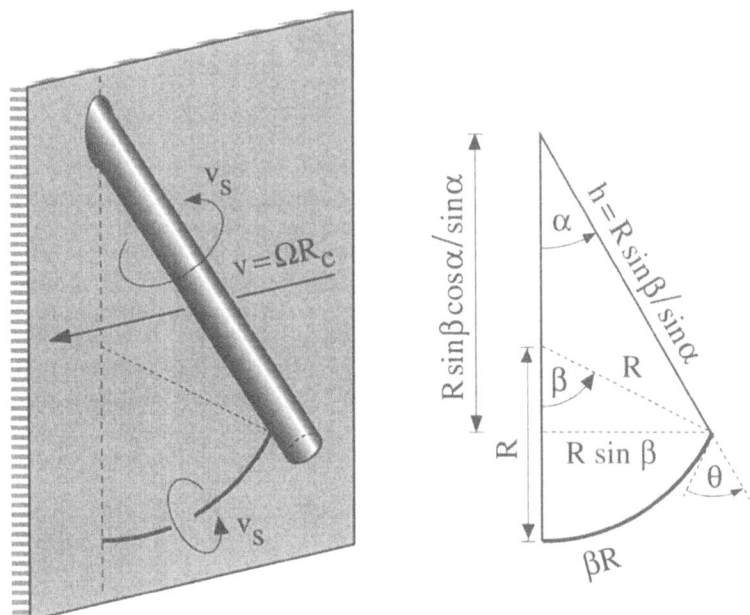

Figure 6: Vortex line partially trapped on a sharp needle, which sticks out from the side wall of a rotating cylinder filled with superfluid in the presence of the counterflow $\vec{v}_s = -\vec{\Omega} \times \vec{R}_c$. Initially the trap is loaded with a remnant vortex during decelerating rotation, when vortices move to the side wall and annihilate. The trapped configuration consists of the circulation around the needle and the free vortex segment between the needle and the side wall. The free segment is a circular arc with the radius $R$. If rotation is later restarted and accelerated in the same direction as originally, then the configuration becomes self-sustained in a state of unstable balance at the velocity $v_s$, which corresponds to the radius $R$ of the free segment. The energy of this configuration represents the barrier against expansion: at higher flow velocities the trapped vortex starts to expands and eventually it forms a rectilinear vortex line in the center of the container. On the right the relations between the different geometrical quantities of the configuration are denoted.

vortex segment. The free vortex segment lies stretched between the wire and the side wall of the container. It forms a circular arc of radius $R$ and covers a sector with the angle $\beta$. In the absence of flow or at low flow velocities such a trapped configuration might persist in a metastable pinned state. However, we neglect pinning here for simplicity. During increasing rotation velocity the trapped vortex will start expanding, when the superfluid counterflow velocity $\vec{v}_s = -\vec{\Omega} \times \vec{R}_c$ exceeds the critical value, which is needed to stabilize the configuration at its self-sustained threshold. Now $\vec{v}_s$ is a velocity in a frame fixed to the rotating vessel, in which $\vec{v}_n = 0$, since the normal fluid corotates with the vessel. What is the value of this flow velocity $v_s$ when this vortex pattern is in a state of unstable balance and its energy corresponds to the threshold value from where expansion starts?

The angle $\theta$ between the vortex line and the wire at the disconnection point can be determined from the balance of the line-tension forces along the wire and the free vortex filament using Eq. (36). However, to illustrate the principle of mechanical balance we shall rederive the condition in Eq. (36) by minimization of the energy. The vortex line energy in a frame moving with the superfluid consists of the contributions from the free vortex segment of length $L = R\beta$, with $\beta = \frac{\pi}{2} + \alpha - \theta$, and from the trapped segment of length $h = R\sin\beta/\sin\alpha$:

$$E = \varepsilon L + \varepsilon' h = R(\varepsilon\beta + \varepsilon'\sin\beta/\sin\alpha) . \tag{43}$$

Here $\varepsilon$ and $\varepsilon'$ are the line tensions for the free and trapped segments, respectively. As usual, the momentum in a frame moving with the superfluid is determined by the area which is encircled by the vortex line, ie. in this case by the free and trapped segments:

$$P = \rho_s \kappa S = \frac{1}{2}\rho_s \kappa R^2(\beta + \sin^2\beta\frac{\cos\alpha}{\sin\alpha} - \sin\beta\cos\beta) = \frac{1}{2}\rho_s \kappa R^2\left(\beta + \frac{\sin\beta\sin(\beta-\alpha)}{\sin\alpha}\right) . \tag{44}$$

By varying the energy $E$ with respect to $\beta$ at fixed momentum $P$, which means that the radius $R$ varies also in order to keep $P$ constant, we obtain the relation $\sin(\beta - \alpha) = \varepsilon'/\varepsilon$ which is identical to Eq. (36), bearing in mind that $\beta = \frac{\pi}{2} + \alpha - \theta$. Finally the energy and the momentum are:

$$E = \varepsilon R\left(\beta + \frac{\sin\beta\cos\theta}{\sin\alpha}\right) ,$$
$$P = \frac{\rho_s \kappa R^2}{2}\left(\beta + \frac{\sin\beta\cos\theta}{\sin\alpha}\right) . \tag{45}$$

Thus both the energy and the momentum differ from those of the complete ring in Sec. 3.2 by the same factor in the brackets. The energy $\mathcal{E} = E + \vec{v}_s \cdot \vec{P}$ in the frame connected with the rotating vessel obtains its maximum value at a given superflow velocity $v_s$ with the radius $R$ given by Eq. (40). Correspondingly, the barrier height can be written in the form

$$\mathcal{E}_{b\alpha} = \mathcal{E}_b \frac{1}{2\pi}\left(\beta + \frac{\sin\beta\cos\theta}{\sin\alpha}\right) , \tag{46}$$

where $\mathcal{E}_b$ is the energy barrier in the bulk given by Eq. (41). A particular limiting case is the one where the line tension $\varepsilon'$ along the wire becomes negligible. Then $\beta = \alpha$, $\theta = \pi/2$, and $h = R$, ie. the free vortex segment is perpendicular to the wire at the disconnection point. In this case, which also applies to a vortex stretched between two plain surfaces (ie. a vortex trapped in a cavity), the barrier is simply proportional to the sector angle $\alpha$: $\mathcal{E}_{b\alpha} = \frac{\alpha}{2\pi}\mathcal{E}_b$. Of special interest are here a half-ring at a flat wall ($\alpha = \pi$, $\mathcal{E}_{1/2} = \frac{1}{2}\mathcal{E}_b$), or a quarter of a ring in the corner of the container ($\alpha = \pi/2$, $\mathcal{E}_{1/4} = \frac{1}{4}\mathcal{E}_b$). Evidently, the expansion of the trapped vortex segment is most probable when the energy barrier is minimal, ie. for such traps, where the length of the free vortex segment is shortest.

## 3.4 Shape of partially trapped curved vortex line

The classic experiment for demonstrating quantization of circulation in a superfluid is performed with Vinen's vibrating wire cell [8]. It is a cylindrical rotating container with a thin coaxial wire (Fig. 4). When a current pulse is passed through the wire, while it is subject to a transverse magnetic field, the wire is set into a state of decaying vibrational motion. If the surrounding superfluid is at rest, then the two orthogonal plane polarized vibration modes of an ideal wire are degenerate and the vibration frequency is monochromatic. On the other hand, if a quantum of circulation is trapped around the wire by rotating the apparatus, then

210

Figure 7: Measured value of trapped circulation in the vibrating wire experiment of Zieve et al. [22]. The circulation is displayed as a function of time while the free vortex segment precesses around the wire and "unwinds" the persistent supercurrent. The vertical scale is the fraction of circulation relative to its full value with one quantum trapped around the whole length of the wire. The insert is a magnification of the precession with a period of 250 sec, which modulates the signal when the wire is slightly off-axis.

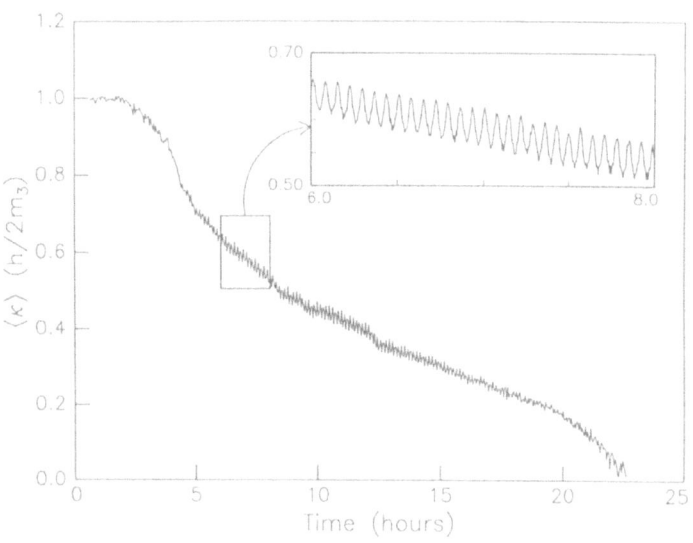

a Magnus force acts on the wire causing its plane of vibration to precess and a beating in the frequency response of the vibrational signal. Vinen originally measured the magnitude of a quantum of circulation with this experiment in $^4$He-II [8].

Now 3 decades later the measurement has been repeated with $^3$He-B. Again quantization of circulation could be demonstrated [22]. In this recent experiment the trapped circulation was observed to disconnect from the wire by nucleating a free vortex segment, as was discussed in Sec. 2.4. The presence of the free vortex filament was concluded from the decay of the measured value of the trapped circulation over a period of several hours (Fig. 7). In addition to the decay the signal for the trapped circulation carried a periodic modulation from the precession of the vortex filament around the wire. In practice the wire is not perfectly centered in the cylindrical container and then the precessional motion, with which the vortex segment moves down along the whole length of the wire and "unwinds" the trapped circulation, brings about the periodic modulation of the measured signal.

The geometry of the experiment is shown in Fig. 4. We use cylindrical coordinates with the $z$-axis along the symmetry axis of the container. The disconnection point of the free vortex segment on the wire is at $z = 0$ and $r = r_w$. The second end point is on the side wall at $r = R_c$. The experiment is conducted in the extreme low temperature limit with $T < 0.2\,T_c$, in order to avoid the large viscosity of the normal fraction in $^3$He-B and to secure a sensitive signal from the vibrating wire. Our approach based on balancing the line-tension forces may now be applied to find the angle at which the free vortex filament meets the wire, when viewed in the large scale limit $r_w \ll r \ll R_c$ (see Sec. 2.4). If dissipation from mutual friction is negligible, which is the case at low temperatures, then the shape of the vortex filament can be solved by minimization of the energy, but now in the coordinate frame rotating around the wire with the angular velocity $\omega_L$ of the precessing vortex, where

$$\mathcal{E} = E - \omega_L J \ . \tag{47}$$

Here $E = \varepsilon' h + \varepsilon L$ and $J = \rho_s \kappa [(R_c^2 - r_w^2)h/2 + \int_{r_w}^{R_c} z(r)r\,dr]$ are the energy in the localized induction approximation and the angular momentum imparted by the vortex to the liquid, $\varepsilon = (\rho_s \kappa^2/4\pi)\ln(R_c/r_c)$ and $\varepsilon' = (\rho_s \kappa^2/4\pi)\ln(R_c/r_w)$ are the line tensions of the free segment of the vortex line and that from the circulation trapped around the wire, $h$ and $L = \int_{r_w}^{R_c} \sqrt{1 + (dz/dr)^2}\,dr$ are the lengths of the trapped and free segments. The energy minimization yields

$$\rho_s \kappa \omega_L r = -\frac{\rho_s \kappa^2}{4\pi} \ln\left(\frac{R_c}{r_c}\right)\frac{1}{\mathcal{R}} \ . \tag{48}$$

This is Eq. (28), but this time written in a cylindrical coordinate system fixed to the container and $\vec{v}_L = d\vec{R}_L/dt$ is the solid-body-rotation velocity $\vec{v}_L = \vec{\omega}_L \times \vec{r}$ of the precessing vortex. The radius of curvature $\mathcal{R}$ of the shape $z(r)$ of the free vortex segment can be expressed in the form

$$\frac{1}{\mathcal{R}} = \frac{d^2 z}{dr^2}\left[1 + \left(\frac{dz}{dr}\right)^2\right]^{-3/2} \ . \tag{49}$$

Eq. (48) is a differential equation for determining the shape $z(r)$. Integrating with respect to $r$ and assuming that $dz/dr = 0$ at $r = R_c$ (ie. the vortex line is normal to the container wall) one obtains:

$$\frac{\rho_s \kappa}{2}\omega_L(R_c^2 - r_w^2) = \frac{\rho_s \kappa^2}{4\pi}\ln\left(\frac{R_c}{r_c}\right)\left[\frac{dz/dr}{\sqrt{1 + (dz/dr)^2}}\right]\Bigg|_{r=r_w} \ . \tag{50}$$

The balance of the line-tension forces at the disconnection point is given by Eq. (36). The inclination angle $\theta$ at the disconnection point in Fig. 4 is thus

$$\cos\theta = \frac{1}{\sqrt{1 + \tan^2\theta}} = \left[\frac{dz/dr}{\sqrt{1 + (dz/dr)^2}}\right]\Bigg|_{r=r_w} = \frac{\varepsilon'}{\varepsilon} = \frac{\ln\frac{R_c}{r_w}}{\ln\frac{R_c}{r_c}} \ , \tag{51}$$

Thus the vortex filament is normal to the wire only if $r_c \to 0$, but the approach to perpendicular disconnection is logarithmicly slow. The opposite extreme is the case when $R_c \to \infty$, the difference between $r_c$ and $r_w$ becomes unimportant, and the vortex filament is tangent to the wire. However, as explained in connection with the derivation of the line-tension balance at the disconnection point, Eq. (36), our approach ignores the fine details of the vortex line shape at distances $r \approx r_w$, since $r_w$ is simply inserted as the core radius for the vortex segment trapped on the wire: while the free vortex segment is actually perpendicular to the wire at the disconnection point it, nevertheless, at distances $r \gg r_w$ approaches the direction as dictated by the line-tension balance.

By substituting Eq. (51) into Eq. (50) one immediately obtains the expression for the angular precession frequency $\omega_L$ of the free vortex segment around the central wire,

$$\omega_L = \frac{\kappa \ln(R_c/r_w)}{2\pi(R_c^2 - r_w^2)} \ . \tag{52}$$

This result may be obtained simply by varying the height $h$ of the trapped segment at fixed shape of the free segment: $\delta E = \varepsilon'\delta h$, $\delta J = \rho_s \kappa(R_c^2 - r_w^2)/2\ \delta h$. The condition $\delta\mathcal{E} = \delta E - \omega_L\delta J = 0$ for a stationary value of the energy in Eq. (47) in the precessing frame gives immediately Eq. (52) for the precession frequency $\omega_L$, which therefore may be found without knowing the shape $z(r)$ of the free vortex segment [22].

The shape $z(r)$ of the free vortex filament is obtained by integrating Eq. (48). Here we restrict to the simplest case when the ratio $\varepsilon'/\varepsilon$ and correspondingly $\cos\theta$ are small. Then $dz/dr$ is also small everywhere along the vortex line and Eq. (48) may then be linearized to yield

$$z(r) = \frac{\ln\frac{R_c}{r_w}}{\ln\frac{R_c}{r_c}} \left( r - \frac{r^3}{3R_c^2} \right) . \tag{53}$$

In the real experimental situation the ratio $\varepsilon'/\varepsilon \approx 1/2$ and thus a more accurate calculation is needed. The exact expression for $z(r)$ within the localized induction approximation has been also obtained [26]. Another approach, in which some nonlocal corrections have been added to the localized approximation, has been developed by Schwarz [27].

# 4 Dynamic single-vortex problems

## 4.1 Kelvin wave

Let us now apply the line-tension approach to dynamic problems by investigating first small oscillations propagating along a straight vortex line aligned parallel to the $z$-axis. These are Kelvin waves, which have been demonstrated in a number of experiments in ${}^4$He-II [1,7,9]. For small amplitude oscillations $dx/dz,\ dy/dz \ll 1$ and the expression in Eq. (29) for the radius of curvature $\vec{\mathcal{R}}$ yields

$$\frac{\vec{\mathcal{R}}}{\mathcal{R}^2} = \frac{d^2\vec{u}}{dz^2} . \tag{54}$$

Vortex motion is now described by the linearized Eq. (28):

$$-\rho_s\vec{\kappa} \times \frac{\partial\vec{u}(z)}{\partial t} = \varepsilon\frac{\partial^2\vec{u}(z)}{\partial z^2} . \tag{55}$$

This is the force-balance equation for a vortex line in the localized induction approximation; it includes the Magnus force and the line-tension force. Eq. (55) has a plane-wave solution $\sim \exp(ipz - i\omega t)$ which corresponds to a circularly polarized Kelvin wave (Fig. 8) propagating along the vortex line with the spectrum

$$\omega = \pm\nu_s p^2 , \tag{56}$$

where

$$\nu_s = \frac{\varepsilon}{\rho_s\kappa} = \frac{\kappa}{4\pi}\ln\frac{2\pi}{pr_c} \tag{57}$$

is the elastic line-tension parameter. It is here assumed that the upper cut-off $r_u$ of the logarithm in the line-tension force in Eq. (9) is equal to the wavelength $2\pi/p$.

To account for mutual friction we should use Eqs. (32) and (33) with the radius of curvature given by Eq. (54). Then the spectrum of the Kelvin wave is

$$\omega = \pm\nu_s p^2 \left( 1 - \frac{\rho_n}{2\rho}B' \mp i\frac{\rho_n}{2\rho}B \right) . \tag{58}$$

Clearly the mutual friction force introduces dissipation. If we use the values of $B$ and $B'$ from Eq. (16) for the Bardeen-Stephen superconductor, we obtain the analog of the Kelvin wave in the superconductor:

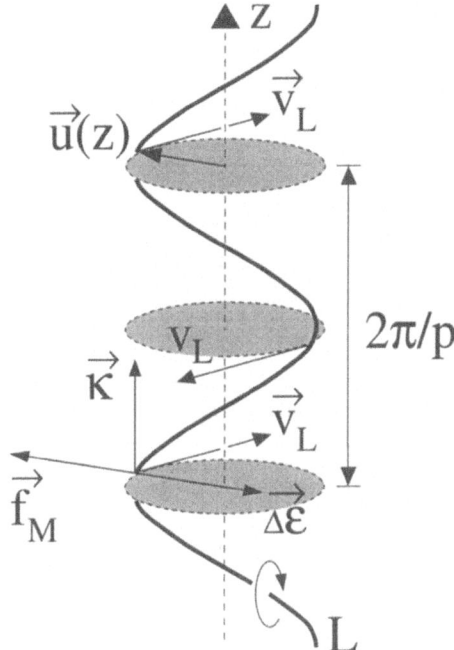

Figure 8: A circularly polarized longitudinal Kelvin wave, which propagates along a solitary vortex line, deforms the line to a helix. The motion along the circular trajectory is in the transverse plane $\perp \vec{\kappa}$, such that $\vec{\kappa}$, $\vec{v}_L$, and $\vec{f}_M$ form an orthogonal triad with the line-tension force $\overrightarrow{\Delta\varepsilon}$ balancing the Magnus force $\vec{f}_M$.

$$\omega = -\frac{i\rho_s\kappa\nu_s p^2}{\eta} \, . \tag{59}$$

The even mode of the standing Kelvin wave in a layer between two planes $z = \pm L/2$ is given by $u(z) \sim \cos(pz)$. Thus the wave number $p$ is determined by the boundary condition Eq. (34),

$$p\tan(pL/2) = b \, . \tag{60}$$

The fundamental or lowest-frequency standing Kelvin mode is pictured in Fig. 9. In the case of weak pinning, when the vortex line remains nearly straight during the oscillation, $pL \ll 1$ and $p = \sqrt{2b/L}$. This yields for the fundamental frequency of a single vortex in a layer with a width $L$:

$$\omega = \pm\frac{2b\nu_s}{L}\left(1 - \frac{\rho_n}{2\rho}B' \mp i\frac{\rho_n}{2\rho}B\right) \, . \tag{61}$$

For this result to be valid the condition for weak pinning has to be met, which for a solitary vortex line is $b \ll 1/L$. In practice this requirement is difficult to fulfill, since it requires that the radius of curvature $1/b$ associated with the pinning site is much larger than the length of the vortex line.

## 4.2  Vortex ring expansion resisted by mutual friction

In a smooth-walled rotating cylinder filled with $^3$He-B, vortices are believed to nucleate as half-rings at a critical counterflow velocity, which depends on the surface structure of the side wall at the nucleation center. The critical counterflow velocity appears to correspond closely to the Landau limit for the breaking of Cooper pairs, when corrected for the enhancement expected from a local rough spot at the side wall. A demonstration of such single quantum

214

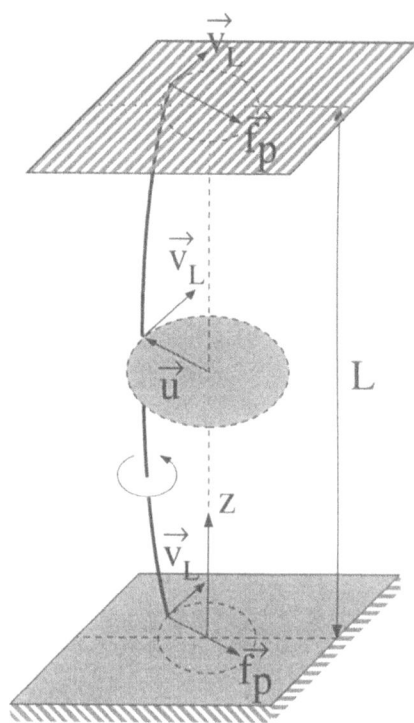

Figure 9: Fundamental mode of the standing Kelvin wave for a solitary vortex line. The oscillation takes place along a circular trajectory (shown as the dashed circles) with respect to the equilibrium position (shown as the rectilinear dashed line). The $z$ dependence of the contour of the vortex line is denoted with the displacement vector $\vec{u}(z,t)$ which is contained in the transverse plane. The boundary condition at the top and bottom surfaces is determined by the pinning force $d\vec{u}/dz = \mp b\vec{u}$ from Eq. (34). Here an idealized case is pictured when two identical pinning sites at the top and the bottom lie on the same vertical line.

nucleation events is shown in Fig. 10, measured during rotation at a constant rate of acceleration. When the counterflow velocity reaches the critical value in the course of the acceleration, which first happens locally at the nucleation center on the side wall, a vortex half-ring is nucleated. Next the ring expands to become a rectilinear vortex line. The arrival of the vortex line in the center of the container is marked by the abrupt step-like increase in the measured signal in Fig. 10. This process reduces the counterflow in the container by the equivalent of one circulation quantum. An interesting single-vortex problem in this experiment is the evolution of the nucleated half-ring to a rectilinear vortex line. We thus want to address the question: How fast does the newly nucleated vortex ring expand and finally end up as a rectilinear vortex line in the center of the container?

In a cylindrical container with radius $R_c$ superflow is created by uniform rotation, which at the angular velocity $\Omega$ in the vortex-free state results in a counterflow of the normal and superfluid components with the relative velocity $\vec{v} = \vec{v}_s - \vec{v}_n = -\vec{\Omega} \times \vec{r}$. As usual for $^3$He superfluids, the normal component can be considered to be clamped to solid-body rotation with the container ($\vec{v}_n = 0$). The velocity of a line element of the half-ring is described by Eq. (33), where the local superfluid velocity incorporates the counterflow velocity and the self-induced velocity from the line curvature. Both contributions have only a component normal to the plane of the half-ring, ie. parallel to the wall:

$$v_{sl} = v - \frac{\nu_s}{R} \ . \tag{62}$$

For simplicity here we treat $v$ as a constant, ie. we follow the early part of the expansion when $v$ can be regarded to be independent of $R$. This means that the form of a half-ring is preserved during this phase of the expansion. Otherwise one should start the calculation

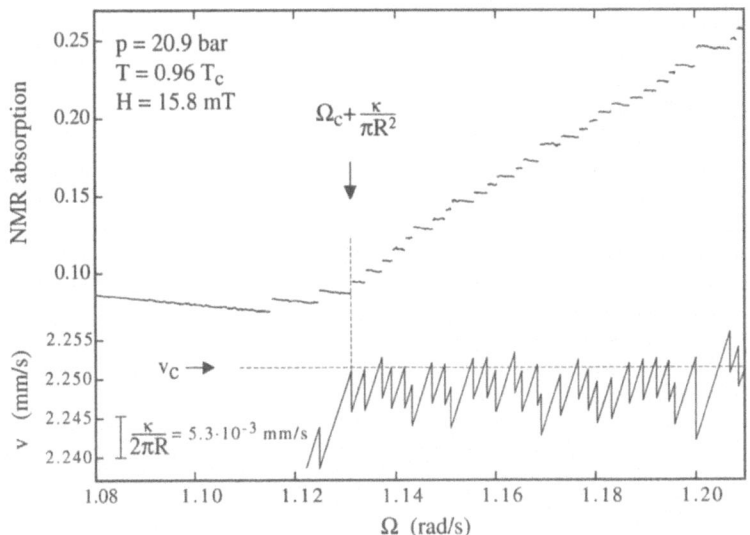

Figure 10: Nucleation of vortices in a cylindrical container filled with $^3$He-B during rotational acceleration with $d\Omega/dt = 2.4 \cdot 10^{-4}$ rad/s$^2$ [28]. In the top part the NMR absorption signal is shown as a function of $\Omega$. It monitors the number of vortices in the container during the acceleration: The steps mark the arrival of one more vortex to the central vortex cluster. In the bottom part the corresponding counterflow velocity $v$ at the cylindrical side wall is shown. It is reduced by the equivalent of one circulation quantum ($\Delta v = \kappa/(2\pi R_c) = 5.3 \cdot 10^{-3}$ mm/s) at each nucleation event. After a nucleation event, $v$ again continues to increase linearly, due to the constant acceleration of $\Omega$, until the next nucleation event comes along and knocks it down by the equivalent of a circulation quantum. Thus $v(\Omega)$ takes the form of a saw tooth pattern and the average of its maxima (except for the first two ones which fall outside the regular behavior) defines the critical velocity $v_c(T, p)$ for an intrinsic nucleation process in this particular sample container.

from Eq. (31). According to Eq. (33), the components of the velocity $\vec{v}_L$ parallel and normal to the central axis of the ring are:

$$v_{L\|} = \left(1 - B'\frac{\rho_n}{2\rho}\right)\left(v - \frac{v_s}{R}\right) , \tag{63}$$

$$v_{L\perp} = \frac{dR}{dt} = B\frac{\rho_n}{2\rho}\left(v - \frac{v_s}{R}\right) . \tag{64}$$

Eq. (63) describes the drift of the half-ring with the azimuthal counterflow while Eq. (64) represents the expansion of the ring in the radial direction of the container. The integration of the latter yields an equation for the radius $R(t)$ of the ring as a function of time during the expansion:

$$R(t) - R_0 - \frac{v_s}{v}\ln\frac{R(t) - v_s/v}{R_0 - v_s/v} = B\frac{\rho_n}{2\rho}vt , \tag{65}$$

where $R_0$ is the radius of the ring at $t = 0$. Thus in its initial phase of expansion the ring drifts azimuthally in the container and simultaneously expands radially towards the center.

Figure 11: Measurement of the fast mutual-friction-resisted motion of vortices in rotating $^3$He-B [21]. On the top the drive $\Omega(t)$ and in the middle its response $P_v(t)$ are plotted during a linear deceleration of $\Omega$ from 2 rad/s in an equilibrium vortex state to a total stop and a final state with no vortices. The rapidly relaxing NMR absorption signal $P_v(t)$ is proportional to the number of vortices $N(t)$ and thus monitors the time dependence in their migration to the side wall, where they annihilate. On the bottom the response is shown once more on a 5 times amplified scale, to illustrate the final phase of the decay of vorticity, after the container already has been stopped. Here the solid curve through the data points represents a fit to the time dependence given by Eq. (99) with the fitted value of $\tau_F = (\Omega_o B \rho_n / \rho)^{-1} = 3.5$ s.

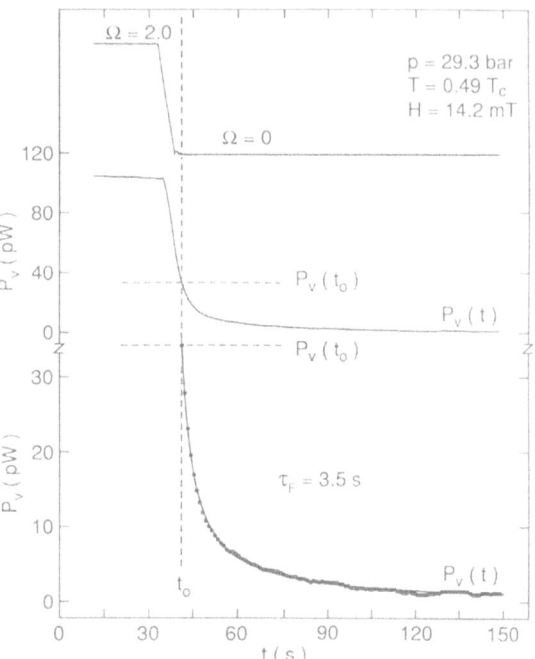

Assuming a perfectly vortex-free counterflow state in the rotating container to start with, then the counterflow velocity near the wall is $v = \Omega R_c$. The logarithmic contribution in Eq. (65) is not important, since $v_s/v$ is small (of order $10^{-3}$ cm). In effect this means that the self induced component in the velocity field in Eq. (62) is neglected. In this case the time for the vortex ring to expand to a size comparable with $R_c$ is

$$t_{exp} = \left(B\frac{\rho_n}{2\rho}\right)^{-1}\frac{R_c}{v} = \left(B\Omega\frac{\rho_n}{2\rho}\right)^{-1}. \tag{66}$$

This expression illustrates the fastest time scale of vortex motion, typical of dynamic processes far from the equilibrium state, which arise from Magnus forces due to a grossly nonequilibrium distribution of the vortex density. In the experiment of Fig. 10 this time is on the order of 1 sec, but it is not distinguishable in the figure. A similar response time governs the motion of vortices in the rotating container, if the rotation velocity is suddenly changed by some amount and the vortex density has to relax towards a new equilibrium value [21]. Experimentally this is most conveniently demonstrated by the decay of vorticity in Fig. 11, where the rotation velocity has suddenly been stopped and all vortices of an initial equilibrium vortex state move to the side wall and annihilate.

# 5 Collective effects in vortex array

## 5.1 Continuum model of vortex array

Regular vortex arrays have been studied in superconductors in the mixed state and in superfluids in rotation. In superconductors the most common technique for imaging the arrays is based on magnetic decoration, where small magnetic particles are allowed to collect on the surface of the superconductor at the sites of high flux concentration, ie. at the end points of the vortex cores. Extended lattices with triangular order have been recorded. In $^4$He-II superfluid vortex cores can be decorated with trapped electrons, which then can be discharged on a screen above the liquid for visualization [29].

It is not tractable to describe analytically the time dependent trajectories of individual vortices in an array formed from discrete vortex lines, as was done above for one solitary vortex line. Instead a continuum approach analogous to the elasticity theory of crystalline lattices can be successfully used. However, the long-range interaction between vortices, which decreases with distance as $1/r$, has pronounced effects on the structure and dynamics of the vortex array. One cannot derive a local description in terms of differential equations for a continuous medium, as is the case in the theory of elasticity, if one uses only the density and the deformations as variables. The effect of other distant vortices on a given vortex line must be incorporated. This was done by Brandt [30] who suggested a non-local theory of the vortex array in which the elastic modulii depended on the wave number.

Another way to incorporate the long-range interaction with distant vortices is to account for it in terms of an average superfluid velocity $\vec{v}_s$, which is induced by all other vortices. Thus the velocity $\vec{v}_s$ (or the magnetic field associated with the supercurrent $\vec{j}_s = e n_s \vec{v}_s$ via Maxwell's equation) has to be included among the variables of the continuum theory. A similar concept is applied when dealing with electrical charges in a crystalline lattice: In order to incorporate the long-range Coulomb interaction between distant charges one introduces a local average electric field, which is derived as a solution to the Poisson equation.

These features have a number of important consequences on the structure and dynamics of the vortex array, as has been concluded from studies on vorticity in $^4$He-II [18]: An additional length scale on the order of the intervortex distance $r_v$ emerges, which describes the penetration of distortions into the vortex array, caused by perturbations acting on the transverse surfaces, as introduced eg. by surface pinning. In a dense vortex array in the absence of strong bulk pinning vortex lines are straight in the bulk and move in a columnar fashion. Bending is restricted to narrow layers adjacent to the transverse boundaries. These are the superfluid Ekman layers, which have a width $\ell_E$ comparable to the intervortex distance $r_v$. The length scale $\ell_E$ plays a role similar to the Debye screening radius in a plasma with long-range Coulomb interactions. The same scale emerges in rotating superfluid $^3$He [21] and in superconductors [31,32]. This screening is the subject of the next section.

A further characteristic of the vortex array is the property that the elastic tilt modulus, which arises from the line tension $\varepsilon$ and may be defined as

$$C_{44}^* = \varepsilon n_v = \frac{\rho_s \kappa^2 n_v}{4\pi} \ln \frac{r_v}{r_c} , \qquad (67)$$

far exceeds the transverse shear modulus $C_{66} = \rho_s c_T^2 = \rho_s \kappa^2 n_v/(8\pi)$. Here $n_v = 2\Omega/\kappa$ is the density of vortices and $c_T = \sqrt{\kappa\Omega/(8\pi)}$ is the velocity of transverse shear waves in the vortex lattice. We have referred to the tilt modulus $C_{44}^*$, which is different from the Labusch modulus $C_{44}$ of the superconductivity theory: $C_{44}^*$ incorporates the line-tension effect, but not the energy of a uniform magnetic field (see Eq. (82) and the discussion following it). The shear modulus $C_{66}$ is associated with small transverse relative shifts of the vortex lines with respect to each other, without change in the average density $n_v$ of the lines or their longitudinal alignment. We use here Voigt's notation of elastic modulii which is traditional in the theory of superconductivity [33]. The shear modulus does not contain the large logarithm as opposed to the tilt modulus. Thus in many applications of the phenomenological theory one can neglect the shear modulus. Such an approach has been widely used in the hydrodynamics of rotating superfluids [7,9,10,18], as well as in the

218

Figure 12: Vortex array in a rotating cylindrical container [36]. Here the state is metastable and the array does not fill the entire container, but forms a cluster, or a bundle of vortices in the center of the container. The graph below the container shows the radial distributions of the normal and superfluid velocities, $\vec{v}_n$ and $\vec{v}_s$, in the laboratory frame. Inside the vortex cluster, ie., at $r < R_0$, the average superfluid velocity $< \vec{v}_s >$ equals the solid body rotation of the normal component with $\vec{v}_n = \vec{\Omega} \times \vec{r}$. Outside the cluster, at $R_0 < r < R_c$, the azimuthal vortex-free counterflow is given by $|\vec{v}| = |\vec{v}_s - \vec{v}_n| = \Omega r - \kappa N/(2\pi r)$, where $N = \pi n_v R_0^2$ is the number of vortices in the cluster.

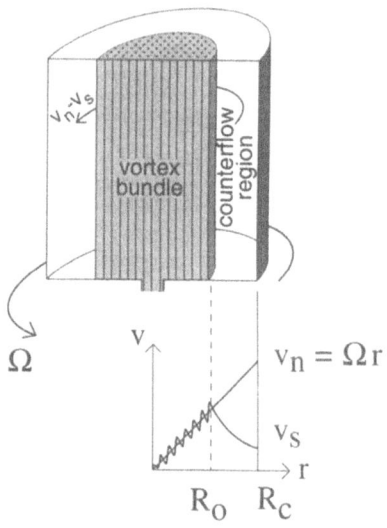

theory of type-II superconductors [34,31,25]. Though there are problems for which the shear rigidity of the vortex lattice is of principal importance, eg. in collective pinning or the phase transition from a vortex crystal to a vortex liquid, these are not discussed in the present article, and the shear rigidity will be neglected. Here we are not going to present a thorough discussion of the continuum theory of vortex arrays and its validity [18,35], but instead we shall apply it to a few simple cases. Also instead of the tilt modulus $C_{44}^*$ we shall make use of a simpler notation in terms of the elastic line-tension parameter

$$\nu_s = \frac{\varepsilon}{\rho_s \kappa} = \frac{\kappa}{4\pi} \ln \frac{r_v}{r_c} \; . \tag{68}$$

## 5.2   Screening of longitudinal distortions in vortex array

Let us generalize Eq. (2) for many vortex lines:

$$\vec{\nabla} \times \vec{v}_s = \vec{\kappa} \sum_j \delta(\vec{r} - \vec{r}_j) \; , \tag{69}$$

where $\vec{r}_j$ is the 2 dimensional position vector of the $j$th vortex line. In the continuum description this relation is averaged over a length scale, which exceeds the intervortex distance, and thus

$$\vec{\nabla} \times \vec{v}_s = \vec{\kappa} n_v \; , \tag{70}$$

where $n_v$ is the vortex density. In a state of steady rotation with constant angular velocity $\Omega$ the requirement of solid body rotation leads to a vortex density $n_v = 2\Omega/\kappa$. The vortex array fills the entire container except for the annular vortex-free counterflow layer near the wall with a width slightly more than one intervortex distance $r_v$ [7]. But metastable states with the counterflow region of any size can be prepared and maintained in rotating experiments with $^3$He-B, since its critical rotation velocity for nucleating vortices is often relatively high

(Fig. 12). For small deviations from the solid body rotation, when the vortex lines are only slightly tilted, we have in the rotating frame:

$$\vec{\nabla} \times \vec{v}_s = \delta\vec{\kappa}\, n_v \,, \tag{71}$$

where $\delta\vec{\kappa} = \kappa d\vec{u}(z)/dz$ is the deviation of the orientation of $\vec{\kappa}$ from the $z$ axis and $\vec{u}(z)$ is the displacement of the vortex lines as a function of $z$ from their equilibrium positions. If the perturbation depends only on $z$ and one is allowed to neglect a small $z$-component in $\vec{v}_s$ of second order in nature, then Eq. (71) may be rewritten as

$$\frac{\partial \vec{v}_s}{\partial z} = -2\vec{\Omega} \times \frac{\partial \vec{u}(z)}{\partial z} \,. \tag{72}$$

This relation shows that for any $z$ dependent mode the average superfluid velocity, which incorporates the long-range interactions within the vortex array, is given by

$$\vec{v}_s(z) = -2\vec{\Omega} \times \vec{u}(z) \,. \tag{73}$$

The same result may be derived also from the Euler equation

$$\frac{\partial \vec{v}_s}{\partial t} + 2\vec{\Omega} \times \vec{v}_L = -\vec{\nabla}\mu \,, \tag{74}$$

which describes superfluid motion in the rotating frame. Here $\mu$ is the chemical potential, which does not contribute if the variation is only along the rotation axis (the axis $z$). To obtain Eq. (73) one should integrate Eq. (74) with respect to time, instead of $z$ in Eq. (72).

The average superfluid velocity in the vortex array produces a Magnus force $-\rho_s\vec{\kappa} \times \vec{v}_s$ which acts on a vortex and should be added to the right-hand side of Eq. (55):

$$-\rho_s\vec{\kappa} \times \frac{\partial \vec{u}(z)}{\partial t} = -\rho_s\vec{\kappa} \times \vec{v}_s + \varepsilon\frac{\partial^2 \vec{u}(z)}{\partial z^2} \,. \tag{75}$$

Eqs. (73) and (75) provide a complete description of the dynamics of the vortex array when mutual friction and thermal effects are neglected. By excluding $\vec{v}_s$ from the equations, one obtains an equation for the vortex displacement $\vec{u}(z)$:

$$-\rho_s\vec{\kappa} \times \frac{\partial \vec{u}(z)}{\partial t} = -2\rho_s\kappa\Omega\vec{u}(z) + \varepsilon\frac{\partial^2 \vec{u}(z)}{\partial z^2} \,. \tag{76}$$

A plane-wave solution of Eq. (76) corresponds to a wave propagating along the rotation axis with the spectrum

$$\omega = \pm(2\Omega + \nu_s p^2) = \pm(\kappa n_v + \nu_s p^2) \,. \tag{77}$$

Thus a gap $2\Omega = \kappa n_v$ appears in the spectrum of Kelvin waves in a vortex array. The gap results from the long-range interaction between vortices in an array of finite density, similar to a finite plasma frequency in an electron liquid due to the long-range Coulomb interaction. The Kelvin wave does not propagate, if $\omega < 2\Omega$, as is seen from its spectrum in Eq. (77). In the limit of very slow motion $\omega \ll \Omega$ the imaginary wave-number of the Kelvin mode is given by

$$p^2 = -\frac{2\Omega}{\nu_s} \,. \tag{78}$$

Figure 13: Fundamental mode of a dense vortex array where the screening of deformations becomes important. The transverse solid surface on the bottom depicts one of two plates in a pile of parallel disks, which is driven to execute torsional oscillations around the vertical axis inside a rotating container filled with a superfluid. The vortices in the array between two disks are oscillating along elliptical trajectories, which are highly eccentric, since the slow motion is primarily azimuthal (see Sec. 6). The displacement $\vec{u}(z,t)$ is constant over most of the length of a vortex line as the bending from the pinning forces $\vec{f}_p$, which act at the transverse surfaces, is confined to the Ekman layers with a width $\ell_E$ given by Eq. (79) (cf. motion of a solitary vortex line in Fig. 9). The motions in the slow vortex mode discussed in Sec. 6 are similar except for the damping with the time dependence $\sim \exp\left(-t/\tau_s\right)$, which transforms the elliptical trajectories to elliptical spirals with a monotonously decreasing amplitude.

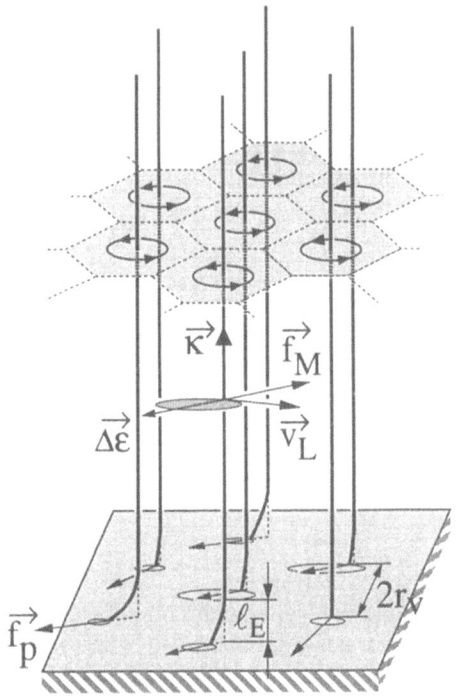

This means that the low-frequency Kelvin mode does not penetrate into the bulk fluid, but is restricted to a layer with the width [18,21]

$$\ell_E = \sqrt{\frac{\nu_s}{2\Omega}} = \sqrt{\frac{\ln(r_v/r_c)}{4\pi n_v}} \, , \qquad (79)$$

which is on the order of the intervortex distance $r_v \sim \sqrt{1/n_v}$. The boundary layer is called the superfluid Ekman layer, in analogy with the classical Ekman layer into which viscous motion penetrates in a rotating fluid. The width of the superfluid Ekman layer involves the quantum parameter $\nu_s = (\kappa/4\pi) \ln\left(r_v/r_c\right)$, which in the classical counterpart is replaced by the kinematic viscosity $\nu = \eta/\rho$ [37,1].

The Ekman layer is a manifestation of the long-range interaction, similar to the Debye screening in a plasma due to the long-range Coulomb interaction: Deviations from quasi-neutrality in a plasma are restricted to the Debye length. Deformations of the vortex array, which are driven by surface forces acting on the ends of the vortex lines, penetrate only into the Ekman layers at the top and bottom surfaces of the container. Like the Debye screening of the charge distribution in a plasma, the Ekman screening in the vortex array strongly influences the dynamics. In Figs. 9 and 13 the oscillation patterns of a single vortex line and of an array are compared, as induced by slow oscillations of the transverse confining borders. The classic measurement for this is conducted in the rotating superfluid with a pile of parallel disks. The pile is supported by a fine torsion fiber and is capacitively driven to perform torsional oscillations at some frequency $\omega$ while the density of vortices is regulated with the rotation velocity $\Omega$. In the case of a single vortex or a very low density of vortices the displacement from the equilibrium configuration is distributed over the whole liquid layer between two adjacent plates in the pile of disks (Fig. 9). In a dense array distortions from rectilinear are confined to the Ekman layers, which coat the transverse surfaces of the disks (Fig. 13). In the following sections we shall discuss some examples where the screening effect from the long-range vortex interaction becomes observable in the properties of the vortex array in superconductors.

## 5.3 Surface impedance of type-II superconductor

Let us consider a type-II superconductor in a dc external magnetic field $\vec{H}$, which is applied along the $z$-axis normal to the surface of the superconductor and coincides in vacuum with the magnetic induction $\vec{B}$. An electromagnetic wave at normal incidence to the surface generates in the superconductor small ac currents $\vec{j}_s$ and fields $\vec{h}$ which give rise to transverse oscillations of the vortex array, similar to those excited in a rotating superfluid by an oscillating disk normal to the rotation axis (cf. Fig. 13).

Here we neglect the current $\vec{j}_n$ from normal quasiparticle excitations as much smaller than the supercurrent $\vec{j}_s$, such that $\vec{j} \approx \vec{j}_s$. In the same manner as we concluded Eq. (72) from Eq. (2), we may average the analogous relation Eq. (5) for superconductors to obtain

$$\vec{h} + \frac{4\pi\lambda^2}{c}\left[\hat{z} \times \frac{\partial \vec{j}_s}{\partial z}\right] = B\frac{\partial \vec{u}}{\partial z} . \tag{80}$$

Besides the notation, the only difference here is the magnetic field, which does not appear in Eq. (72) for the neutral superfluid. The force-balance equation is now written in the form

$$\eta\frac{\partial \vec{u}}{\partial t} = \frac{\Phi_0}{c}[\vec{j}_s \times \hat{z}] + \frac{\Phi_0}{B}C_{44}^*\frac{\partial^2 \vec{u}}{\partial z^2} . \tag{81}$$

Here

$$C_{44}^* = C_{44} - \frac{B^2}{4\pi} = \frac{B(H-B)}{4\pi} = B\,|\,M_0\,| \tag{82}$$

is a renormalized tilt-modulus which is different from the Labusch tilt-modulus $C_{44} = BH/4\pi$ [38], $\vec{M}_0$ is the equilibrium diamagnetic moment of the superconductor in the mixed state and $\vec{H} = \vec{B} - 4\pi\vec{M}_0$ is the thermodynamic magnetic field. Using the known dependence of $B = H - H^*$ on $H$, the explicit expression for the renormalized tilt-modulus becomes:

$$C_{44}^* = \frac{BH^*}{4\pi} . \tag{83}$$

The field $H^* = \left(\Phi_0/(4\pi\lambda^2)\right)\ln\left(r_u/r_c\right)$ coincides with the lower critical field $H_{c1}$, when $B$ is close to $H_{c1}$ and the upper cut-off in the logarithm $r_u = \lambda$. But $r_u = r_v$ at $B \geq H_{c1}$. The tilt-modulus $C_{44}^*$ is directly connected with the line-tension $\varepsilon$ by the relation $C_{44}^* = \varepsilon\sqrt{\Phi_0/B}$. The terms on the right hand side of Eq. (81) are in a one-to-one correspondence to those on the right hand side of Eq. (75). On the left-hand side of Eq. (81) one can see the friction force $\propto \eta$, which is assumed to be in superconductors more important than the Magnus force on the left-hand side of Eq. (75).

The eigenmodes of the set of Eqs. (80) and (81) combined with the Maxwell equation (6) can be found in the plane wave form $u \propto h \propto j_s \propto \exp\left(ipz - i\omega t\right)$, where the wave number $p$ and the frequency $\omega$ fulfill the dispersion equation

$$\frac{i\omega}{\omega_B} = p^2\lambda^2\left[\frac{1}{1+p^2\lambda^2} + \frac{H^*}{B}\right] , \tag{84}$$

with

$$\omega_B = \frac{\Phi_0 B}{4\pi\lambda^2\eta} \ . \tag{85}$$

In the low-frequency limit the wave numbers of the two eigenmodes are:

$$p_1^2 = \frac{4\pi\sigma_f i\omega\mu}{c^2} \ , \qquad p_2^2 = -\frac{1}{\lambda^2}\frac{B+H^*}{H^*} \ . \tag{86}$$

Here the transverse magnetic permeability in the mixed state,

$$\mu = \frac{B}{H} = \frac{B^2}{4\pi C_{44}} = \frac{B}{B+H^*} \tag{87}$$

has been introduced. It incorporates the diamagnetic shielding from circular supercurrents within the vortex array cell. In addition

$$\sigma_f = \frac{c^2\eta}{B\phi_0} \tag{88}$$

denotes the flux-flow conductance. The first wave number $p_1$ is associated with the skin-depth of the conducting medium while the second wave number $p_2$ is a screening length. It is equal to the London penetration depth $\lambda$ at low magnetic induction and low vortex density, when $B \ll H^*$. In the opposite limit of high magnetic field $B \gg H^*$, when the vortex array is dense and behaves similar to that of a neutral superfluid, $p_2$ coincides with the inverse superfluid Ekman layer width $1/\ell_E$ (Eq. (79)). The existence of this second length scale, the screening length for distortions from external perturbations, is crucial for deriving the expression for the surface impedance [39]

$$Z = \sqrt{\frac{-i\omega\mu}{4\pi\sigma_f}} \ , \tag{89}$$

which is well-known in the electrodynamics of continuous magnetic media ($\mu \neq 1$). At the surface of the superconductor the tangential component of the thermodynamic magnetic field $\vec{H} = \vec{B}/\mu$ is continuous, whereas that of the magnetic induction $\vec{B}$ is not. This means that the discontinuity in the tangential component of the magnetic induction occurs on a length scale, which is related to the circular molecular currents responsible for the magnetism of the medium. In a molecular magnetism this length is on the atomic scale. In contrast, in the case of a dense vortex array the length scale is not microscopical, but on the order of the intervortex distance within which the discontinuity in the tangential magnetic field takes place. The distribution of the tangential magnetic field is shown in Fig. 14 in the limit $\omega \to 0$, when the skin-depth becomes infinite. The discontinuity is only accounted for by the second mode of short wave length with the wave number $p_2$. If the mode is neglected and the magnetic field discontinuity is not included then the surface impedance in the low frequency limit becomes [38]

$$Z = \sqrt{\frac{-i\omega}{4\pi\mu\sigma_f}} \ . \tag{90}$$

which differs from Eq. (89) by the factor $\mu$. Recent measurements on a superconductor with very weak bulk pinning [40] have confirmed the former expression Eq. (89) and thus the importance of the screening effect from the vortex array.

Figure 14: Vortices at the surface of a superconductor in a measurement of the surface impedance. On the top the vortex lines are shown at the moment, when the ac component $\vec{h}_e$ of the external magnetic field is not zero and the total field $\vec{H} + \vec{h}_e$ is tilted with respect to the surface. The picture corresponds to the limit when the applied frequency $\omega \to 0$ and the skin-depth becomes infinite. But another screening length appears which accounts for the recovery from the boundary condition: The vortices are perpendicular at the boundary, but curve parallel to the field well inside the superconductor. On the bottom the magnetic field distribution is shown in the vicinity of the surface of the superconductor.

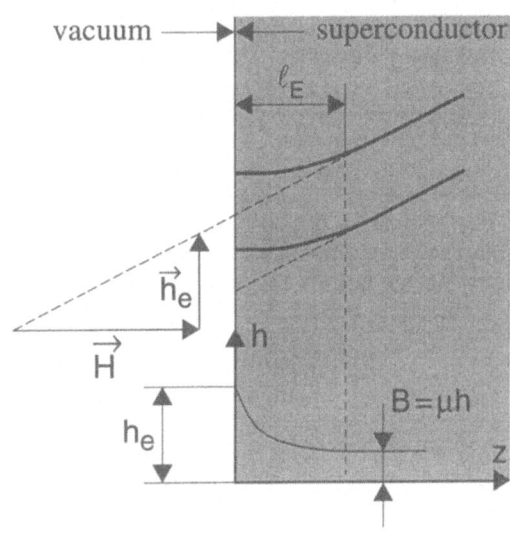

We should mention here that a lot of work has been devoted to the surface impedance of type-II superconductors in the mixed state (see eg. some recent Refs. [41]). However, mostly the surface impedance for very dense vortex arrays has been considered at high magnetic fields, when $\mu \approx 1$, and one may neglect the diamagnetism of the mixed state. Then there is no difference between expressions (89) and (90). Thus all the information on the elastic properties of the flux-line lattice and its diamagnetic screening is only present in the susceptibility $\mu$ of the mixed state.

## 5.4 Collective effects in pinning at twin boundaries

The long-range interaction between vortices and the resulting screening of deformations from surface perturbations has observable consequences also on the pinning of vortex lines at parallel twin boundary planes in a single crystalline YBaCuO superconductor, when a dense array of vortices is involved. In Sec. 3.1 we concluded that if the magnetic field is tilted with respect to the twin-boundary by an angle $\theta_B$, a vortex line may become partially trapped on the boundary. From the line-tension balance it was noted that the vortex line then forms an angle $\theta = \arccos(\varepsilon'/\varepsilon)$ with respect to the plane (Eq. (36)). This angle is small, about 10° according to measurements [23], and provides the upper limit for the inclination of the magnetic field when trapping will still occur, ie. for trapping the requirement $\theta_B < \theta$ has to be fulfilled. In the case of a vortex array the contour of the lines will differ from that of a single line due to the presence of a new length scale, the Ekman layer width $\ell_E$, which at high magnetic fields would be much smaller than the separation between parallel twin-boundary planes. The two cases are illustrated in Figs. 5 and 15.

An analysis in the framework of the continuum theory has been recently employed to analyze the vortex pattern in twinned single crystals of YBaCuO [25]. Above we have calculated the width $\ell_E$ of the screening layer as measured along the vortex line. In the case of trapping on the twin-boundary the width of this layer in the direction normal to the plane is shorter and equal to $\ell^* = \ell_E \sin\theta$. Within this distance $\ell^*$ from the plane the vortex line direction is dictated by the twin boundary. For a single vortex this distance is infinite and the vortex line is straight between two twin-boundaries (Fig. 5). In contrast, in a vortex array,

Figure 15: A vortex array in a YBaCuO single crystal with parallel twin-boundary planes. The screening from the Ekman layers in a dense vortex array forces the vortex lines (thick lines) to align along the external field direction (dashed straight lines) over most of the distance between the planes (vertical straight lines). Only within the Ekman layers with a width $\ell^*$ the tilt angle of the lines is affected by the boundary condition Eq. (36) at the twin plane. In contrast, a single inclined vortex line is constrained to form an angle $\theta = \arccos(\varepsilon'/\varepsilon)$ with respect to the planes, which is largely independent of the magnetic field and its tilt angle $\theta_B$ in Fig. 5.

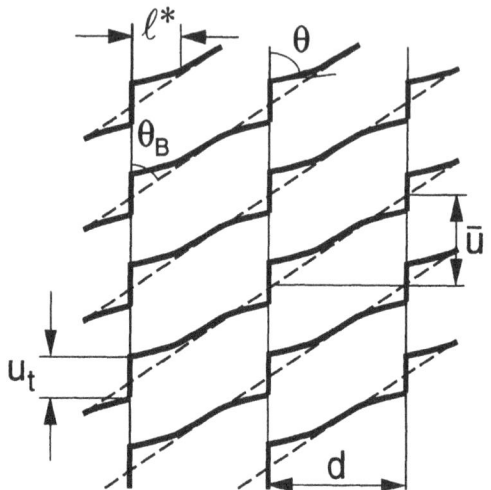

when the Ekman layer width $\ell^*$ is less than the separation $d$ between two twin-boundaries, the tilt angle of the vortex lines outside the boundary layers is governed by the external magnetic field and is equal to $\theta_B$ (Fig. 15). The existence of these boundary layers is important for the interpretation of the measurements: The effects from twin-boundary trapping are confined to within the screening layer width $\ell^*$, which decreases when the external magnetic field is increased, while the value of the critical angle $\theta$ only weakly depends on the magnetic field [25].

# 6 Collective dynamic modes of vortex array

As the elastic Kelvin mode does not penetrate into a vortex array beyond the Ekman surface layer (Fig. 13), this property gives vortex array motion a rather special character: Within the Ekman screening layer the motion may be of 3 dimensional nature but well within the bulk it is strictly 2 dimensional and confined into the transverse plane. This is called *columnar motion*, in which the rectilinear vortex lines move like solid straight sticks. Columnar motion is well-known from rotating classical liquids [37]. In superconductors the situation is more complicated because of the presence of other length scales, the skin-depth or that arising from bulk pinning, whichever is shorter. Thus columnar motion of vortices would be possible in superconducting layers which are thinner than these other length scales.

In the following we shall briefly characterize the principal modes of columnar motion in a rotating superfluid and point out the role of collective effects [21]. In the boundary layer the motion may have a $z$ dependent component due to surface pinning, but it decays away exponentially with the Ekman screening length $\ell_E$ and we may simply focus on the remaining motions in the transverse xy plane well within the array. Such motion is excited by a rotation drive, which includes a time dependent component $\Omega(t)$. Suppose $\Omega$ is abruptly changed by a small amount in a step-like manner. If the array is an isolated cluster, as in Fig. 12, and the step change in $\Omega$ is sufficiently small, then no existing vortices are annihilated or new ones created. In the response two distinct modes of motion can be distinguished: an initial fast mode and a final slow mode.

The analysis of the dynamic response of a vortex array is based on Eq. (33), which accounts for the motion of a vortex line in the presence of damping from mutual friction.

The local superfluid velocity $\vec{v}_{sl}$ near the vortex line differs from the average superfluid velocity $\vec{v}_s$ due to the elastic deformation of the vortex array:

$$\vec{v}_{sl} = \vec{v}_s + \nu_s \left[ \hat{z} \times \frac{d^2 \vec{u}}{dz^2} \right] . \tag{91}$$

We average Eq. (91) over the $z$-axis by assuming that the bending of the vortex lines is small:

$$\vec{v}_{sl} = \vec{v}_s + \frac{\nu_s}{L} \left[ \hat{z} \times \left( \frac{d\vec{u}}{dz} \bigg|_{z=L/2} - \frac{d\vec{u}}{dz} \bigg|_{z=-L/2} \right) \right] . \tag{92}$$

Here it is assumed that the liquid is confined between two transverse planes located at $z = \pm L/2$. The boundary condition for linearized pinning expressed in Eq. (34) is used to define the slope of a vortex line at its end points. If pinning is weak and these slopes are small (see Fig. 13), then the displacements at the transverse boundaries in Eq. (34) do not greatly differ from their value within the bulk fluid. Thus the difference between the average and local superfluid velocities is

$$\vec{v}_{sl} - \vec{v}_s = -\frac{2\nu_s b}{L} \left[ \hat{z} \times \vec{u} \right] . \tag{93}$$

Now all the $z$-variation in the fluid parameters has been removed and the problem has become two-dimensional. Next we substitute Eq. (93) into the expression for the vortex velocity $\vec{v}_L = d\vec{u}/dt$ in Eq. (33) and solve it together with the Euler equation (74). The solution depends on what type of motion we are interested in. If we neglect the chemical potential $\mu$ in the Euler equation we obtain a circularly polarized Kelvin mode, which propagates along the $z$-axis without any preferable direction in the $xy$-plane. However, such motion can exist only in the Ekman layer, as was pointed out above. Another type of solution is azimuthal motion, described by the azimuthal component $v_{s\phi}$ of the superfluid velocity $\vec{v}_s$. The radial component vanishes, $v_{sr} = 0$, due to the incompressibility condition $\vec{\nabla} \cdot \vec{v}_s = 0$. Then only one component of the Euler equation Eq. (74) is needed which now connects $v_{s\phi}$ with the radial component $v_{Lr}$ of the vortex line velocity $\vec{v}_L$: $i\omega v_{s\phi} = 2\Omega v_{Lr}$. The other component of the Euler equation is not used in the analysis: it connects the azimuthal component $v_{L\phi}$ with the gradient of the chemical potential: in the present case $\vec{\nabla}\mu$ is not constant, in contrast to the Kelvin mode.

If we exclude the average and the local superfluid velocities $v_s$ and $v_{sl}$ from the system of equations we are left with two equations written in terms of the components of the vortex line velocity:

$$v_{Lr}\left(i\omega - \frac{\rho_n}{\rho} B\Omega\right) + v_{L\phi}\omega_\Sigma \left(1 - \frac{\rho_n}{2\rho} B'\right) = 0 , \tag{94}$$

$$-v_{Lr} 2\Omega \left(1 - \frac{\rho_n}{2\rho} B'\right) + v_{L\phi}\left(i\omega - \frac{\rho_n}{2\rho} B\omega_\Sigma\right) = 0 . \tag{95}$$

Here the frequency

$$\omega_\Sigma = \frac{2\nu_s}{L} b \tag{96}$$

depends on the line-tension parameter $\nu_s$ of the vortex line and its elastic pinning parameter $b$. On equating the determinant of this set of equations to zero we obtain the dispersion equation

$$\omega^2 + i\omega \frac{\rho_n}{2\rho} B(2\Omega + \omega_\Sigma) - 2\Omega\omega_\Sigma \left[ \left(1 - \frac{\rho_n}{2\rho} B'\right)^2 + \left(\frac{\rho_n}{2\rho} B\right)^2 \right] = 0 . \tag{97}$$

Usually $\omega_\Sigma \ll \Omega$ and if we neglect it in the dispersion equation, we obtain the overdamped fast mode with an imaginary frequency

$$i\omega = \frac{1}{\tau_F} = \frac{\rho_n}{\rho} B\Omega \ . \tag{98}$$

This motion thus corresponds to pure exponential relaxation, with a time dependence $\propto \exp(-t/t_F)$. Here the damping originates from mutual friction, since pinning can be omitted when the ends of the straight vortex lines are sliding over the transverse surfaces. The trajectories of the vortices in the $xy$ plane are dominated by a large radial component, but are not purely radial because of a smaller azimuthal component from the reactive part $1 - B'\rho_n/(2\rho)$ of mutual friction, as can be derived from Eq. (95). This is the *fast mutual-friction-resisted mode* which is responsible for restoring an average value for the vortex line density, which is close to the appropriate equilibrium at the current rotation velocity. The characteristic time $\tau_F$ of the fast mode is typically measured in a rapid spin-down experiment, as shown in Fig. 11. Here no measurable energy barrier has been detected against the annihilation of vortices at the side wall, in contrast to the nucleation barrier which is involved in a spin-up measurement. The time dependence of the monotonous relaxation in Fig. 11 is dominated by the time it takes for the vortices to move to the wall and for the vortex density to decay to zero, but since the angular velocity variation drops down to zero during the experiment, the time dependence does not follow a simple exponential law, but should be derived from the nonlinear Euler equation [21] and is given by

$$n_v(t) = \frac{n_v(t_0)}{1 + (t - t_0)/\tau_F} \ , \tag{99}$$

where $n_v(t_0)$ is the density at some reference point in time $t = t_0$ after the rotation has been brought to a complete stop, ie. $\Omega = 0$, and $\tau_F = [\kappa n_v(t_0) B \rho_n/(2\rho)]^{-1}$.

In order to identify the low frequency branch of the spectrum in Eq. (97) one can use the inequality $\omega_\Sigma \ll 2\Omega$ and neglect the term with $\omega^2$. Then

$$\frac{1}{2\tau_s} = i\omega = \frac{\omega_\Sigma}{\beta} = \frac{2\nu_s b}{\beta L} \ , \quad \beta = \frac{\frac{\rho_n}{2\rho} B}{\left(1 - \frac{\rho_n}{2\rho} B'\right)^2 + \left(\frac{\rho_n}{2\rho} B\right)^2} \ . \tag{100}$$

This is an exponentially relaxing overdamped mode, where the time constant for convenience has been denoted with $2\tau_s$. (The experimental signal in Fig. 16 records directly $\tau_s$ [21]). It is called the *slow vortex mode* which involves predominantly azimuthal motion of vortices at nearly constant vortex density, since according to Eq. (94) the ratio between the radial and azimuthal components of the vortex velocity is small:

$$\frac{v_{Lr}}{v_{L\phi}} = \frac{\omega_\Sigma}{2\Omega} \frac{1 - \frac{\rho_n}{2\rho} B'}{\frac{\rho_n}{2\rho} B} \ . \tag{101}$$

The slow mode is responsible for the final approach to the equilibrium configuration in the vortex array restoring the local triangular equilibrium pattern. In the case of $^3$He-B there are typically two orders of magnitude difference between the two time constants in Eqs. (98) and (100).

The slow mode is governed not only by pinning, but also by the collective effects from the long-range interaction between vortices. This property is not evident from the expression for the exponential relaxation time in Eq. (100): it does not depend on vortex density. This

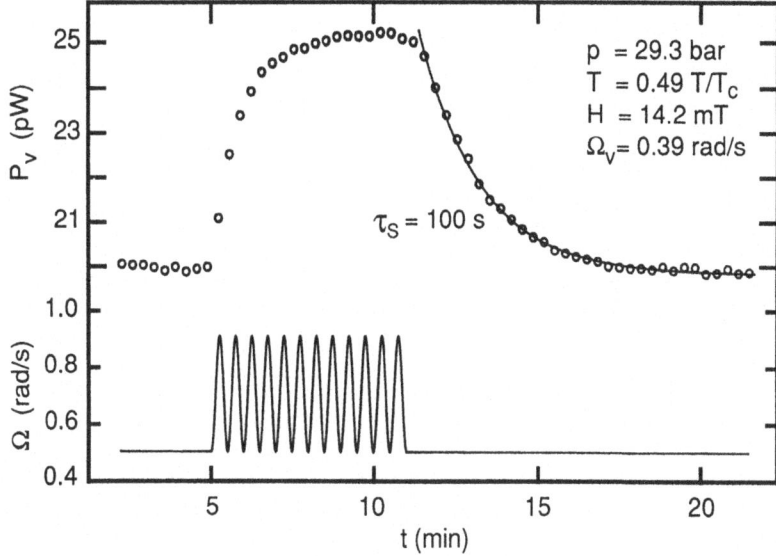

Figure 16: Slow mode response to a sinusoidal rotation drive in rotating $^3$He-B [21]. On the top the behaviour of the NMR absorption signal $P_v(t)$ is recorded when a sinusoidal component is superimposed on the steady rotation at $\Omega = 0.5$ rad/s, as shown in the bottom part. After switching off the sinusoidal drive component the signal relaxes exponentially with a time constant of 100 sec to the equilibrium state at constant $\Omega$. The initial rise time after switching on the harmonic drive represents more complicated forced motion. The measurement is performed on an isolated vortex cluster which is not in contact with container (Fig. 12) and which contains $N = 2\pi\Omega_v R_c^2/\kappa = 450$ vortices.

is a consequence from the fact that we are using the continuum model of the vortex array for the derivation and here, in first order, the relaxation is independent of $n_v$. However, if pinning becomes a collective phenomenon such that the vortex motion is correlated over several intervortex distances even in the Ekman layer, then correlated regions are formed, "Larkin-Ovchinnikov domains", within which the vortices are pinned in unison. The size of the domains depends on the vortex density, thus the pinning parameter $b$ and also the time constant in Eq. (100) become explicitly density dependent. This has been observed to be the case in $^3$He-B [21]. Collective pinning is possible only, if pinning is weak. The condition for weak pinning is different for the slow collective mode and the single-vortex Kelvin mode. Pinning is weak if the vortex remains nearly straight: the difference between the displacements of the ends of the vortex and its center, as measured from the equilibrium position, is much smaller than the displacement itself. For the single-vortex Kelvin mode this gives the condition $Ldu/dz \approx Lbu \ll u$ or $bL \ll 1$, as was mentioned at the very end of Sec. 4.1. However, in the collective slow mode vortex bending is confined to the superfluid Ekman layer and the condition of weak pinning is $bl_E \ll 1$, which is much less stringent than that for the single-vortex mode. The condition of weak pinning for the slow mode also coincides with the condition for pinning to become collective; thus weak pinning in this sense can always be collective. To summarize we remind that although the existence of the Ekman layer is not explicitly present in our result for the exponential response of the slow mode, nevertheless, in the derivation of the mode the limit of weak pining is phrased explicitly in terms of the screening length $\ell_E$.

# 7 Concluding remarks

Our analysis of the hydrodynamics of a solitary single vortex line has been based on a few central concepts which include the elastic line tension, the equation of motion in the localized induction approximation and the boundary condition for elastic vortex line pinning. The basic principles and the underlying physical interpretation, such as the quantized vortex line itself, its nucleation and the processes by which its motion leads to dissipation in superflow, were all worked out in the pioneering decades of the fifties and sixties. In recent experimental work convincing new verification of these ideas has been demonstrated, either due to more advanced technology such as the possibility to fabricate sub-micron-size orifices and to perform measurements with single vortex sensitivity on the superflow through such orifices, or by diversifying through studies on new systems such as superfluid $^3$He or single-crystal high-$T_c$ superconductors.

Here we have dealt with a few examples on solitary vortex lines, where the lines are curved. One may avoid numerical calculations in determining the shape of the line by using the localized induction approximation, which is valid if the logarithm in the line tension is large. Example cases include a vortex half-ring at the boundary of the superfluid system, the configuration which is believed to approximate a vortex immediately after nucleation. A further example is that of superfluid circulation trapped around a thin wire. In superfluid $^3$He-B with weak pinning a vortex has been observed to peel off a vibrating wire with trapped circulation. This filament of free vortex precesses around the wire and simultaneously moves down along the wire, whereby the trapped circulation is unwound and removed from the wire. Such linear traps for quantized circulation can be realized also in superconductors with columnar pins. These are constructed by shooting accelerated heavy ions into the crystal.

A large assembly of vortices in a superfluid can be either a disordered tangle or an ordered array of rectilinear lines. Both states have been under thorough investigation for the last 4 decades. Here we have only dealt with the regular array configuration. Especially in a rotating superfluid in the absence of bulk pinning the vortex lines in the array remain straight and approximately regularly spaced even in a state of motion. This leads to columnar motion, in which distortions due to pinning are restricted to the narrow Ekman layers at the transverse solid boundaries. If pinning is weak at these surfaces, then the motion of vortex lines tends to become coupled even within the Ekman layer and collective effects influence pinning by creating collectively pinned domains, known as Larkin-Ovchinnikov domains. Vortex array formation and columnar motion are thus a result from the vortex line tension and the long range interactions between vortices, which together act to screen distortions due to pinning from the bulk volume.

These features control the dynamic properties of the vortex array, in particular the slow mode. It represents the asymptotically slowest motion with exponential time dependence, which restores the equilibrium state in the vortex array after a disturbance. It has recently been studied in rotating $^3$He-B [21]. In superconductors vortex motion corresponds to an electrodynamic mode, which is restricted to the surface by the skin depth or by some other length characteristic of bulk pinning. However, both of these length scales may be arranged to be very large compared to the width of the Ekman layer. Then the screening effect is observable also in superconductors as diamagnetism in the surface impedance. We have noted that the columnar structure of the flux-line array and the formation of the Ekman layers can be studied continuously as a function of the angle between the external field orientation with respect to twin-boundaries in YBaCuO single crystals. These examples emphasize the central theme of the present volume: The fundamental properties of quantized vortices are common to both neutral and charged superfluids and give rise to very similar behavior, even though finer details may vary.

# References

[1] R.J. Donnelly, *Quantized Vortices in Helium II* (Cambridge University Press, Cambridge, UK, 1991).

[2] M. Tinkham, *Introduction to Superconductivity* (R.E. Krieger Publ. Co., Florida, 1975).

[3] A.G. Lyne and F. Graham-Smith, *Pulsar Astronomy* (Cambridge University Press, Cambridge, UK, 1990); G. Baym, R.I. Epstein, and B. Link, Physica B **178**, 1 (1992).

[4] F. London, *Superfluids*, vols I, II (Wiley, New York, 1954; reprinted Dover, New York, 1964).

[5] D.R. Tilley and J. Tilley, *Superfluidity and Superconductivity* (Adam Hilger, Bristol, 1990).

[6] H. Lamb, *Hydrodynamics* (Dover Publ. Co., New York, 1945).

[7] H.E. Hall, Adv. Phys. **9**, 89 (1960).

[8] W.F. Vinen, in *Progress in Low Temperature Physics*, ed. C.J. Gorter, Vol. **III**, Chap. 1 (North-Holland Publ. Co., Amsterdam, 1961).

[9] E.L. Andronikashvili and Yu.G. Mamaladze, in *Progress in Low Temperature Physics*, ed. C.J. Gorter, Vol. **V**, Chap. 3 (North-Holland Publ. Co., Amsterdam, 1967).

[10] I.L. Bekarevich and I.M. Khalatnikov, Zh. Eksp. Teor. Fiz. **40**, 920 (1961) [Sov. Phys. JETP **13**, 643 (1961)]; I.M. Khalatnikov, *Introduction to the Theory of Superfluidity* (Benjamin Inc., New York, 1965).

[11] O. Avenel, G.G. Ihas, and E. Varoquaux, J. Low Temp. Phys., in press (1993).

[12] D. Vollhardt and P. Wölfle, *The Superfluid Phases of Helium 3* (Taylor & Francis Publ. Co., London, 1990).

[13] M. Krusius, J. Low Temp. Phys. **91**, 233 (1993); P.J. Hakonen, O.V. Lounasmaa, and J.T. Simola, Physica B **160**, 1 (1989); M.M. Salomaa and G.E. Volovik, Rev. Mod. Phys. **59**, 533 (1987).

[14] L. Onsager, Nuovo Cimento Suppl. **6**, 249 (1949).

[15] R.P. Feynman, in *Progress in Low Temperature Physics*, ed. C.J. Gorter, Vol. **I**, Chap. 2 (North-Holland Publ. Co., Amsterdam, 1955).

[16] A.A. Abrikosov, Zh. Eksp. Teor. Fiz. **32**, 1442 (1957) [Sov. Phys. JETP **5**, 1174 (1957)].

[17] H.E. Hall and W.F. Vinen, Proc. Royal Soc. London Ser. A **238**, 204 (1956).

[18] E.B. Sonin, Rev. Mod. Phys. **59**, 87 (1987); Physica B **178**, 106 (1992).

[19] A.L. Fetter, in *The Physics of Liquid and Solid Helium*, eds. K.H. Bennemann and J.B. Ketterson, Pt. I, p. 207 (John Wiley and Sons, New York, 1976).

[20] R.J. Arms and F.R. Hama, Physics of Fluids **8**, 553 (1965).

[21] M. Krusius, Y. Kondo, J.S. Korhonen, and E.B. Sonin, Phys. Rev. B, **47**, 15113 (1993); Europhys. Lett. **22**, 125 (1993).

230

[22] J.C. Davis, J.D. Close, R. Zieve, and R.E. Packard, Phys. Rev. Lett. **66**, 329 (1991); R.J. Zieve, Yu. Mukharsky, J.D. Close, J.C. Davis and R.E. Packard, Phys. Rev. Lett. **68**, 1327 (1992); J. Low Temp. Phys. **90**, 243 (1993); ibid. **91**, 315 (1993).

[23] L.J. Swartzendruber, A. Roitburd, D.L. Kaiser, F.W. Gayle, and L.H. Bennett, Phys. Rev. Lett. **64**, 483 (1990); ibid. **64**, 2962 (1990); W.K. Kwok, U. Welp, G.W. Crabtree, K.G. Vandervoort, R. Hultscher, and J.Z. Liu, Phys. Rev. Lett. **64**, 966 (1990).

[24] G. Blatter, J. Rhyner, and V.M. Vinokur, Phys. Rev. B **43**, 7826 (1990).

[25] E.B. Sonin, Phys. Rev. **48**, 10487 (1993).

[26] E.B. Sonin, to be published.

[27] K.W. Schwarz, Phys. Rev B **47**, 12030 (1993).

[28] Ü. Parts, J.H. Koivuniemi, M. Krusius, V.M. Ruutu, and S.R. Zakazov, Physica B, in press (1993).

[29] G.A. Williams and R.E. Packard, Phys. Rev. Lett. **33**, 280 (1974); E.J. Yarmchuck and R.E. Packard, J. Low Temp. Phys. **46**, 479 (1982).

[30] E.H. Brandt, J. Low Temp. Phys. **26**, 709 (1979).

[31] P. Mathieu and Y. Simon, Europhys. Lett. **5**, 67 (1988); see also T. Hocquet, P. Mathieu, and Y. Simon, Phys. Rev. B **46**, 1061 (1992).

[32] E.B. Sonin, A.K. Tagantsev, and K.B. Traito, Phys. Rev. B **46**, 5830 (1992).

[33] A.M. Campbell and J.E. Evetts, *Critical Currents in Superconductors* (Taylor & Francis Ltd, London, 1972).

[34] A.A. Abrikosov, M.P. Kemoklidze, and I.M. Khalatnikov, Zh. Eksp. Teor. Fiz. **48**, 765 (1965) [Sov. Phys. JETP **21**, 506 (1965)].

[35] G. Baym and E. Chandler, J. Low Temp. Phys. **50**, 57 (1983); ibid. **62**, 119 (1986).

[36] M. Krusius, E.V. Thuneberg, and Ü. Parts, Physica B, in press (1993).

[37] H.P. Greenspan,*The Theory of Rotating Superfluids* (Cambridge University Press, Cambridge, UK, 1968).

[38] L.P. Gorkov and N.B. Kopnin, Usp. Fiz. Nauk **116**, 413 (1975) [Sov. Phys. Usp. **18**, 496 (1975)].

[39] L.D. Landau and E.M. Lifshitz, *Electrodynamics of Continuous Media* (Addison-Wesley, Reading, MA, 1969).

[40] V.A. Berezin, E.V. Il'ichev, V.A. Tulin, E.B. Sonin, A.K. Tagantsev, and K.B. Traito, Phys. Rev. B, in press (1993).

[41] A.M. Portis, K.W. Blazey, K.A. Muller, and J.G. Bednorz, Europhys. Lett. **5**, 467 (1988); M.W. Coffey and J.R. Clem, Phys. Rev. Lett. **67**, 386 (1991); IEEE Trans. Magn. **27**. 2136 (1991); E.H. Brandt, Phys. Rev. Lett. **67**, 2219 (1991).

# EXPERIMENTAL OBSERVATION OF THE KOSTERLITZ-THOULESS TRANSITION

J.E. MOOIJ
*Department of Applied Physics*
*Delft University of Technology*
*P.O.Box 5046, 2600 GA Delft*
*The Netherlands*

ABSTRACT. In this chapter it is discussed how the Kosterlitz-Thouless transition can be observed in helium films and in superconducting thin films. Experimental results are given. In helium films and in films of low-temperature superconductors the transition has been observed. For films of superconductors with high critical temperature the data are still incomplete.

## 1. Introduction

For two-dimensional superfluids the phase transition between the ordered ground state at zero temperature and the high-temperature regime where excitations dominate is of a very special nature. At zero temperature the phase of the order parameter is constant over the system. With increasing temperature long-wavelength phase oscillations, spin waves, start to occur. Also vortex-antivortex pairs are formed, initially with a very small distance between the partners. With increasing temperature more and more pairs with larger and larger separation occur. The smaller pairs screen the interaction for larger pairs. Although there is no phase-phase correlation over long distances, topological order is preserved in the sense that the line integral of the phase gradient along a large closed contour is always zero. At a certain higher temperature the first unbinding of a vortex-antivortex pair takes place [1]. This temperature, above which free vortices are present in the infinite 2D superfluid system in thermodynamic equilibrium, is called the Kosterlitz-Thouless or Kosterlitz-Thouless-Berezinskii (KTB) temperature. The theory of the 2D topological KTB phase transition is discussed in the chapter by Young of these proceedings [2]. Here we will address the experimental observation of the KTB transition in helium films and in superconducting thin films.

It is not so easy to observe the specific features of the true KTB transition. In all 2D superfluid films, however disordered or inhomogeneous, one will find reasonably strong phase coherence at the lowest temperatures and dissipative behavior at high temperatures. Very likely in all these systems vortex-antivortex pairs and free vortices will be generated by thermal excitation, which have a strong influence on many properties. One might take the liberal view that any gradual, temperature-stimulated cross-over involving vortex pairs is a KTB transition. In particular with films of high-temperature superconductors many authors appear to take this attitude. However, the theory for true Kosterlitz-Thouless transitions makes specific, quantitative predictions. We will in this chapter take the stricter view that observation requires a detailed quantitative comparison of measured data with theory leading to consistent results. More detailed discussions are to be found in previous review papers by Mooij [3] and Minnhagen [4].

231

*N. Bontemps et al. (eds.), The Vortex State,* 231–243.

## 2. theoretical predictions

In a two-dimensional superfluid the free energy of a single vortex-antivortex pair depends logarithmically on the distance r between the vortex cores:

$$U(r) = 2\mu_c + A \ln (r/\xi) \tag{1}$$

Here $2\mu_c$ is the free energy for the pair at the smallest separation $\xi$, containing the formation energy for the two cores but also an entropy term that is determined by the number of independent ways to realize such a small pair. A is a prefactor that depends on film properties. For helium and superconducting films A is temperature dependent, in contrast with the usual assumption in statistical mechanics where the constant A sets the energy scale to compare temperature with. In the presence of other pairs, the interaction is modified. This is expressed in a scale-dependent "dielectric constant" $\varepsilon(r)$. The interaction for separation r is modified by the pairs with separation smaller than r, so that $\varepsilon(\xi)=1$ and $\varepsilon(r)$ increases to values above 1 for $r>\xi$. $\varepsilon$ is larger when $\mu_c$ is small, as then the pair density is higher. $\varepsilon$ obviously also increases with temperature. The approximations of the theory are only valid for small densities of pairs. As a consequence $\varepsilon$ should never be much larger than 1. At the transition temperature $T_c$, the value of for infinite separation is called $\varepsilon_c$. Even for an ideal superfluid film theory makes no prediction for the exact values of $\mu_c$ or $\varepsilon_c$.

The transition temperature is given by:

$$k_B T_c = A(T_c) / 4\varepsilon_c \tag{2}$$

If the theoretical description of the superfluid (e.g. Ginzburg-Landau theory for the superconducting film) allows the calculation of the "bare" interaction energy A for the sample, the transition temperature is only known approximately.

Various scales play an important role. It is best to express the scale in the logarithm of the length divided by the smallest pair separation: $\lambda = \ln(r/\xi)$. The smallest pair is on the scale $\lambda=0$. The sample size W sets an upper scale for the practical sample: $\lambda_W = \ln(W/\xi)$. The dynamical theory of Ambegaokar, Halperin, Nelson and Siggia (AHNS, [5]) gives typical scales for the average pair separation below $T_c$ and the average distance between free vortices above $T_c$. In an approximation that is only valid in a very narrow range near $T_c$, these scales are:

$$\lambda^- \approx 0.5 \, ( b / |\tau| )^{1/2} \qquad \tau < 0, \ |\tau| << 1 \tag{3a}$$
$$\lambda^+ \approx ( b / \tau )^{1/2} \qquad \tau > 0, \ \tau << 1 \tag{3b}$$

Here $\tau \equiv T/T_c - 1$ and b is a constant of order unity. The free-vortex density is coupled to $\lambda^+$ so that:

$$n_f \approx \xi^{-2} \exp\{-2b/\tau )^{1/2} \} \tag{3c}$$

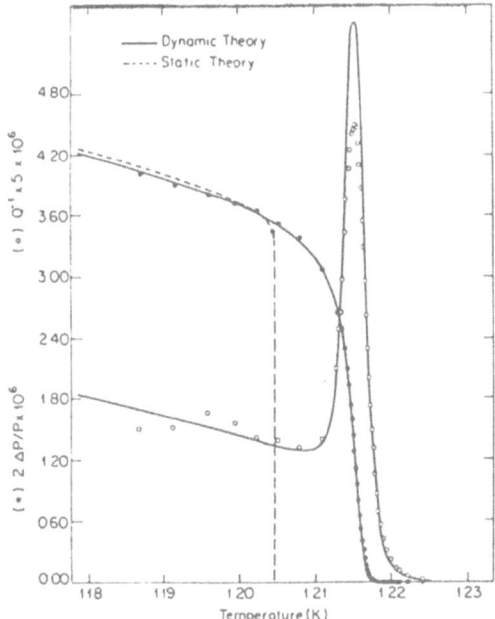

Figure 1. Period shift $\Delta P/P$ (solid dots) and dissipation $Q^{-1}$ (open dots) for torsion oscillator experiment on a helium film by Bishop and Reppy [7]. The drawn line is a fit to the dynamical theory of ref. 5. The dashed line indicates the transition for infinite scale.

All derivatives of the density are zero at the transition. The vortex density increases only very slowly above $T_c$.

AHNS calculated the response to AC excitation of the superfluid medium near the KTB transition. The calculations were made for helium films, but were later adapted to superconductors by Halperin and Nelson [6]. Bound pairs respond to oscillatory motion for frequencies below a cut-off value that depends on the pair separation. When D is the diffusion constant for vortices, the limiting scale for frequency $\omega$ is:

$$\lambda_\omega = \ln (14D/\omega) \tag{3d}$$

For a given frequency $\lambda_\omega$ sets the maximum scale to where pairs are probed. To detect the real phase transition, one wants to address pairs with infinite separation, so frequencies should be as low as possible.

Other predictions from the dynamical AHNS theory, for the non-linear response at high superfluid velocities, will be discussed later in the section on superconducting films.

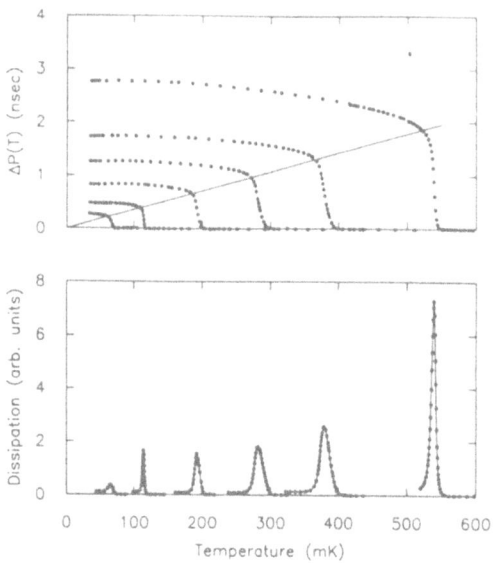

Figure 2. Period shift (upper plot) and dissipation (lower plot) for a set of different films with widely varying $T_C$, as measured by McQueeney et al.[10]. The drawn line is the prediction for the "universal jump of the superfluid density" by Nelson and Kosterlitz [9].

## 3. helium films

In helium films, the KTB transition was detected by Bishop and Reppy [7] in their torsion-oscillator experiments on thin superfluid films. They used a version of the Andronikashvili experiment as discussed by Vinen [8]. A sheet of mylar was rolled into a "jelly roll" and rotated around the axis in the center of the roll. Helium was condensed onto the mylar, the areal density known by calibration of cell volume and film area. With the torsional oscillator arrangement, they could measure the shift in period of the oscillator as well as the damping. A typical result is given in figure 1. At the highest temperatures the film is locked to the oscillator. The resonance frequency in this regime is used as the reference. The dissipation is zero here because oscillator and film move as a whole. At the lower temperatures the superfluid is disconnected and the resonance frequency is higher. The presence of bound pairs shows up as a gradual decrease of the frequency shift. The dissipation sharply rises when the density of bound pairs increases near $T_C$, the drop of the resonance frequency clearly coincides with the onset of dissipation (the dissipation level for the lowest temperatures is probably not connected with the film). The dissipation peak is not a resonance peak: the onset is due to the increasing number of pairs, the fall-off due to the presence of free vortices. Figure 1 demonstrates the beautiful agreement between the experiment and the dynamical AHNS theory with reasonable fitting parameters.

The position of the dissipation peak and the frequency drop is not quite at the KTB transition temperature. For a finite oscillator frequency, the active probing scale as given be Eq. 3d is finite. Only when the density of free vortices is high enough that their average distance is equal to the probing scale ($\lambda^+ = \lambda_\omega$) does the experimental transition show up. The higher the frequency, the higher the offset of the experimental peak. Bishop and Reppy varied the film thickness and found a clear linear relation between the observed transition temperature and the molar two-dimensional density of the film. The fact that the linear plot extrapolates to zero transition temperature at a relatively high value of the molar density indicates that not all the helium on the mylar substrate behaves as a superfluid. The torsion experiment by definition measures the "superfluid density" of the film. This is not connected with the bare coefficient A of Eq. 1, but with the renormalized value for the scale set by the oscillator frequency at the given temperature $A/\varepsilon(\lambda,T)$. This quantity is equal to $\rho_s h^2/2m^2$. If the measurement is performed at a low frequency (large scale), at $T_c$ the measured superfluid density should be directly proportional to $A/\varepsilon_c$ and related to the value of the transition temperature by Eq. 2. This universal value of the superfluid density as measured in the experiment at the transition temperature was predicted by Nelson and Kosterlitz [9]. McQueeney et al.[10] performed oscillation measurements on films with a large variation of thickness. In figure 2 a set of results for several films is given. In the upper plot the universal Nelson-Kosterlitz prediction, transferred to the cell parameters, is indicated with a drawn line. The results are clearly in excellent agreement. A more detailed description of the experiments is found in reference 11.

The transition manifests itself also in the thermal conduction and in the propagation of third sound. The thermal conductance is inversely proportional to the density of free vortices and therefore decreases sharply above $T_c$. Agnolet et al. [12] have observed this drop and have shown that the temperature coincides with the transition temperature as measured with the torsion oscillator. Rudnick [13] showed that also the results from measurements of third sound are in good agreement with the predicted position of the transition.

## 4. superconducting films

### 4.1. REQUIREMENTS

The superfluidity of a superconducting film has many aspects in common with the properties of a superfluid helium film. However, there are very significant differences that have a strong influence on the observability of the Kosterlitz-Thouless transition. It was originally supposed that the charged nature of the carriers in a superconductor would prohibit the KTB transition, as the interaction between vortices is only logarithmic over distances below the screening length. Beasley et al. [14] showed that this is in practice not significant as the screening length is of the order of centimeters at the transition, larger than the typical sample size. More serious are the limitations of fabricated superconducting films. As will be discussed further on, there is a simple relation between the BCS transition temperature for the bulk material and the KTB temperature that involves only the sheet resistance (the resistance of a film with equal length and width). Unless this sheet resistance is of the order of the quantum resistance, 5 to 10 k$\Omega$, the KTB temperature lies well within the fluctuation region near the BCS transition. Fluctuations of the  magnitude of the order parameter above $T_{BCS}$ are difficult to separate from the phase fluctuations that lead to the KTB transition. It is extremely difficult to fabricate a homogeneous, well-defined superconducting film with such a high sheet resistance. In a granular film, where the resistance is dominated by the grain boundaries, transport is strongly influenced by Coulomb charging effects. Even worse, percolative behavior may lead to an effective one-dimensionality. Extremely thin amorphous alloy films seem the only good solution. Superconductors do provide an advantage: electrical measurements allow a wider range of experimental approach.

The equivalent of the "superfluid density" concept that is used with helium films is not the density of Cooper pairs but rather the effective screening or the inverse effective kinetic inductance. It is influenced by the presence of pairs and scale-dependent. The bare interaction constant for vortices in superconductors is related to the two-dimensional penetration length $\Lambda = 2\lambda^2/d$ for films with thickness d below the 3D penetration length $\lambda$. In S.I. units the relation between A and $\Lambda$ is:

$$A = \Phi_0^2 / \pi\mu_0\Lambda \qquad (4)$$

where $\Phi_0$ is the flux quantum h/2e. According to Eq. 2 at $T_c$ the value of $\Lambda$ is:

$$\Lambda(T_c) = \Phi_0^2/\pi\mu_0 A(T_c) = \Phi_0^2 / 4\varepsilon_c\pi\mu_0 k_B T_c \qquad (5)$$

When $T_c$ is expressed in Kelvin, $\Lambda(T_c)$ in meters is equal to 0.018 $(\varepsilon_c T_c)^{-1}$. Clearly the penetration length near the transition is very long.

The transition temperature can be calculated from the penetration length. For dirty films of a BCS superconductor in the Ginzburg-Landau regime, the 2D penetration length is directly proportional to the sheet resistance $R_n$:

$$\Lambda = 0.021 \,\mu_0^{-1}\{ \, h / k_B(T_{co}-T)\} \, R_n \qquad (6a)$$

$$T_c = T_{co} \{ \, 1 + 0.27 \, \varepsilon_c \, R_n/R_q \,)^{-1} \qquad (6b)$$

here $T_{co}$ is the BCS transition temperature and $R_q = h/4e^2 = 6.5$ k$\Omega$. It is not practically possible with a BCS superconductor to reach the regime where $T_c$ is far below $T_{co}$.

The best example of a thorough analysis of the Kosterlitz-Thouless transition in superconducting thin films has been given by Hebard, Fiory and Glaberson [15,16] in their measurements on amorphous InO$_x$ films. They performed a full set of measurements with different techniques on the same set of samples. Examples of their data will be quoted here. References to other experimental data on low-$T_{co}$ superconducting films can be found in reference 3.

## 4.2. MEASUREMENTS

AC measurements can be performed on superconducting films by means of a small magnetic coil. The AC field is applied at low frequency, the kinetic response of the film is detected either with the same coil or with a second coil. The most sensitive arrangement uses a SQUID as the detector. One can measure both the in-phase and the out-of-phase response. The first yields a dissipation peak, similar to the dissipation peak for the helium films. The out-of-phase component is the inverse kinetic inductance of the superfluid. It decreases gradually with increasing temperature but drops down sharply at the transition.

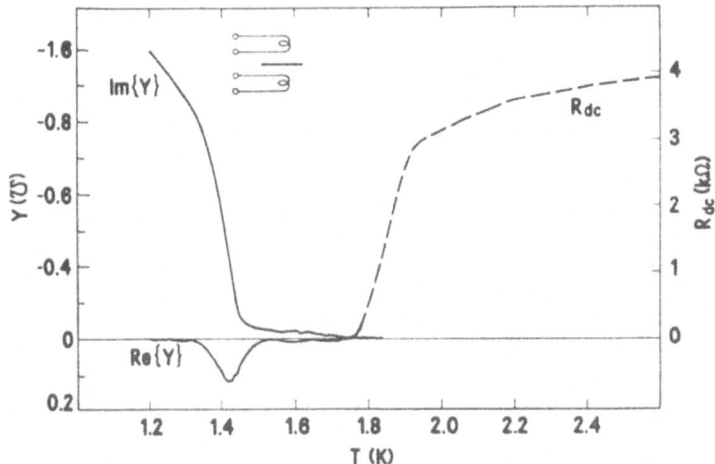

Figure 3. Measured values of the real and imaginary components of the admittance of a granular aluminum film with a sheet resistance of 4.1 kΩ [17].

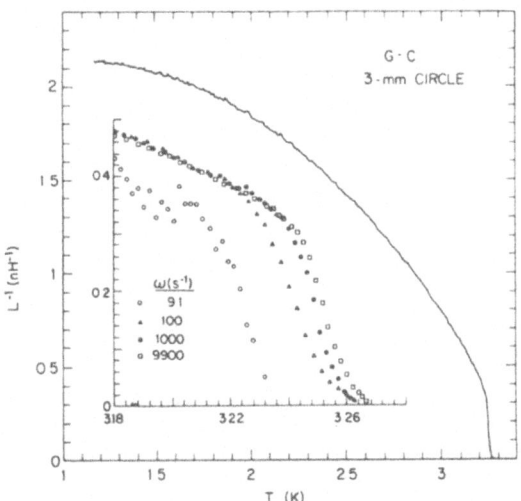

Figure 4. Measured inverse kinetic inductance of an InO$_x$ film at 160 Hz. Inset: values for different frequencies [15].

The gradual decrease is the Ginzburg-Landau behavior in the temperature region near $T_{co}$, where all measurements take place. In figure 3 results are shown for a granular Al film with a sheet resistance of 4.1 k$\Omega$. In figure 4 the inverse kinetic inductance is shown over a wider temperature range for a film of InO$_x$ with $R_n$=1.25 k$\Omega$ and a $T_{co}$ of 3.234 K. $T_{co}$ was determined independently from the resistance at high temperatures, by fitting to the theoretical prediction for paraconductivity above $T_{co}$.

For this type of measurement it is relatively easy to change the frequency, unlike for the helium torsion oscillator. The inset of figure 4 shows results at four different frequencies. The observed transition is at the lowest temperature for the lowest frequency. To find the "real" KTB temperature one would need, as discussed before, the infinite probing scale that is connected with zero frequency. Obviously the sensitivity for the AC measurement decreases for lower frequencies.

DC measurements of the resistance for small currents are certainly possible, but do not yield much significant information on the nature of the phase transition. One observes a broad transition, in particular for high-resistance films. One can fit the observed data against the predicted density of free vortices (Eq. 3c) above $T_c$ and usually obtains excellent coverage. However, the temperature range where the theoretical prediction is valid is extremely small [3]. Given the large number of unknown parameters and the wide range of resultant values for different films, the DC linear resistance cannot by itself be considered a significant test for Kosterlitz-Thouless behavior. In combination with other results, it adds to the overall picture. In figure 5a an example is shown.

A more specific test is provided by the non-linear resistance. The partners of a bound pair experience opposite forces from a tranport current. So the pair experiences a moment. In the orientation with the smallest energy the pair is pulled apart by the current. The stronger the current, the easier pairs will unbind. The current leads to a saddle point in the pair energy at a finite separation given by the scale $\lambda_j = \ln(0.4 J_{GL}/J)$, where J is the current density and $J_{GL}$ is the Ginzburg-Landau critical depairing current density. The theory for the current-induced vortex-pair unbinding has been worked out by Halperin and Nelson [6]. In the region near $T_c$, for not too large currents, the theory predicts a power-law dependence of the voltage on the current $V=I^a$. The exponent a is temperature dependent. Just below Tc, in the regime where $\lambda_j >> \lambda^-$, it should approach the universal value 3:

$$a(T) = 3 + \pi \, (b/\tau')^{-1/2} \tag{8}$$

$\tau'$ is equal to $T_{co}(T_c-T)/T_c(T_{co}-T)$; b is the usual unknown constant of order one. At $T_c$ the exponent drops to one.

To measure the exponent, one should determine the current and voltage over many orders of magnitude and plot the values for each temperature on a logarithmic scale. The sign for $T_c$ is the straightness of the plot for the lower currents. The highest temperature where straight lines are seen is $T_c$. The slope for this line should be 3. In figure 5b a set of I-V characteristics is shown for an InO$_x$ film. The higher temperatures are to the left, starting with curve a. Curves a, b and c are not straight, the ones at lower temperatures are. The temperature of curve d is therefore the critical one. Indeed, when the slopes are determined for curves m to d, they decrease linearly with increasing temperature. For curve d the slope is just above 3.

From the many measurements on high-resistance films of low-temperature superconductors a mixed picture emerges. One might hope that all 2D superconductors would fall in

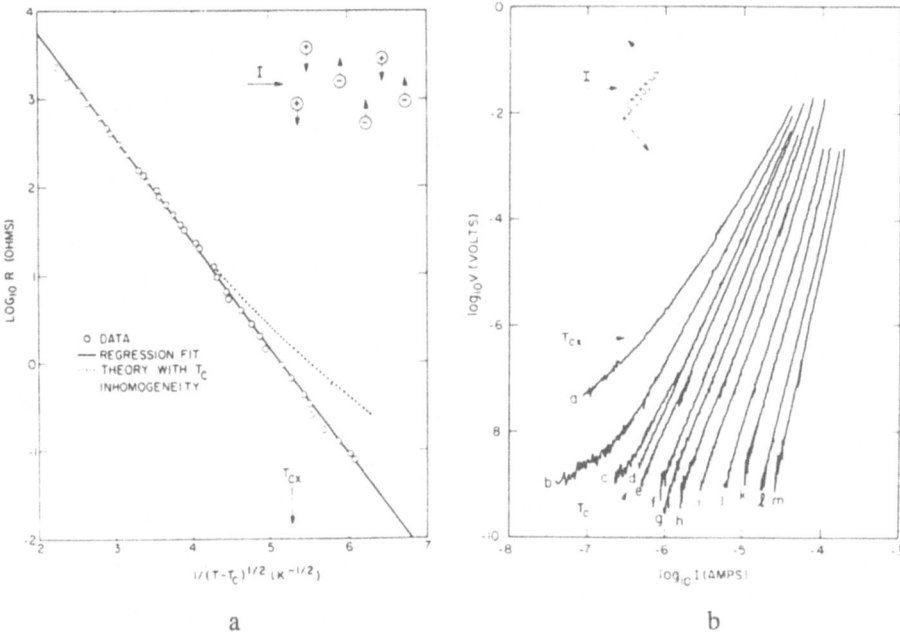

Figure 5. DC measurements on InO$_x$ film [15].

a. Linear resistance plotted against $\tau^{1/2}$. The straight line is a fit
   to the predicted behavior just above T$_c$, Eq. 3c.
b. Nonlinear resistance: plot of log V against log I at different
   temperatures. Curve d is at T$_c$.

one group with the same value of the non-universal quantities like the core energy and the b-parameter. The value of $\varepsilon_c$ should certainly be the same for all homogeneous films. However, deduced values are widely different. One could understand granular films to be different from amorphous films and (partial) granularity may be responsible for variation. Inhomogeneity will also strongly influence the results.

The careful analysis by Hebard, Fiory and Glaberson is the most encouraging. For two of the three films they investigated, a very consistent fit to the same parameters could be obtained from a complete series of AC and DC measurements. They yield values for b of about 8, $\varepsilon_c$ is found to be 1.20. This latter value is low enough to be consistent with the assumptions of the theory that was used. The T$_c$ is also in very good agreement with the calculated value from the sheet resistance and T$_{co}$. Unfortunately the third film yields deviating parameter values that are inconsistent with original assumptions ($\varepsilon_c$=1.8). Given the excellent detailed agreement with theory for the two "best" films, one feels justified to say that the Kosterlitz-Thouless transition has been observed in high resistance superconducting films.

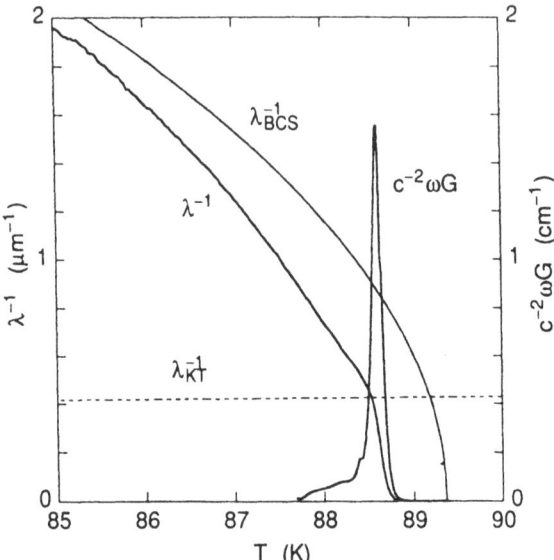

Figure 6. Real ($c^{-2}\omega G$) and imaginary ($\lambda^{-1}$) response of a YBCO film [22].

A different type of superconducting KTB systems are 2D arrays of superconducting islands that are weakly coupled by Josephson junctions. They are fabricated using lithographic techniques. On the one hand they represent the controlled model of a granular film, on the other hand they are a physical realization of the 2D X-Y model that is used extensively in statistical mechanics. The 2D X-Y model is the mother system for the Kosterlitz-Thouless transition. Indeed, these arrays clearly exhibit the characteristics that have been discussed previously. A translation of the KTB concepts to Josephson junction arrays has been given by Lobb et al. [18]. Examples of experimental studies are the dynamical measurements on superconductor-normal metal-superconductor junctions by Leeman et al. [19] and the linear and non-linear measurements on tunnel junction arrays by Van der Zant et al. [20]. Beautiful agreement between theory and experiment is obtained for these systems.

## 4.3 HIGH-TEMPERATURE SUPERCONDUCTORS

Over the last years, much attention has been paid to the resistive transition and other properties of high-temperature superconductors. By many authors, references have been made to a Kosterlitz-Thouless transition. Claims to experimental observation have been laid. Among these authors, only few have paid detailed attention to the precise aspects that accompany a KTB transition. When the available data are reviewed, it is certainly quite plausible that the transition in high-$T_c$ films contains related features. However, when numbers are extracted from the data, they seem to be inconsistent with the straightforward interpretation of KTB transitions as previously discussed for helium films and for films of low-$T_c$ supeconductors. In this section a few examples will be discussed. The possibility of a KTB-like transition for pancake vortices in single copper oxide layers of strongly anisotropic compounds will be ignored here.

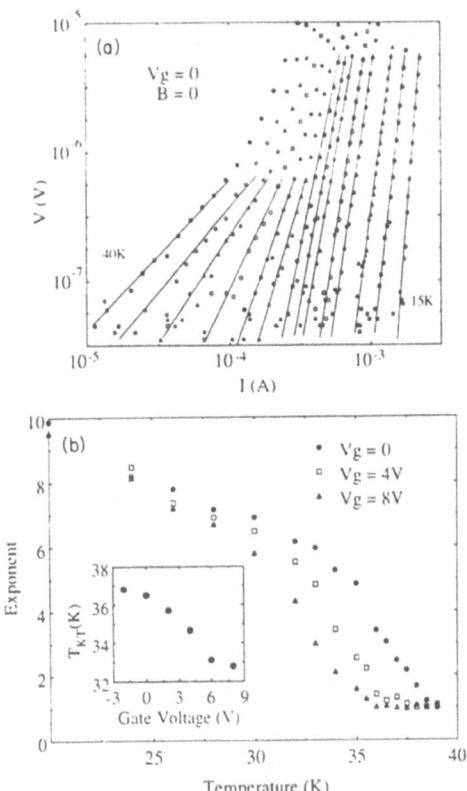

Figure 7. Nonlinear resistance measurement for ultrathin YBCO film [23]. Top: I-V characteristics; bottom: exponent (a from V=I$^a$) as a function of temperature.

One of the most detailed studies of the transition of thin YBCO films has been made by the same group that has been quoted extensively for low-$T_c$ measurements: Fiory et al. [21]. In figure 6 the real and imaginary components of the dynamic magnetic response (dissipation and inverse kinetic inductance, here indicated as the inverse penetration length) are plotted. The film is 50 nm thick, the measuring frequency is 13 kHz. The BCS prediction for the inverse penetration length is also indicated. One notices the familiar drop of the inverse inductance at the same temperature where the dissipation peak is found, similar to the data for helium films and low-$T_c$ films. Fiory et al. analyze the data and conclude from the renormalization of the inductance at the transition that $\varepsilon_c$ is 4.6. This value lies far outside the range where the Kosterlitz-Thouless renormalization scheme can be applied. The approximations fail for vortex pair densities that are not very small. For consistency, $\varepsilon$ should always be close to 1. So, although the picture of figure 6 looks very convincing, the numbers are wrong. Davis et al. [22] in their discussion of the same data come to a similar conclusion by different reasoning: they show that the temperature dependence is too weak compared with the KTB predictions. A more detailed study would be necessary to establish the nature of the transition, but it seems likely that it will be intermediate, not easily categorizable.

More recent studies focus on extremely thin films and on multiple well-separated single layers of YBCO or other 123 compounds between other materials. One finds transition temperatures that are far below the 90 K regime of the bulk material These depressions are sometimes interpreted as due to a Kosterlitz-Thouless transition of the very thin layers. The field effect can be used to depress the carrier density of thin films. If the KTB transition is relevant, such a depression should immediately lead to a lower transition temperature. In figure 7 a set of data on an ultrathin (5 nm) YBCO film is shown, as measured by Walkenhorst et al. [23]. The top shows I-V characteristics at zero gate voltage, the bottom plot gives the values of the exponent (the a from $V=I^a$) derived from the slopes for three different values of the gate voltage. The data show again a kink like the one for KTB transitions, that shifts to lower temperatures for lower carrier density. The authors rightly state that the transition is Kosterlitz-Thouless-like. When a stricter definition is used, the numbers are wrong from the pure KTB point of view. The jump should occur at a value for the exponent of 3, instead of 6.

It is interesting to estimate the value of $T_c$ for a film of YBCO with a thickness of a single cell. Lacking a microscopic theory, the BCS expressions are used. The first approach would be to assume the extremely thin film to be dirty. Then the regular relation between the sheet resistance and the ratio of the "BCS" and KTB transition temperatures should apply. The sheet resistance is typically of order 1 k$\Omega$. This corresponds to a ratio $T_c/T_{c0}$ of 0.93 (Eq. 7b) when a value of 1.6 is taken for $\varepsilon_c$. A higher value is not consistent with the original assumptions. If the superconductor is "clean" one should estimate the areal density of Cooper pairs. For one cell, the pair density is expected to be $3\ 10^{14}$ cm$^{-2}$ at low temperatures. Near $T_{c0}$ it is $2(1-T/T_{c0})$ times this value. The Nelson-Kosterlitz relation [9] with the same value of 1.6 for $\varepsilon_c$ leads to $T_c/T_{c0}=0.94$ when the pair mass is taken to be 4 times the bare electron mass. If the pair mass is taken as $2m_0$, the transition is even higher. From the previous numbers it follows that a direct application of the theoretical framework for Kosterlitz-Thouless transitions cannot lead to a depression of $T_c$ by more than 10 or 20 %. New assumptions may be needed, but are hard to justify without a better understanding of the nature of the superfluid carriers in the high-$T_c$ materials.

## 5. conclusions

Kosterlitz-Thouless transitions have been observed in helium films and in thin films of low-temperature superconductors. Detailed quantitative agreement between theory and experiment is obtained for a number of properties. In helium the universal value of the superfluid density is observed at the transition. In superconductors, the detailed knowledge of the background parameters should in principle allow the determination of the non-universal constants for the transition in the homogeneous superfluid system. Probably due to non-ideality of samples values for these parameters differ. In thin films of high-temperature superconductors transitions have been observed that show qualitative similarity with the characteristics for Kosterlitz-Thouless behavior. However, quantitative differences lead to inconsistency with theoretical assumptions. It is not yet clear whether these deviations from the pure Kosterlitz-Thouless transition are due to sample inhomogeneities or that the nature of the transition is intrinsically different.

## 6. references

[1] J.M. Kosterlitz and D.J. Thouless, J. Phys. C **6**, 1181 (1973)

[2] A.P. Young, these proceedings

[3] J.E. Mooij, Proceedings NATO Advanced Study Institute on Percolation, Localization and Superconductivity, editors A.M. Goldman and S.A. Wolf, Plenum, New York 1983, p. 325

[4] P. Minnhagen, Rev. Mod. Phys. **59**, 1001 (1987)

[5] V. Ambegaokar, B.I. Halperin, D.R. Nelson and E.D. Siggia, Phys. Rev. B **21**, 1806 (1980)

[6] B.I. Halperin, D.R. Nelson, J. Low Temp. Phys. 36, **599** (1979)

[7] D.J. Bishop and J.D. Reppy, Phys. Rev. Lett. **40**, 1727 (1978)

[8] V.W. Vinen, these proceedings

[9] D. Nelson and J.M. Kosterlitz, Phys. Rev. Lett. **39**, 1201 (1977)

[10] D. McQueeney, G. Agnolet and J.D. Reppy, Phys. Rev. Lett. **52**, 1325 (1984)

[11] D.J. Bishop and J.D. Reppy, Phys. Rev. B **22**, 5171 (1980)

[12] G. Agnolet, S.L. Teitel and J.D. Reppy, Phys. Rev. Lett. **47**, 1537 (1981)

[13] I. Rudnick, Phys. Rev. Lett. **40**, 1454 (1978)

[14] M.R. Beasley, J.E. Mooij and T.P. Orlando, Phys. Rev. Lett. **42**, 1165 (1979)

[15] A.F. Hebard and A.T. Fiory, Phys. Rev. Lett. **50**, 1603 (1983)

[16] A.T. Fiory, A.F. Hebard and W.I. Glaberson, Phys. Rev. B **28**, 5075 (1983)

[17] A.F. Hebard and A.T. Fiory, Phys. Rev. Lett. **44**, 291 (1980)

[18] C.J. Lobb, D.W. Abraham and M. Tinkham, Phys. Rev. B **27**, 150 (1983)

[19] C. Leeman, P. Lerch, G.A. Racine and P. Martinoli, Phys. Rev. Lett. **56**, 1291 (1986)

[20] H.S.J. van der Zant, H.A. Rijken and J.E. Mooij, J. Low Temp. Phys. **79**, 289 (1990)

[21] A.T. Fiory, A.F. Hebard, P.M. Mankiewich and R.E. Howard, Phys. Rev. Lett. **61**, 1419 (1988)

[22] L.C. Davis, M.R. Beasley and D.J. Scalapino, Phys. Rev. B **42**, 99 (1990)

[23] A. Walkenhorst, C. Doughty, X.X. Xi, Qi Li, C.J. Lobb, S.N. Mao and T. Venkatesan, Phys. Rev. Lett. **69**, 2709 (1992)

# FERMI-LIQUID THEORY of NON-S-WAVE SUPERCONDUCTIVITY

P. Muzikar[a], D. Rainer[b], J.A. Sauls[c]

[a] *Department of Physics, Purdue University, West Lafayette, IN 47907 USA*
[b] *Physikalisches Institut, Universität Bayreuth, D-95440 Bayreuth, Germany*
[c] *Northwestern University, Evanston, IL 60208 USA\* & Nordita, DK-2100 Copenhagen ØDenmark*

**Abstract**

These lectures present the Fermi-liquid theory of superconductivity, which is applicable to a broad range of systems that are candidates for non-s-wave pairing, *e.g.* the heavy fermions, organic metals and the CuO superconductors. Ginzburg-Landau (GL) theory provides an important link between experimental properties of non-s-wave superconductors and the more general Fermi-liquid theory. The multiple superconducting phases of UPt$_3$ provide an ideal example of the role that is played by the GL theory for non-s-wave superconductors. The difference between non-s-wave superconductivity and conventional anisotropic superconductivity is illustrated here by the unique effects that impurities are predicted to have on the properties of non-s-wave superconductors.

## I. INTRODUCTION

Historically, "non-s-wave pairing" began with the publication "Generalized Bardeen-Cooper-Schrieffer States and the Proposed Low-Temperature Phase of $^3$He" by Anderson and Morel.[1] They considered the possibility of BCS pairing with non-zero angular momentum, and studied its physical consequences. When superfluidity was discovered in 1972 by Osheroff et al.[2] in $^3$He it was immediately clear that this was not a conventional s-wave BCS superfluid because there was more than one superfluid phase. Increasing evidence for non-s-wave pairing came from many experimental results and within about a year after their discovery the three superfluid phases of $^3$He were undisputedly identified as p-wave (as proposed by Anderson and Morel) spin-triplet superfluids.

The search for "non-s-wave superconductivity", the metallic analog to superfluidity in $^3$He, was unsuccessful for more than a decade. In 1979 Steglich[3] discovered superconductivity in CeCu$_2$Si$_2$. This was the first in a new class of heavy-fermion superconductors, which now include the U-based compounds UBe$_{13}$, UPt$_3$, URu$_2$Si$_2$, UNi$_2$Al$_3$, and UPd$_2$Al$_3$. Unusual temperature dependences of heat capacity, penetration depth, and sound absorption led to conjectures that these materials were non-s-wave superconductors.[4] Much more experimental information is now available,[5,6] and there is consensus that some heavy-fermion superconductors (if not all of them) show non-s-wave pairing. The interest in non-s-wave superconductivity reached new levels recently with the reports of several experiments on cuprate superconductors that supported earlier predictions[7] of d-wave pairing. However, there is not yet a generally accepted identification of the specific type of pairing in any of these superconductors.

For metals, the popular notion of "non-s-wave superconductivity" should not be taken literally. A rigorous classification of superconductors by the angular momentum of the Cooper pairs (e.g. s-wave, p-wave, d-wave pairing, etc.) fails in crystalline materials because angular momentum is not a good quantum number. There is no ideally isotropic superconductor in nature, and any superconductor is, in this sense, "non-s-wave". However, we will use the term "non-s-wave pairing" interchangeably with "unconventional pairing" or "reduced symmetry superconductivity" for superconductors in which the superconducting state spontaneously breaks one or more symmetries of the crystalline phase.

245

*N. Bontemps et al. (eds.), The Vortex State, 245–264.*
© *1994 Kluwer Academic Publishers.*

Liquid $^3$He, heavy-fermion metals, and high-$T_c$ cuprates are systems of strongly correlated fermions, for which we do not have a practical microscopic theory. One is forced
to rely on phenomenological theories in order to describe superfluidity or superconductivity in these systems. This lecture gives an introduction to the two most powerful
phenomenological theories of correlated fermions with non-s-wave pairing: the Ginzburg-
Landau theory and Fermi liquid theory. The Ginzburg-Landau theory is summarized
in section [II] where we discuss the order parameter for non-s-wave pairing, introduce
the basic concepts and notations, and apply the theory to UPt$_3$, the best studied candidate for non-s-wave pairing in the heavy-fermion superconductors. Section [III] gives
an introduction to the Fermi-liquid theory of superconductivity. This theory includes the
Ginzburg-Landau theory for $T \sim T_c$, but is more general; it covers the entire temperature
and field range of interest and has proven to be very powerful in describing the superfluid
phases of $^3$He, the paradigm of a Fermi liquid. Fermi-liquid theory is also very useful
in describing the heavy-fermion metals at low temperatures; i.e. below the "coherence
temperature" which marks the cross over to the Fermi-liquid state. Special versions of
Fermi-liquid theory (e.g. nearly antiferromagnetic Fermi liquid, marginal Fermi liquid,
Eliashberg's strong-coupling model, etc.) seem to be promising steps toward a theory of
high-$T_c$ superconductivity. Finally, in section [IV] we discuss the effects of impurities on
the properties of non-s-wave superconductors. Impurity scattering leads to several novel
effects in many non-s-wave superconductors; these examples demonstrate the predictive
power of the Fermi-liquid theory for unconventional superconductors.

This lecture presents selected aspects of the theory of non-s-wave superconductors, and
does not cover the full spectrum of interesting and promising developments in the theory
of unconventional pairing. We refer the interested reader to the reviews by Lee et al.,[8]
Gorkov,[9] and Sigrist and Ueda[10] for a broader discussion.

## II. GINZBURG-LANDAU THEORY

One of the most important phenomenological theories available for investigating the properties of superconductors is the Ginzburg-Landau (GL) theory. This theory is applicable
to a wide class of materials and superconducting phenomena, but is limited to temperatures near $T_c$ where the order parameter is small. Here we introduce the order parameter
for non-s-wave superconductors, develop the GL theory from symmetry considerations
and apply the GL theory to UPt$_3$.

The basic quantity that describes all BCS-type superconductors is the equal-time pair
amplitude,

$$f_{\alpha\beta}(\vec{k}_f) \sim \left\langle a_{\vec{k}_f \alpha} a_{-\vec{k}_f \beta} \right\rangle , \tag{1}$$

where $\vec{k}_f$ is the momentum of a quasiparticle on the Fermi surface and $\alpha, \beta$ are the spin
labels of the quasiparticles.* Fermion statistics requires that the pair amplitude obey

---

*In the heavy-fermion materials it is generally assumed that spin-orbit coupling is strong;[11,12,8] thus, the labels
characterizing the quasiparticle states are not eigenvalues of the spin operator for electrons. Nevertheless, in zero-
field the Kramers degeneracy guarantees that each $\vec{k}$ state is two-fold degenerate, and thus, may be labeled by a
'pseudo-spin' quantum number $\alpha$, which can take on two possible values. We use the term 'spin' interchangeably
with 'pseudo-spin' in this article.

the anti-symmetry condition, $f_{\alpha\beta}(\vec{k}_f) = -f_{\beta\alpha}(-\vec{k}_f)$. Thus, the most general pairing amplitude can be written in terms of a sum of spin-singlet and spin-triplet amplitudes,

$$f_{\alpha\beta}(\vec{k}_f) = f_0(\vec{k}_f)\,(i\sigma_y)_{\alpha\beta} + \vec{f}(\vec{k}_f) \cdot (i\vec{\sigma}\sigma_y)_{\alpha\beta}\,, \tag{2}$$

where $f_0(\vec{k}_f) = f_0(-\vec{k}_f)$ $(\vec{f}(\vec{k}_f) = -\vec{f}(-\vec{k}_f))$ has even (odd) parity.

The actual realization of superconductivity in a given material will be described by a pair amplitude of this general form which minimizes the free energy. The GL theory is formulated in terms of a stationary free energy functional of the pair amplitude, or order parameter, of eq.(1). The central assumptions of the GL theory are (i) that the free energy functional can be expanded in powers of the order parameter and (ii) that the GL functional has the full symmetry of the normal state. The GL functional can then be constructed from basic symmetry considerations.

Essentially all of the candidates for non-s-wave superconductivity, including the heavy-fermion and cuprate superconductors, have inversion symmetry. This has an important consequence; the pairing interaction that drives the superconducting transition decomposes into even- and odd-parity sectors.[11,13] Thus, $f_{\alpha\beta}(\vec{k}_f)$ necessarily has even or odd parity unless there is a second superconducting instability into a state with different parity. Furthermore, the pairing interaction separates into a sum over invariant bilinear products of basis functions, $\mathcal{Y}_{\Gamma,i}(\vec{k}_f)$, for each irreducible representation $\Gamma$ of the crystal point group, for both even- and odd-parity sectors. Representative basis functions for the group $D_{6h}$, appropriate for UPt$_3$ with strong spin-orbit coupling, are given in Table I (see Ref. ( 14) for a complete discussion of the allowed basis functions). The general form of the order parameter is then,

$$f_0(\vec{k}_f) = \overset{even}{\sum_{\Gamma}} \eta_i^{(\Gamma)} \mathcal{Y}_{\Gamma,i}(\vec{k}_f) \quad, \quad \vec{f}(\vec{k}_f) = \overset{odd}{\sum_{\Gamma}} \eta_i^{(\Gamma)} \vec{\mathcal{Y}}_{\Gamma,i}(\vec{k}_f)\,. \tag{3}$$

There is a single quadratic invariant for each irreducible representation, so the leading terms in the GL free energy are of the form,

$$\mathcal{F} = \int d^3x \sum_{\Gamma} \alpha_{\Gamma}(T) \sum_{i=1}^{d_{\Gamma}} |\eta_i^{(\Gamma)}|^2 + \dots. \tag{4}$$

The coefficients $\alpha_{\Gamma}(T)$ are material parameters that depend on temperature and pressure. Above $T_c$ all the coefficients $\alpha_{\Gamma}(T) > 0$. The instability to the superconducting state is then the point at which one of the coefficients vanishes, e.g. $\alpha_{\Gamma^*}(T_c) = 0$. Thus, near $T_c$ $\alpha_{\Gamma^*}(T) \simeq \alpha'(T - T_c)$ and $\alpha_{\Gamma} > 0$ for $\Gamma \neq \Gamma^*$. At $T_c$ the system is unstable to the development of all the amplitudes $\{\eta_i^{(\Gamma^*)}\}$, however, the higher order terms in the GL functional which stabilize the system, also select the ground state order parameter from the manifold of degenerate states at $T_c$. In most superconductors the instability is in the even-parity, $A_{1g}$ channel. This is the crystalline analog of 's-wave superconductivity' in which only gauge symmetry is spontaneously broken. An instability in any other channel is a particular realization of non-s-wave superconductivity.

### Application to UPt$_3$

Considerable evidence in support of non-s-wave superconductivity state in the heavy-fermion materials has accumulated from specific heat, upper critical field and various

TABLE I. Basis functions for $D_{6h}$

| | Even parity | | | Odd parity |
|---|---|---|---|---|
| $A_{1g}$ | $1$ | | $A_{1u}$ | $\hat{z}\,k_z$ |
| $A_{2g}$ | $Im(k_x + ik_y)^6$ | | $A_{2u}$ | $\hat{z}\,k_z\,Im(k_x + ik_y)^6$ |
| $B_{1g}$ | $k_z\,Im(k_x + ik_y)^3$ | | $B_{1u}$ | $\hat{z}\,Im(k_x + ik_y)^3$ |
| $B_{2g}$ | $k_z\,Re(k_x + ik_y)^3$ | | $B_{2u}$ | $\hat{z}\,Re(k_x + ik_y)^3$ |
| $E_{1g}$ | $k_z \begin{pmatrix} k_x \\ k_y \end{pmatrix}$ | | $E_{1u}$ | $\hat{z} \begin{pmatrix} k_x \\ k_y \end{pmatrix}$ |
| $E_{2g}$ | $\begin{pmatrix} k_x^2 - k_y^2 \\ 2k_x k_y \end{pmatrix}$ | | $E_{2u}$ | $\hat{z}\,k_z \begin{pmatrix} k_x^2 - k_y^2 \\ 2k_x k_y \end{pmatrix}$ |

transport measurements, all of which show anomalous properties compared to those of conventional superconductors (see Refs. ( 4–6) for original references). However, the strongest evidence for unconventional superconductivity comes from the multiple super-conducting phases of UPt$_3$.[15-19] The important features of the H-T phase diagram are: (i) There are two zero-field superconducting phases with a difference in transition temperatures of $\Delta T_c/T_c \simeq 0.1$. (ii) A change in slope of the upper critical field (a 'kink' in $H_{c2}^{\perp}$) is observed for $\vec{H} \perp \hat{c}$, but not for $\vec{H} \| \hat{c}$, or at least it is much smaller. (iii) There are three flux phases, and the phase transition lines separating the flux phases appear to meet at a tetracritical point for all orientations of $\vec{H}$ relative to $\hat{c}$.

There are two basic types of models that have been proposed to explain the phase diagram: (i) theories based on a *single* primary order parameter belonging to a higher dimensional representation of the symmetry group of the normal state, and (ii) theories based on *two* primary order parameters belonging to different irreducible representations which are nearly degenerate. Below we construct a particular GL theory for non-s-wave superconductivity which describes the superconducting phase diagram of UPt$_3$. We start from the assumption that the superconducting instability occurs in one of the 2D representations appropriate to strong spin-orbit coupling, then show how the presence of weak perturbations can give rise to multiple superconducting phases.[20,21] A detailed discussion of this model for UPt$_3$ is given in Ref. ( 22).

Consider one of the 2D representations, *e.g.* the $E_{2u}$ representation of Table I. Near $T_c$ all other order parameter amplitudes vanish. The GL order parameter is then a complex two-component vector, $\vec{\eta} = (\eta_1, \eta_2)$, transforming according to the relevant 2D representation. The terms in the GL functional must be invariant under the full symmetry group of point rotations, time-reversal and gauge transformations. The form of $\mathcal{F}$ is governed by the linearly independent invariants that can be constructed from fourth-order products, $\sum b_{ijkl}\,\eta_i\eta_j\eta_k^*\eta_l^*$, and second-order gradient terms, $\sum \kappa_{ijkl}(D_i\eta_j)(D_k\eta_l)^*$. For the 2D representations there are only two independent fourth-order invariants and four independent second-order gradients; the GL free energy functional has the general form,

$$\mathcal{F}\left[\vec{\eta}, \vec{A}\right] = \int d^3R \left\{ \alpha(T)\,\vec{\eta} \cdot \vec{\eta}^* + \beta_1 (\vec{\eta} \cdot \vec{\eta}^*)^2 + \beta_2 |\vec{\eta} \cdot \vec{\eta}|^2 + \frac{1}{8\pi}|\vec{\nabla} \times \vec{A}|^2 \right.$$
$$\left. + \kappa_1 (D_i\eta_j)(D_i\eta_j)^* + \kappa_2 (D_i\eta_i)(D_j\eta_j)^* + \kappa_3 (D_i\eta_j)(D_j\eta_i)^* + \kappa_4 (D_z\eta_j)(D_z\eta_j)^* \right\}. \tag{5}$$

The equilibrium order parameter and current distribution are determined by the stationarity conditions for variations of $\mathcal{F}$ with respect to $\vec{\eta}$ and the vector potential, $\vec{A}$. This functional provides an important connection between experiment and the Fermi-liquid theory of superconductivity because the material parameters, $[\alpha(T), \beta_1, \beta_2, \kappa_1, \kappa_2, \kappa_3, \kappa_4]$, can be determined from comparison of GL theory with experiment, and calculated from

Fermi-liquid theory (see below).

There are two possible homogeneous equilibrium states of this GL theory. For $-\beta_1 < \beta_2 < 0$ the equilibrium order parameter, $\vec{\eta} = \eta_0 \hat{x}$ (or any of the six degenerate states obtained by rotation), breaks rotational symmetry in the basal plane, but preserves time-reversal symmetry. However, for $\beta_2 > 0$ the order parameter retains the full rotational symmetry (provided each rotation is combined with an appropriately chosen gauge transformation), but spontaneously breaks time-reversal symmetry. The equilibrium state is doubly-degenerate with an order parameter of the form $\vec{\eta}_+ = (\eta_0/\sqrt{2})(\hat{x} + i\hat{y})$ [or $\vec{\eta}_- = \vec{\eta}_+^*$], where $\eta_0 = (|\alpha|/2\beta_1)^{1/2}$. The broken time-reversal symmetry of the two solutions, $\vec{\eta}_\pm$, is exhibited by the two possible orientations of the internal orbital angular momentum, $\vec{l} \sim i(\vec{\eta} \times \vec{\eta}^*) \sim \pm\hat{z}$, or spontaneous magnetic moment of the Cooper pairs.

The case $\beta_2 > 0$ is relevant for the 2D models of the double transition of UPt$_3$. However, the 2D theory has only one phase transition in zero field, and by itself cannot explain the double transition. The small splitting of the double transition in UPt$_3$ ($\Delta T_c/T_c \simeq 0.1$) suggests the presence of a small symmetry breaking energy scale and an associated lifting of the degeneracy of the possible superconducting states belonging to the 2D representation. The second zero-field transition just below $T_c$ in UPt$_3$, as well as the anomalies observed in the upper and lower critical fields, have been explained in terms of a weak symmetry breaking field (SBF) that lowers the crystal symmetry from hexagonal to orthorhombic, and consequently reduces the 2D E$_2$ (or E$_1$) representation to two 1D representations with slightly different transition temperatures.[20,21] The key point is that right at $T_c$ all states belonging to the 2D representation are degenerate, thus any SBF that couples second-order in $\vec{\eta}$ and prefers a particular state will dominate very near $T_c$. At lower temperatures the SBF energy scale, $\Delta T_c$, is a small perturbation compared to the fourth order terms in the fully developed superconducting state and one recovers the results of the GL theory for the 2D representation, albeit with small perturbations to the order parameter.

In UPt$_3$ there appears to be a natural candidate for a SBF,[23] the AFM order in the basal plane reported by Aeppli, et al.[24] In this case a contribution to the GL functional corresponding to the coupling of the AFM order parameter to the superconducting order parameter is included; $\mathcal{F}_{\text{SBF}}[\vec{\eta}] = \varepsilon M_s^2 \int d^3R \, (|\eta_1|^2 - |\eta_2|^2)$, where $\vec{M}_s$ is the AFM order parameter and the coupling parameter $\varepsilon M_s^2$ determines the magnitude of the splitting of the superconducting transition. The analysis of this GL theory, including the SBF, is given in Ref. ( 20); the main results are:

1. A double transition occurs only if $\beta_2 > 0$. The splitting of the transition temperature is $\Delta T_c \propto M_s^2$.

2. The relative magnitudes of the two heat capacity anomalies, $\Delta C_*/T_{c*} > \Delta C/T_c$, are consistent with the stability condition $\beta_2 > 0$ required by the double transition in the 2D model.

3. The low temperature phase ($T < T_{c*}$) has broken time-reversal symmetry, and is doubly degenerate: $\vec{\eta}_\pm \sim (a(T), \pm i\, b(T))$, reflecting the two orientations of the internal angular momentum of the ground state.

4. The upper critical field exhibits a change in slope, a 'kink' at high temperature for $\vec{H} \perp \hat{c}$, but not for $\vec{H} \| \hat{c}$. The kink in $H_{c2}^\perp$ is isotropic in the basal plane provided the in-plane magnetic anisotropy energy is weak compared to the Zeeman energy acting on $\vec{M}_s$, in which case $\vec{M}_s$ rotates to maintain $\vec{M}_s \perp \vec{H}$. Recent magnetoresistance experiments support this interpretation of weak magnetic anisotropy energy in the

basal plane.[25]

5. Kinks in $H_{c1}$, for all field orientations, are predicted to occur at the second zero-field transition temperature, $T_{c*}$. This was confirmed by several groups.[26–28,25] The increase in $H_{c1} \propto \eta^2$ at $T_{c*}$ is also a strong indication of the onset of a second superconducting order parameter.

6. Additional evidence for a SBF model of the double transition comes from pressure studies of the superconducting and AFM phase transitions. Heat capacity measurements by Trappmann, et al.[29] show that both zero-field transitions are suppressed under hydrostatic pressure, and that the double transition disappears at $p_* \simeq 4\,kbar$. Neutron scattering experiments reported by Hayden, et al.[30] show that AFM order disappears on the same pressure scale, at $p_c \simeq 3.2\,kbar$.

The phase diagram determined by ultrasound velocity measurements indicates that the phase boundary lines meet at a tetracritical point for both $\vec{H}\|\hat{c}$ and $\vec{H} \perp \hat{c}$. This has been argued to contradict the GL theory based on a 2D order parameter.[31–33] The difficulty arises from gradient terms of the form, $\kappa_{23}[(D_x\eta_1)(D_y\eta_2)^* + (D_x\eta_2)(D_y\eta_1)^* + c.c.]$, that couple the two components of the order parameter. These terms lead to 'level repulsion' effects in the linearized GL differential equations which prevent the crossing of two $H_{c2}(T)$ curves, corresponding to different superconducting phases. This feature of the 2D model has spawned alternative theories, designed specifically to eliminate the 'level repulsion' effect,[32,33] and recently to a more specific version of the 2D model coupled to a SBF.[22] We discuss this latter model below and direct readers to Refs. ( 31–33, 22) for discussions of alternative GL models for UPt₃.

Although the GL functionals are formally the same for any of the 2D orbital representations, the predictions for the material parameters from Fermi-liquid theory differ substantially depending on the symmetry properties of the Cooper pair basis functions. For example, the interpretation of the unusual anisotropy of $H_{c2}$ in UPt₃ in terms of anisotropic Pauli limiting requires an odd-parity, spin-triplet representation with the $\vec{d}$-vector parallel to the $\hat{c}$ direction.[34] This restricts one to the $E_{2u}$ or $E_{1u}$ basis functions among the four possible 2D representations. The case for an odd-parity order parameter with $\vec{d}\|\hat{c}$ is discussed in detail in Refs. ( 34, 22).

There are several other important predictions from Fermi-liquid theory for the 2D representations. For any of the four 2D representations, and independent of the geometry of the Fermi surface or the presence of non-magnetic, s-wave impurity scattering, the fourth-order free energy coefficients have the ratio, $\frac{\beta_2}{\beta_1} = \frac{1}{2}$.[35] This ensures that the coupling to the SBF will produce a double transition in zero field for any of the 2D orbital representations. Differences between the 2D models appear for the in-plane gradient terms in the GL functional. These gradient coefficients can be calculated from quasiclassical theory; in the clean limit the results are[35]

$$\kappa_1 = \kappa_0 \left\langle \mathcal{Y}_1(\vec{k}_f)\, v_{fy}\, v_{fy}\, \mathcal{Y}_1(\vec{k}_f) \right\rangle \quad , \quad \kappa_2 = \kappa_3 = \kappa_0 \left\langle \mathcal{Y}_1(\vec{k}_f)\, v_{fx}\, v_{fy}\, \mathcal{Y}_2(\vec{k}_f) \right\rangle , \quad (6)$$

where $\mathcal{Y}_i(\vec{k}_f)$ are the basis functions, $\kappa_0 = \frac{7\zeta(3)}{16\pi^2 T_c^2} N_f$, and $N_f$ is the density of states at the Fermi level. There are important differences between the $E_1$ and $E_2$ representations when we evaluate these averages. Using the basis functions in Table I and a Fermi surface with weak hexagonal anisotropy, the $E_1$ model gives $\kappa_2 = \kappa_3 \simeq \kappa_1$, while for the $E_2$ model, $\kappa_2 = \kappa_3 \ll \kappa_1 \sim N_f(v_f^\perp/\pi T_c)^2$. In fact, the three in-plane coefficients are identical for $E_{1u}$ in the limit where the in-plane hexagonal anisotropy of the Fermi surface vanishes. In contrast, the coefficients $\kappa_2$ and $\kappa_3$ for the $E_{2u}$ model both vanish when the hexagonal anisotropy of the Fermi surface is neglected. This latter result follows directly from the approximation of a cylindrically symmetric Fermi surface and Fermi velocity, $\vec{v}_f = v_f^\perp(\hat{k}_x\,\vec{x} + \hat{k}_y\,\vec{y}) + v_f^\parallel \hat{k}_z \vec{z}$, and the higher angular momentum components of the $E_2$ basis functions,

$$\kappa_2(E_{2u}) \propto \left\langle \hat{k}_z(\hat{k}_x^2 - \hat{k}_y^2) v_{fx} v_{fy} (2\hat{k}_x\hat{k}_y)\hat{k}_z \right\rangle \equiv 0. \tag{7}$$

If we have weak hexagonal anisotropy of the Fermi surface then $\kappa_{23} \ll \kappa_1$ only for $E_2$. Thus, there is a natural explanation for the absence (or at least the smallness) of the 'level repulsion' terms in the orbital 2D model if there is weak hexagonal anisotropy in the basal plane, but we are required to select the $E_2$ representation. There is support for the assumption of weak hexagonal anisotropy; if the hexagonal anisotropy of $\vec{v}_f$ were significant it should be observable at low temperature as an in-plane anisotropy of $H_{c2}^\perp(T)$. The angular dependence of $H_{c2}^\perp$ was investigated, but no in-plane anisotropy was observed.[36]

In order to account for the discontinuities in the slopes of the transition lines near the tetracritical point an additional ingredient in the GL theory for the $E_{2u}$ model is needed that is not contained in the theory of Hess, et al.[20] For $E_{2u}$ with $\kappa_{23} = 0$ the gradient energy reduces to $\mathcal{F}_{grad} = \int d^3x \{\kappa_1(\,|\vec{D}_\perp\eta_1|^2 + |\vec{D}_\perp\eta_2|^2) + \kappa_4(\,|D_z\eta_1|^2 + |D_z\eta_2|^2)\}$. Because both order parameter components appear with the same coefficients there is no crossing of different $H_{c2}(T)$ curves corresponding to different eigenfunctions, and therefore no tetracritical point. However, the slopes of the transition lines near the tetracritical point suggest that the difference in the gradient energies associated with the two components of the order parameter are finite, but small, e.g. $|\Delta\kappa/2\kappa_1| \lesssim 0.2$. This suggests that the SBF may be responsible for a splitting in the gradient coefficients as well as the transition temperature.

In the model of Hess, et al.[20] the coupling to the SBF was included through second order in both the superconducting order parameter, $\vec{\eta}$, and the AFM order parameter, $\vec{M}_s$, but only for the homogeneous terms in the free energy. However, there is also a contribution to the gradient energy that is second-order in $M_s$. These terms are as essential for describing a double transition as a function of field, as the homogeneous term is for the double transition in zero field. The relevant invariants are of the form $M_s^2(|\vec{D}_\perp\eta_1|^2 - |\vec{D}_\perp\eta_2|^2)$ and $M_s^2(|D_z\eta_1|^2 - |D_z\eta_2|^2)$, and lead to a gradient energy

$$\mathcal{F}_{grad} = \int d^3x \left\{ (\kappa_1^+|\vec{D}_\perp\eta_1|^2 + \kappa_1^-|\vec{D}_\perp\eta_2|^2) + (\kappa_4^+|D_z\eta_1|^2 + \kappa_4^-|D_z\eta_2|^2) \right\}, \tag{8}$$

where $\kappa_1^\pm = \kappa_1(1 \pm \epsilon_\perp M_s^2)$ and $\kappa_4^\pm = \kappa_4(1 \pm \epsilon_\parallel M_s^2)$. Dimensional analysis implies that the coupling coefficients, $\epsilon$, $\epsilon_\perp$, $\epsilon_\parallel$, for the homogeneous term, the in-plane gradient energies and the $\hat{c}$-axis gradient energies are formally the same order of magnitude, in which case we conclude that the splittings in the gradient coefficients are relatively small,

$$\left|\frac{\kappa_{1,4}^+ - \kappa_{1,4}^-}{2\kappa_{1,4}}\right| = \left|\epsilon_{\perp,\parallel}M_s^2\right| \sim \left|\frac{\Delta T_c}{T_c}\right|. \tag{9}$$

Thus, in this GL theory the SBF is essential for producing an apparent tetracritical point, and at this semi-quantitative level, can account for the magnitudes of the slopes near the tetracritical point.[22] As this model for UPt$_3$ illustrates, the GL theory provides a central link between experiments, in this case the phase diagram, and the more microscopic Fermi-liquid theory which we now discuss.

### III. FERMI-LIQUID THEORY

Conduction electrons in metals interact strongly with each other and with the lattice. These interactions lead to correlations among the electrons, and we have to view conduction electrons, in general, as a system of correlated Fermions. On the other hand, the model of non-correlated electrons occupying single-particle states in conduction bands describes most properties of traditional metals very well. This simple behaviour of an intrinsically complicated system was explained by Landau in a series of papers in which he established his "Fermi-liquid theory." Landau argued that interacting Fermions can form a "Fermi-liquid state," which for many properties has the appearance of a non-correlated state. The reason, according to Landau, is that the physical properties of a Fermi liquid at low temperatures are dominated by low-lying excitations (quasiparticles) which are complex objects but have basic features (e.g. charge, spin, fermion number) in common with non-interacting electrons. Landau showed that an ensemble of quasiparticles is described by a classical distribution function $g(\vec{k}_f, \vec{R}; \epsilon, t)$,[†] and that this distribution function obeys a classical transport equation (Boltzmann-Landau transport equation).

Derivations of the Boltzmann-Landau transport equation from first principles[38,39] use many-body Green's function techniques, and lead to explicit expressions for the various terms of the transport equation in terms of self-energies. The self-energies describe the effects of electron-electron and electron-phonon interactions as well as impurity scattering. The complete set of diagrams for the leading order self-energies are shown in Fig. (I); the full circles are vertices representing quasiparticle-quasiparticle, quasiparticle-phonon, and quasiparticle-impurity interactions. We follow Landau and consider these vertices together with the Fermi surface properties (e.g. shape of the Fermi surface and Fermi velocities) as phenomenological parameters of the Fermi-liquid model, but in principle they can be obtained from the full many-body theory.

The present notion of a normal Fermi liquid is somewhat more general than standard definitions. For the conduction electrons of a metal to form a Fermi liquid we require that

- the dominant charge-carrying excitations have the signatures of electrons or holes; they are fermions with charge $\pm e$, spin 1/2, *large* velocities $\vec{v}_f$ and *large* momenta $\vec{k}_f$ near a Fermi surface.[‡]

- The physical properties of the ensemble of these excitations can be described by a Boltzmann-Landau transport equation.

In the following we call the electronic quasiparticles simply "electrons", but emphasize that the physical properties of such an electron are, in general, significantly affected by many-body effects.

---

[†]We use the "energy representation" in which the traditional variables $\vec{k}$, $\vec{R}$, $t$ of the distribution function are replaced by the equivalent set $\vec{k}_f$, $\epsilon$, $\vec{R}$, $t$. This amounts to a transformation to new coordinates in momentum space. The 3-dimensional momentum variable ($\vec{k}$) is replaced by the 2-dimensional momentum variable $\vec{k}_f$, which is the point on the Fermi surface nearest to $\vec{k}$, and the 1-dimensional energy variable $\epsilon = E(\vec{k}, \vec{R}, t) - E_f$. The energy representation of the Boltzmann-Landau equation has a wide range of validity. It also describes overdamped excitations whose lifetime is comparable to or shorter than its oscillation period $\hbar/\epsilon$. For a review on transport theory in energy representation see Ref. ( 37).

[‡]The term *large* means that the corresponding energies ($k_f^2/2m^* \sim m^* v_f^2/2 \sim E_f$) are much greater than typical excitation energies $\epsilon \sim k_B T_c$.

Landau's Fermi-liquid model predicts a number of universal results for temperature and magnetic field dependences of thermodynamic and transport properties at very low temperatures. The universal laws can be used as experimental signatures for Fermi-liquid behaviour, provided the low-temperature limit is reached before the superconducting transition. A detailed discussion of these normal-state properties can be found in various textbooks and review articles.[40] The systems with a potential interest for non-s-wave superconductivity, i.e. heavy-fermion superconductors and high-$T_c$ superconductors, often have not yet reached their ideal low-temperature limit when superconductivity sets in. In such systems the investigation of Fermi-liquid behaviour requires an analysis of measurements in both the normal and superconducting phases. We focus on the Fermi-liquid model of the superconducting phases.

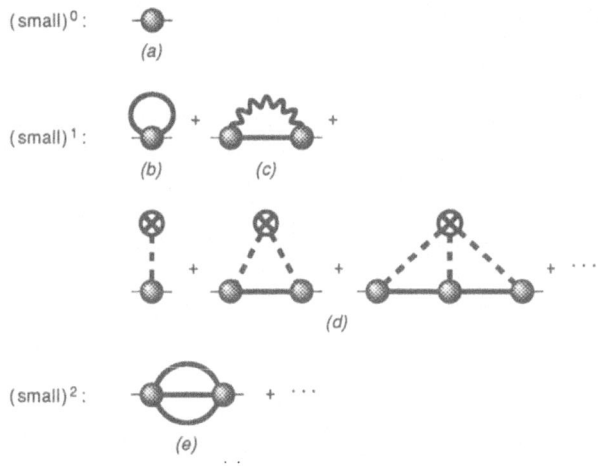

Fig. I  Leading order self-energy diagrams of Fermi-liquid theory. The vertices (shaded circles) represent the sum of all high-energy processes and give rise to interactions between the quasiparticles (smooth propagator lines), phonons (wiggly propagator lines) and impurities (dashed lines). The order in the parameter 'small' is indicated for each diagram.

### Superconducting Fermi Liquids

It took more than 10 years after the breakthrough in the theory of superconductivity by BCS to establish a complete Fermi-liquid theory of superconductivity. Earlier theories of superconductivity were formulated in terms of Bogolyubov's equations, Gorkov's equations, or other fully quantum mechanical schemes. They lacked the quasiclassical aspects of Landau's Fermi-liquid theory. The first complete quasiclassical theory of superconductivity was formulated in a series of publications by Eilenberger,[41] Larkin and Ovchinnikov,[42,43] and Eliashberg.[44] It is presented and discussed in several review articles.[45-47,37] The quasiclassical theory allows one to calculate all superconducting phenomena of interest, including transition temperatures, excitation spectra, Josephson effects, vortex structures, the electromagnetic response, etc. In this theory the dynamical degrees of freedom of electronic quasiparticles are described partly by classical statistical mechanics, and partly by quantum statistics. The classical degrees of freedom are the motion in $\vec{k}$-$\vec{R}$ phase space; i.e. quasiparticles move along classical trajectories. Of special importance for understanding superconducting phenomena are the quantum-mechanical "internal degrees of freedom", the *spin* and the *particle-hole* degrees of freedom of an electron. Quantum coherence between particle excitations and hole excitations is a key feature of the BCS theory of superconductivity and the origin of all non-classical effects in superconductors (e.g. supercurrents, coherence factors in transition amplitudes, Andreev reflection, etc.). Particle-hole coherence is taken care of in the quasiclassical theory by grouping particle excitations (occupied one-electron states with energy above the Fermi energy ($\epsilon > 0$)) and hole excitations (empty one-electron states with $\epsilon < 0$) into a doublet. The spin and particle-hole doublets form a 4-dimensional Hilbert space of internal degrees of freedom for quasiparticle excitations. In quantum statistical mechanics the internal state of quasiparticle excitations is described by a $4 \times 4$ density matrix. Hence, the distribution function of the quasiclassical theory is a $4 \times 4$-matrix whose elements are functions of $\vec{k}_f$, $\vec{R}$, $\epsilon$, and $t$.

The central object of the quasiclassical theory is the "Keldysh quasiclassical propagator" $\hat{g}^K(\vec{k}_f, \vec{R}; \epsilon, t)$. It is a $4 \times 4$-matrix which generalizes the classical Boltzmann-Landau distribution function to the superconducting state, and carries the information on both the classical and quantum degrees of freedom. We denote $4 \times 4$-matrices by a "hat". In addition, we use standard notation and introduce a superscript on the quasiclassical propagators to identify their microscopic origin. Thus, $\hat{g}^K$ denotes the Keldysh propagator which is obtained by $\xi$-integration[46] from the microscopic Keldysh Green's function. The 16 matrix elements of $\hat{g}^K$ can be expressed in terms of 4 spin-scalars ($g^K$, $\underline{g}^K$, $f^K$, $\underline{f}^K$) and 4 spin-vectors ($\vec{g}^K$, $\underline{\vec{g}}^K$, $\vec{f}^K$, $\underline{\vec{f}}^K$):

$$\hat{g}^K = \begin{pmatrix} g^K + \vec{g}^K \cdot \vec{\sigma} & \left( f^K + \vec{f}^K \cdot \vec{\sigma} \right) i\sigma_y \\ i\sigma_y \left( \underline{f}^K + \underline{\vec{f}}^K \cdot \vec{\sigma} \right) & \underline{g}^K - \sigma_y \underline{\vec{g}}^K \cdot \vec{\sigma} \sigma_y \end{pmatrix}. \tag{10}$$

Retarded and advanced quasiclassical propagators, $\hat{g}^{R,A}(\vec{k}_f, \vec{R}, ; \epsilon, t)$, are defined analogously by $\xi$-integration of the retarded and advanced microscopic Green's functions.[46] The diagonal components of (10) are directly related to scalar distribution functions for particles, $n_p(\vec{k}_f, \vec{R}; \epsilon, t)$, and holes, $n_h(\vec{k}_f, \vec{R}; \epsilon, t)$:[§]

$$n_p = \frac{-i}{(2\pi)^4 \mid \vec{v}_f \mid} \left( g^K - (g^R - g^A) \right), \quad n_h = \frac{+i}{(2\pi)^4 \mid \vec{v}_f \mid} \left( \underline{g}^K - (\underline{g}^R - \underline{g}^A) \right) \tag{11}$$

Similarly, $\vec{g}^K$ and $\underline{\vec{g}}^K$ are related to spin-distribution functions. The off-diagonal components $f^K(\vec{k}_f, \vec{R}; \epsilon, t)$, $\vec{f}^K(\vec{k}_f, \vec{R}; \epsilon, t)$ in (10) can be interpreted as pair amplitudes. [**] They carry the information on particle-hole coherence. Non-zero pair amplitudes indicate particle-hole coherence and thus superconductivity. More specifically, a non-vanishing scalar pair amplitude, $f^K$, (vector pair amplitude $\vec{f}^K$) describes even parity, spin-singlet pairing (odd parity, spin-triplet pairing).

---

[§] $n_{p(h)}(\vec{k}_f, \vec{R}; \epsilon, t) d^2 k_f d^3 R d\epsilon$ is the number of particle excitations (hole excitations) with momentum $\vec{k}_f$, position $\vec{R}$, and energy $\epsilon$ in the phase space element $d^2 k_f d^3 R d\epsilon$.

[**] The amplitudes $\underline{f}^K$ and $\underline{\vec{f}}^K$ are redundant since they are related to $f^K$ and $\vec{f}^K$ by fundamental symmetry relations.[46]

Measurable quantities such as the charge density $n(\vec{R}, t)$, current density $\vec{j}(\vec{R}, t)$, etc. can be calculated from the diagonal components of the quasiclassical propagator, $\hat{g}^K$, or, equivalently, from the distribution functions $n_p$ and $n_h$. If we ignore the Landau parameters (contained in diagram Ib), then

$$
\begin{aligned}
n(\vec{R}, t) &= n_0 + 2e^2 N_f \Phi(\vec{R}, t) + 2e \int_{fs} \frac{d^2 k_f}{(2\pi)^3 |\vec{v}_f|} \int_{-\infty}^{\infty} \frac{d\epsilon}{4\pi i} g^K(\vec{k}_f, \vec{R}; \epsilon, t) \\
&= n_0 + 2e^2 N_f \Phi(\vec{R}, t) + e \int_{fs} d^2 k_f \int_0^{\infty} d\epsilon \left( n_p - n_h + \frac{(g^R - g^A) + (g^R - \underline{g}^A)}{(2\pi)^4 |\vec{v}_f| i} \right) ,
\end{aligned}
\tag{12}
$$

$$
\begin{aligned}
\vec{j}(\vec{R}, t) &= 2e \int_{fs} \frac{d^2 k_f}{(2\pi)^3 |\vec{v}_f|} \int_{-\infty}^{\infty} \frac{d\epsilon}{4\pi i} \vec{v}_f \, g^K(\vec{k}_f, \vec{R}; \epsilon, t) \\
&= e \int_{fs} d^2 k_f \int_0^{\infty} d\epsilon \vec{v}_f \left( n_p + n_h + \frac{(g^R - g^A) - (g^R - \underline{g}^A)}{(2\pi)^4 |\vec{v}_f| i} \right) .
\end{aligned}
\tag{13}
$$

In eq. (12), $n_0 + 2e^2 N_f \Phi(\vec{R}, t)$ is the local equilibrium charge density of electrons in the presence of an electro-chemical potential $\Phi$. The terms involving $n_p$, $n_h$ in eqs.(12, 13) represent the contributions of thermally excited particles and holes, whereas the terms involving $g^{R,A}$, $\underline{g}^{R,A}$ are contributions from the superconducting condensate. Note that the hole-distribution function enters the density and the current with different signs, because electrons and holes with the same momentum $\vec{k}_f$ have opposite charge but carry the same current.

The central equation of the quasiclassical theory is the quasiclassical transport equation for $\hat{g}^K$, and is the generalization of the Landau-Boltzmann equation to the superconducting state. It is a set of coupled partial differential equations of first order in the spatial derivatives and, in general, of infinite order in the time derivative. In a compact notation, the quasiclassical transport equation reads

$$
\left( \epsilon \hat{\tau}_3 - \hat{v} - \hat{\sigma}^R \right) \otimes \hat{g}^K - \hat{g}^K \otimes \left( \epsilon \hat{\tau}_3 - \hat{v} - \hat{\sigma}^A \right) - \hat{\sigma}^K \otimes \hat{g}^A + \hat{g}^R \otimes \hat{\sigma}^K + i \vec{v}_f \cdot \vec{\nabla} \hat{g}^K = 0, \tag{14}
$$

where the $\otimes$-product stands for the usual $4 \times 4$-matrix product and a product in the energy-time variables defined by $\hat{a} \otimes \hat{b}(...; \epsilon, t) = exp \left( i(\partial_\epsilon^a \partial_t^b - \partial_t^a \partial_\epsilon^b)/2 \right) \hat{a}(...; \epsilon, t) \hat{b}(...; \epsilon, t)$. The superscripts $a$ $(b)$ on the partial derivatives indicate derivatives with respect to the arguments of $\hat{a}$ $(\hat{b})$. A determination of $\hat{g}^K$ from the transport eq. (14) requires knowledge on the external potentials $\hat{v}(\vec{k}_f, \vec{R}, t)$, the advanced, retarded and Keldysh self-energies $\hat{\sigma}^{R,A,K}(\vec{k}_f, \vec{R}; \epsilon, t)$, and the advanced and retarded quasiclassical propagators $\hat{g}^A(\vec{k}_f, \vec{R}; \epsilon, t)$ and $\hat{g}^R(\vec{k}_f, \vec{R}; \epsilon, t)$. These propagators are auxiliary quantities which in general have no direct physical interpretation, except in the adiabatic limit[46] where they determine the local quasiparticle density of states $N(\vec{k}_f, \vec{R}; \epsilon, t)$ via the relation $N(\vec{k}_f, \vec{R}; \epsilon, t) = -N_f \Im m \left( g^R(\vec{k}_f, \vec{R}; \epsilon, t) \right) / \pi$. Finally, in equilibrium $\hat{g}^K$ is given in terms of $\hat{g}^{R,A}$ by $\hat{g}^K = \left( \hat{g}^R - \hat{g}^A \right) \tanh(\epsilon/2T)$.

The solutions to the quasiclassical equations for $\hat{g}^{R,A}$,

$$
\left[ \epsilon \hat{\tau}_3 - \hat{v} - \hat{\sigma}^{R,A}, \hat{g}^{R,A} \right]_\otimes + i \vec{v}_f \cdot \vec{\nabla} \hat{g}^{R,A} = 0 , \tag{15}
$$

are necessary inputs to the transport eq. (14), and the set of physically relevant solutions of eqs. (14, 15) must satisfy the normalization conditions,

$$
\hat{g}^{R,A} \otimes \hat{g}^{R,A} = -\pi^2 \hat{1} , \quad \hat{g}^R \otimes \hat{g}^K + \hat{g}^K \otimes \hat{g}^A = 0 . \tag{16}
$$

The quasiclassical equations (14, 15) need to be complemented by self-consistency equations for the the quasiclassical self-energies $\hat{\sigma}^{R,A,K}$. They give $\hat{\sigma}^{R,A}$ in terms of $\hat{g}^{R,A}$, and $\hat{\sigma}^K$ in terms of $\hat{g}^R$, $\hat{g}^A$ and $\hat{g}^K$. The self-consistency equations are shown, in a diagrammatic notation, in Fig. (I). For example, these equations include the "gap equation", which is the self-consistency equation for the off-diagonal self-energy. Explicit forms of the quasiclassical self-consistency equations can be found in the reviews.[45–47,37]

This completes our list of equations of the theory of Fermi-liquid superconductivity. The quasiclassical equations, normalization conditions, and self-consistency equations (Fig. (I)) form a set of non-linear integro-differential equations. They are exact to leading orders in the expansion parameters of Fermi-liquid theory (e.g. $k_B T/E_f$, $1/k_f \xi_0$, $1/k_f \ell$, $\omega_D/E_f$, ...), and cover equilibrium phenomena as well as non-equilibrium superconductivity.

### Non-s-Wave Pairing

The quasiclassical theory is especially well suited for studying non-s-wave superconductors. Many calculational steps of the theory are identical for s-wave and non-s-wave superconductors, and many s-wave results can be simply generalized to anisotropic, non-s-wave wave superconductors. The reason is that a quasiparticle at point $\vec{R}$ moving with momentum $\vec{k}_f$ senses only the gap function $\Delta(\vec{k}_f, \vec{R})$. Since it does not sample the full momentum dependence of the gap function it cannot distinguish an anisotropic gap from an isotropic one. For example, the density of states $N(\vec{k}_f; \epsilon)$ of a homogeneous superconductor in equilibrium has the traditional BCS form, $N(\epsilon, \vec{k}_f) = N_f \Re e(\epsilon/\sqrt{\epsilon^2 - |\Delta|^2})$, but with the isotropic BCS gap $|\Delta|$ replaced by an anisotropic local gap $|\Delta(\vec{k}_f)|$ for singlet pairing or $|\vec{\Delta}(\vec{k}_f)|$ for unitary triplet pairing.

Several measurable properties of a non-s-wave superconductor can be calculated in a rather simple way by using standard s-wave formulas locally at any point $\vec{k}_f$ on the Fermi surface, and then averaging over the Fermi surface. Typical examples for such averaged quantities are the free energy of a superconductor (and related quantities such as the critical field $H_c(T)$ and the specific heat $c_s(T)$), the stripped penetration depth tensor, $\lambda_{ij}$,[tt]

$$\frac{H_c^2(T)}{8\pi} = N_f \left\langle |\Delta(\vec{k}_f)|^2 \left( \ln(T_c/T) - \pi T \sum_{n=-\infty}^{+\infty} \frac{|\Delta(\vec{k}_f)|^2}{|\epsilon_n| \left( |\epsilon_n| + \sqrt{\epsilon_n^2 + |\Delta(\vec{k}_f)|^2} \right)^2} \right) \right\rangle_{fs} \quad (17)$$

$$(\lambda^{-2})_{ij} = N_f \frac{4\pi e^2}{c^2} \pi T \sum_{n=-\infty}^{+\infty} \left\langle \frac{v_{f_i}(\vec{k}_f) v_{f_j}(\vec{k}_f) |\Delta(\vec{k}_f)|^2}{\left[ \epsilon_n^2 + |\Delta(\vec{k}_f)|^2 \right]^{3/2}} \right\rangle_{fs}, \quad (18)$$

where $\epsilon_n = (2n+1)\pi T$ are the Matsubara energies and the Fermi surface average is,

$$\left\langle \cdots \right\rangle_{fs} = \frac{1}{N_f} \int \frac{d^2 k_f}{(2\pi)^3 |\vec{v}_f|} \cdots . \quad (19)$$

Eq. (18) is the clean limit of a general formula derived in Ref. ( 49, 50).

---

[tt]The measured penetration depth is affected (dressed) by quasiparticle interactions.[48] We follow the terminology of superfluid $^3$He and use the adjective "stripped" for properties of non-interacting quasiparticles.

For these averaged properties there is no fundamental difference between anisotropic conventional superconductors and genuine unconventional superconductors. One significant feature are here the zeros of the energy gap on the Fermi surface. These zeros (points or lines of zeros) determine the phase space for quasiparticle excitations, and lead to characteristic temperature dependences for $T \to 0$ which might be used to identify unconventional pairing. Zeros of the gap in unconventional superconductors are often a direct consequence of the symmetry of the order parameter, and thus robust features. They lead to power laws in the $T$-dependence of Fermi-surface averaged quantities, whereas superconductors with a finite minimum gap on the Fermi surface show an exponential $T$-dependence at low temperatures. The various power laws predicted by theory for unconventional superconductors are discussed in the reviews.[9,10]

The most striking differences between s-wave and non-s-wave superconductors, which are intimately connected with unconventional superconducting order parameters, are (i) complex phase diagrams due to the existence of a variety of different superconducting phases (section II), (ii) new types of vortices and other defects in the order parameter field,[51–56] (iii) new collective modes which affect the electromagnetic response and other dynamical response functions,[57,58] (iv) a significant effect on the order parameter due to impurities (section IV), surfaces and interfaces,[59,60] (v) anomalous Josephson effects.[61–64] All these anomalous collective effects have been observed in superfluid $^3$He, which is the best studied unconventional "superconductor". By contrast, only a few of these indicators for non-s-wave superconductivity have been observed in superconducting metals. For example, the $H$-$T$ phase diagram of UPt$_3$ shows a variety of superconducting phases and provides evidence for unconventional superconducting order parameter (see section II). And in high-$T_c$ superconductors (YBCO) anomalous Josephson effects have recently been reported.[65]

Collective effects are different in origin than the "averaged effects" discussed above. They probe the coherence of quasiparticles with different momenta $\vec{k}_f$, i.e., quasiparticles moving along different classical trajectories. The coupling of quasiparticles with different momenta is a consequence of quasiparticle interactions and self-energy terms in the quasiclassical equations. The self-consistency equations for the self-energies are integrals over the Fermi surface which sample, in general, the full anisotropy and residual symmetries of the order parameter. New superconducting phases and novel defects in the order parameter field, e.g. new types of vortices, are special self-consistent solutions for the off-diagonal self-energies (gap functions) (see section II). Collective modes in systems with unconventional pairing are discussed most thoroughly for triplet pairing in $^3$He; we refer to the extensive literature in this field.[66,67] In the following section we focus on collective effects induced by impurity scattering.

## IV. IMPURITY SCATTERING

Scattering by ordinary, nonmagnetic impurities is expected to have many dramatic effects on the properties of unconventional superconductors.[9] In this section we discuss the theory of such effects, using the framework of the Fermi liquid theory of superconductivity.

In unconventional superconductors, it is useful to distinguish two types of effects. Impurities affect global properties of superconductors, such as thermodynamic properties ($T_c$, critical fields, heat capacity,...) and long wavelength response coefficients (conductivity, sound absorption,...). Measurements of these macroscopic quantities average over the positions of the impurities and give coarse-grained, "impurity averaged" information. Impurities also cause strong local changes in their environment, out to distances of order $\xi(T)$. Such local effects might be measured by probes such as NMR and high resolution STM. These two different effects require different types of calculations:

- Calculations which average over the positions of the impurities, and so produce coarse-grained, "impurity averaged" quantities such as the transition temperature and density of states.

- Calculations which focus on the local environment of a single impurity, and compute position-dependent quantities such as the current or order parameter.

To start, we note that many of the properties of conventional superconductors are very little affected by the presence of nonmagnetic impurities. For example, quantities such as the transition temperature, specific heat, and density of states are quite insensitive to such scattering. On the other hand, transport response functions such as the superfluid density are affected rather strongly.

In contrast, in unconventional superconductors, all of the quantities mentioned in the preceding paragraph are strongly affected by nonmagnetic impurities. This opens up interesting areas of research in two ways- novel effects in superconductivity can be studied, and conversely, superconductivity can be used as a probe of the properties of such impurities. We give some illustrative examples:

1. Transition Temperature– The value of $T_c$ is strongly reduced by impurity scattering. The precise value of the reduction depends on the angular dependence, in $\vec{k}$ space, of both the order parameter and the scattering matrix.[9] If we let $\tau$ be the normal state transport time for the impurity scattering in question, then we can make a rough estimate by saying that the reduction depends on the pair-breaking parameter $1/T_c\tau$ in approximately the same way that the reduction in conventional superconductors depends on $1/T_c\tau_s$, where $\tau_s$ is the spin flip time.

2. Density of States– $N(\omega)$ can be strongly affected. For example, if the energy gap vanishes on a line of nodes on the Fermi surface, for a pure superconductor we have $N(\omega) \sim \omega$ for small energies. The addition of an arbitrarily small concentration of impurities makes the superconductor gapless;[9,68] this means that $N(\omega = 0)$ does not vanish.

3. Order Parameter– The question here is: how is the order parameter $\Delta(\vec{k}_f, \vec{R})$ affected in the vicinity of a nonmagnetic impurity? In general $\Delta(\vec{k}_f, \vec{R})$ is distorted away from its pure state value $\Delta_0(\vec{k}_f)$. However, in a conventional superconductor this distortion is important only out to distances of order $1/k_f$ from the impurity.[69] For an unconventional superconductor this distortion has a much larger spatial extent, going out to distances of order $\xi(T)$.[70] In addition, if the order parameter is a complex function of $\vec{k}_f$, a pattern of supercurrents is set up near the impurity, leading to the existence of a magnetic field;[70,71] these currents will be discussed later in this section. Thus, what we are calling a "nonmagnetic" impurity can develop magnetic properties in the superconducting state, if the order parameter breaks time reversal symmetry. These properties can be used as an experimental check of such an order parameter.

4. Superfluid Density Tensor– In general, impurity scattering reduces the elements of $\bar{\bar{\rho}}_s$. One surprising effect can occur in unconventional superconductors: impurity scattering can increase the anisotropy in the $\rho_s$ tensor.[49] For example, for an $l =$

$1, m = 0$ order parameter with a line of nodes in the $xy$ plane, and a spherical Fermi surface, we can derive the following simple result close to $T_c$, as the impurity scattering time $\tau$ goes to zero: $\rho_{zz}/\rho_{xx} \to -(10/3)\ln(4\pi\tau T_c)$.

### Impurity Averaging Calculations

To illustrate the use of the impurity-averaged quasiclassical equation, we review one particular calculation, a calculation which is perhaps somewhat novel. The general question to be addressed is: if the magnitude of the energy gap varies through space, do supercurrents arise?[72,73] For weak-coupling theory in the absence of impurities, the answer is no, if particle-hole symmetry is assumed.[74] That is, supercurrents are produced by spatial variations in the phase and orientation of the order parameter, not by variations in its magnitude.

Note that if the gap is a complex function of $\vec{k}$, and so breaks time reversal symmetry, a term in the supercurrent proportional to the gradient of the magnitude of the gap is allowed by symmetry. Gorkov[9] first pointed out that impurity scattering could give such a term a nonzero coefficient, even in the context of a weak-coupling calculation.

For definiteness, we consider a singlet gap transforming according to the $E_{1g}$ representation of the hexagonal group $D_{6h}$. This particular case has been discussed as a possible model for the heavy-fermion superconductor $UPt_3$. [‡‡] Thus we write the gap as $\Delta(\vec{k}_f) = \eta_1 \mathcal{Y}_1(\vec{k}_f) + \eta_2 \mathcal{Y}_2(\vec{k}_f)$. Here, the $\mathcal{Y}_i(\vec{k}_f)$ are the basis functions, and the coefficient $\vec{\eta} = (\eta_1, \eta_2)$ transforms as a vector in the $xy$ plane. In the Ginzburg-Landau region, the supercurrent is given, in terms of the gauge invariant derivative $\vec{D} = \vec{\nabla} + \frac{2ie}{\hbar c}\vec{A}$, by[72]

$$\vec{J} = \frac{4e}{\hbar}Im[\kappa_1\eta_j\vec{D}^*\eta_j^* + \frac{1}{2}\kappa_{23}(\vec{\eta}D_j^*\eta_j^* + \eta_jD_j^*\vec{\eta}^*) + \hat{z}\kappa_4\eta_jD_z^*\eta_j^*] - \kappa_a\frac{ie}{\hbar}[\vec{\nabla} \times (\vec{\eta} \times \vec{\eta}^*)]. \quad (20)$$

This expression is obtained as the stationarity condition of the GL functional of eq. (5); the coefficients $\kappa_1$, $\kappa_{23}$, $\kappa_4$, and $\kappa_a$ can be calculated from the quasiclassical theory. We focus our attention on the coefficient $\kappa_a$; this is the one which vanishes in weak-coupling theory unless impurity scattering is included. Let us choose a particular form for the order parameter: $\vec{\eta}(\vec{R}) = \eta_0(\vec{R})\vec{\phi}$, where $\vec{\phi}$ is a fixed two dimensional vector and $\eta_0(\vec{R})$ is a spatially varying, real amplitude. If we make the choice $\vec{\phi} = (\hat{x} + i\hat{y})/\sqrt{2}$ we then have $\vec{J} = (e\kappa_a/\hbar)\hat{z} \times \vec{\nabla}\eta_0^2$.

To calculate $\kappa_a$ we use the following strategy. We assume the order parameter has the form given above, with an $\eta_0(\vec{R})$ which varies slowly in space. We then compute the Matsubara propagator[46] $\hat{g}$ in a gradient expansion($\hat{g} = \hat{g}^{(0)} + \hat{g}^{(1)}$), use this $\hat{g}$ to compute the current, and then read off $\kappa_a$ from our answer. For simplicity we take our Fermi surface to be spherical ($\vec{k}_f = k_f\hat{k}$), and choose the following basis functions: $\mathcal{Y}_1(\hat{k}) = \sqrt{15}\hat{k}_x\hat{k}_z$, $\mathcal{Y}_2(\hat{k}) = \sqrt{15}\hat{k}_y\hat{k}_z$. We evaluate the impurity self energy in the Born approximation, and consider spherically symmetric impurities. This gives

$$\hat{\sigma}^{(j)}(\hat{k}, \vec{R}; \epsilon_n) = cN_f \int \frac{d^2\hat{k}'}{4\pi} W(\hat{k}, \hat{k}')\hat{g}^{(j)}(\hat{k}', \vec{R}; \epsilon_n). \quad (21)$$

---

[‡‡] Similar calculations to those in this section can be carried out for the $E_{2u}$ representation, which in our opinion is the most likely candidate for the phases of $UPt_3$.

Here, $c$ is the concentration of impurities. We can expand $W$ in spherical harmonics as follows:

$$W(\hat{k},\hat{k}') = 4\pi \sum_l W_l \sum_m Y_{lm}^*(\hat{k}) Y_{lm}(\hat{k}') . \tag{22}$$

Note that we are keeping all terms in the spherical harmonic expansion of $W$; very often, calculations make the so-called s-wave approximation, and keep only the $W_0$ term. As we shall shortly see, the s-wave approximation is not good enough for our purposes.

If we solve for $\hat{g}^{(1)}$, and then compute the current, we find the following formula for $\kappa_a$[72,73]

$$\kappa_a = \frac{N_f \xi_0^2}{80} \sum_{n \geq 0} \frac{1}{n_+(n_+ - \alpha_2)^2} \times \left\{ \frac{\alpha_1}{(n_+ - \alpha_1)} - \frac{\alpha_3}{(n_+ - \alpha_3)} \right\} . \tag{23}$$

Here, $n_+ = n + 1/2 + \alpha_0$, the pair breaking parameters $\alpha_l$ are defined by $\alpha_l = cN_fW_l/2T_c$, and the correlation length is given by $\xi_0 = v_f/\pi T_c$.

There are several noteworthy features about this result for $\kappa_a$. First, we see that it vanishes in the absence of impurity scattering. In fact, if we had only kept the s-wave piece of $W$, we would still have found that $\kappa_a$ was zero. Second, we can see that $\kappa_a$ depends on the details of the impurity scattering in a quite intricate fashion. The $l = 0, 1, 2, 3$ pieces of $W$ all enter into the answer. Finally, the sign of $\kappa_a$ can go either way. Note that $\alpha_0$ must be positive, but the other $\alpha_l$'s can be either positive or negative.

We also remark that the coefficient $\kappa_a$ plays a role in the discussion of whether or not Cooper pairs have angular momentum. For a discussion of this point we refer to the literature.[72,74,75]

### Single Impurity Calculations

Calculations concerning the local environment of a single impurity are of interest for several reasons. On the one hand, experimentally measurable quantities, such as local magnetic field, can be investigated. These quantities can be important probes of the type of order parameter present in the material. It is also true, however, that these local types of calculation can be important in increasing our physical insight;[76] the impurity-averaging technique is quite powerful, but the microscopic origin of interesting effects may be hard to recover after the averaging procedure.

In an unconventional superconductor, much interesting structure develops in the neighborhood of a nonmagnetic impurity. In contrast to the conventional case, this structure is significant out to quite long distances from the impurity, distances of order $\xi(T)$.[70] If the order parameter breaks time reversal symmetry, these local perturbations are particularly interesting: a local pattern of equilibrium supercurrents is set up, leading to the existence of a spatially varying magnetic field.[71]

We consider a single, nonmagnetic impurity, located at $\vec{R} = 0$; the interaction of the electronic quasiparticles with the impurity can be described by the potential $v(\vec{k}_f, \vec{k}_f')$. The quasiclassical approach[77,78] then leads to coupled equations for the Matsubara propagator $\hat{g}(\vec{k}_f, \vec{R}; \epsilon_n)$, the order parameter $\Delta(\vec{k}_f, \vec{R})$, and the t-matrix $\hat{t}(\vec{k}_f, \vec{k}_f'; \epsilon_n)$. The equations for the propagator are

$$[i\epsilon_n \hat{\tau}_3 - \hat{\Delta}, \hat{g}] + i\vec{v}_f \cdot \vec{\nabla}\hat{g} = [\hat{t}, \hat{g}_{imt}]\delta(\vec{R}) , \qquad \hat{g}\hat{g} = -\pi^2 \hat{1} . \tag{24}$$

Here, the intermediate propagator $\hat{g}_{imt}$ satisfies the following equation: $[i\epsilon_n\tau_3 - \hat{\Delta}, \hat{g}_{imt}] + i\vec{v}_f \cdot \vec{\nabla}\hat{g}_{imt} = 0$, $\hat{g}_{imt}\hat{g}_{imt} = -\pi^2\hat{1}$. Note that the equations for both $\hat{g}$ and $\hat{g}_{imt}$ involve the order parameter $\Delta(\vec{k}_f, \vec{R})$, which must be found from the gap equation; this gap equation needs $\hat{g}$ as input.

The final ingredient is an equation for the t-matrix:

$$\hat{t}(\vec{k}_f, \vec{k}_f'; \epsilon_n) = v(\vec{k}_f, \vec{k}_f')\hat{1} + \int \frac{d^2\vec{k}_f''}{(2\pi)^3 \mid \vec{v}_f'' \mid} v(\vec{k}_f, \vec{k}_f'')\hat{g}_{imt}(\vec{k}_f'', \vec{R} = 0; \epsilon_n)\hat{t}(\vec{k}_f'', \vec{k}_f'; \epsilon_n). \quad (25)$$

So in general we face the difficult task of solving self-consistently for all of the involved quantities. However, for a small impurity, a certain simplification is possible.[70,77] By small, we mean an impurity with a cross section much smaller than the square of the coherence length: $\sigma \ll \xi_0^2$. We may then work to first order in the small parameter $\sigma/\xi_0^2$ and write

$$\hat{g} = \hat{g}_0 + \delta\hat{g}, \quad \hat{\Delta} = \hat{\Delta}_0 + \delta\hat{\Delta}. \quad (26)$$

Here, $\hat{g}_0$ and $\hat{\Delta}_0$ are the spatially uniform, unperturbed propagator and order parameter, and we work to first order in the small quantities $\delta\hat{g}$ and $\delta\hat{\Delta}$. We may then take $\hat{g}_0 = \hat{g}_{imt} = -\pi(i\epsilon_n\hat{\tau}_3 - \hat{\Delta}_0)/(\epsilon_n^2 + |\Delta_0|^2)^{1/2}$. The leading order answer for the Fourier transform $\delta\hat{g}(\vec{k}_f, \vec{q}; \epsilon_n)$ is then[79]

$$\delta\hat{g}(\vec{k}_f, \vec{q}; \epsilon_n) = \frac{E}{2\pi(E^2 + Q^2)}(\hat{g}_0 - \pi Q/E)([\delta\hat{\Delta}, \hat{g}_0] + [\hat{t}, \hat{g}_0]). \quad (27)$$

Here, $Q = \frac{1}{2}\vec{v}_f \cdot \vec{q}$, and $E = (\epsilon_n^2 + |\Delta_0|^2)^{1/2}$. This solution for $\delta\hat{g}$ can be used to calculate quantities of direct physical interest, such as the supercurrent or density of states. The off-diagonal, in particle-hole space, elements of $\delta\hat{g}$ can be substituted into the gap equation to obtain a closed equation for $\delta\Delta(\vec{k}_f, \vec{q})$.

We investigate in further detail the supercurrent. For any singlet, or separable unitary triplet gap, we may write $\delta\hat{\Delta}(\vec{k}_f, \vec{q}) = i\delta\Delta_1(\vec{k}_f, \vec{q})\hat{\tau}_1 + i\delta\Delta_2(\vec{k}_f, \vec{q})\hat{\tau}_2$. We then obtain

$$\vec{J}(\vec{q}) = -\frac{eT}{\pi}\sum_{\epsilon_n}\int\frac{d^2\vec{k}_f}{(2\pi)^3 \mid \vec{v}_f \mid}\vec{v}_f\frac{E}{E^2 + Q^2}I(\vec{k}_f, \vec{q}; \epsilon_n), \quad (28)$$

$$I(\vec{k}_f, \vec{q}; \epsilon_n) = \frac{2\pi^2 iQ}{E^2}(\Delta_2\delta\Delta_1 - \Delta_1\delta\Delta_2) + \frac{1}{2}Tr\left(\hat{\tau}_3\hat{g}_0[\hat{t}, \hat{g}_0]\right) - \frac{\pi Q}{2E}Tr\left(\hat{\tau}_3[\hat{t}, \hat{g}_0]\right). \quad (29)$$

We can now discuss the magnetic field, $B(\vec{R})$, and the total magnetic moment $\vec{M}$, produced by these currents. We stress that these currents vanish unless the unperturbed order parameter breaks time reversal symmetry. We ignore the Meissner screening current; this is a good approximation for extreme type-II superconductors, since the magnetic screening length is then much longer than the spatial extent of the currents.

- Magnetic Field $B$– To estimate the overall size of the effect, we consider the magnetic field produced at the impurity site, $\vec{R} = 0$. This is given by

$$\vec{B}(0) = \frac{1}{c}\int d^3R\frac{\vec{R}\times\vec{J}(\vec{R})}{R^3} = \frac{4\pi i}{c}\int\frac{d^3q}{(2\pi)^3}\frac{\vec{q}\times\vec{J}(\vec{q})}{q^2}. \quad (30)$$

To make further progress in our estimates, we need the expression for $\delta\Delta(\vec{k}_f, \vec{q})$. As mentioned above, this is obtained by substituting $\delta\hat{g}$ into the gap equation. We refer to the literature for these details,[70,71] and simply quote the final result:[71]

$$|\vec{B}(0)| \approx \beta(T)(ek_f^2)(\frac{T_c}{T_f})^2(\frac{v_f}{c})(\sigma k_f^2). \tag{31}$$

In this formula, $\beta(T)$ is a temperature dependent function, which vanishes at $T_c$, and is of order unity at lower temperatures. The quantity $ek_f^2$ is a magnetic field, and can be quite large, of order $10^6 G$; however, it is reduced by the small factors $(T_c/T_f)^2$ and $(v_f/c)$.

- Total Magnetic Moment $\vec{M}$ – The magnetic moment of the currents around the impurity is given by

$$\vec{M} = \frac{1}{2c}\int d^3R\, \vec{R}\times \vec{J}(\vec{R}) = \frac{i}{2c}[\vec{\nabla}_q \times \vec{J}(\vec{q})]_{q=0}. \tag{32}$$

The above formula for $\vec{J}(\vec{q})$ then allows us to deduce a surprisingly general result:[70,79] $\vec{M} = 0$, unless $\delta\Delta(\vec{k}_f, \vec{q})$ is singular as $\vec{q} \to 0$. This result holds for any $n(\vec{k}_f)$, $v(\vec{k}_f, \vec{k}_f'), \Delta_0(\vec{k}_f)$, and $\vec{v}_f$.

The subtle point behind the $\vec{M} = 0$ answer is the requirement of nonsingular behavior at small $q$ of the perturbation in the order parameter. This means that $\delta\Delta(\vec{k}_f, \vec{R})$ should go to zero quickly enough at large $R$. Now, the dangerous distortions are the soft modes, those that leave the bulk free energy unchanged. In a crystalline superconductor, however, the orientation of the order parameter is pinned by the crystal; thus, the only soft mode available is the phase mode, which for real $\delta\theta(\vec{R})$ is described by $\delta\Delta(\vec{k}_f, \vec{R}) = i\delta\theta(\vec{R})\Delta_0$.

So, we must study the gap equation for $\delta\Delta$ to see if the impurity couples to the phase mode. No general proof exists, but in all the cases studied so far,[79] a long range disturbance in the phase has not been found.

We hope that the preceding discussion gives a good idea of the interesting spatial structure which should develop in the neighborhood of an impurity in an unconventional superconductor. It should be stressed that the length scale of this structure, of order $\xi(T)$, is quite long by microscopic standards.

## V. ACKNOWLEDGEMENTS

Part of this manuscript was written at the Institute for Scientific Interchange in Torino, Italy. The authors DR and JAS thank the ISI and the organizers for the workshop on 'Phenomenological Theories of Superconductivity' for their hospitality and support. The work of JAS was supported in part by NSF grant no. DMR 91-20521 through the Materials Research Center at Northwestern University, NSF grant no. DMR 91-20000 through the Science and Technology Center for Superconductivity, and the Nordic Institute for Theoretical Physics in Copenhagen (NORDITA).

* permanent address.
[1] P. W. Anderson and P. Morel, Phys. Rev. **123**, 1911 (1961).
[2] D. D. Osheroff, R. C. Richardson, and D. M. Lee, Phys. Rev. Lett. **28**, 885 (1972).
[3] F. Steglich *et al.*, Phys. Rev. Lett. **43**, 1892 (1979).
[4] H. Ott *et al.*, in *18th International Conference on Low Temperature Physics, Kyoto* (Japanese Journal of Applied Physics, Tokyo, 1987).
[5] B. Sarma, M. Levy, S. Adenwalla, and J. Ketterson, in *Physical Acoustics* (Academic Press, New York, 1992), Vol. XX.
[6] F. Steglich *et al.*, Frontiers Sol. State Sciences **1**, 527 (1993).
[7] N. E. Bickers, D. J. Scalapino, and R. T. Scalettar, Int. Journ. Mod. Phys. **B1**, 687 (1987).
[8] P. A. Lee *et al.*, Comments on Cond. Mat. Phys. **12**, 99 (1986).
[9] L. Gor'kov, Sov. Sci. Rev. A. **9**, 1 (1987).
[10] M. Sigrist and K. Ueda, Rev. Mod. Phys. **63**, 239 (1991).
[11] P. Anderson, Phys. Rev. **B30**, 4000 (1984).
[12] G. Volovik and L. Gor'kov, Sov. Phys. JETP **61**, 843 (1985).
[13] G. Volovik and L. Gor'kov, Sov. Phys. JETP Lett. **39**, 674 (1984).
[14] S. Yip and A. Garg, Phys. Rev. **B48**, 3304 (1993).
[15] V. Müller *et al.*, Phys. Rev. Lett. **58**, 1224 (1987).
[16] Y. Qian *et al.*, Solid State Commun. **63**, 599 (1987).
[17] A. Schenstrom *et al.*, Phys. Rev. Lett. **62**, 332 (1989).
[18] G. Bruls *et al.*, Phys. Rev. Lett. **65**, 2294 (1990).
[19] S. Adenwalla *et al.*, Phys. Rev. Lett. **65**, 2298 (1990).
[20] D. Hess, T. Tokuyasu, and J. Sauls, J. Phys. Condens. Matter **1**, 8135 (1989).
[21] K. Machida and M. Ozaki, J. Phys. Soc. Jpn. **58**, 2244 (1989).
[22] J. Sauls, submitted to Adv. Phys. (1993).
[23] R. Joynt, Sup. Sci. Tech. **1**, 210 (1988).
[24] G. Aeppli *et al.*, Phys. Rev. Lett. **60**, 615 (1988).
[25] T. Vorenkamp, Ph.D. thesis, University of Amsterdam, 1992.
[26] B. Shivaram, J. Gannon Jr., and D. Hinks, Phys. Rev. Lett. **63**, 1723 (1989).
[27] E. Vincent *et al.*, J. Phys. Cond. Matt. **3**, 3517 (1991).
[28] E. A. Knetch, J. A. Mydosh, T. Vorenkamp, and A. A. Menovsky, J.M.M.M. **108**, 75 (1992).
[29] T. Trappmann, H. v. Löhneysen, and L. Taillefer, Phys. Rev. **B43**, 13714 (1991).
[30] S. Hayden, L. Taillefer, C. Vettier, and J. Flouquet, Phys. Rev. **B46**, 8675 (1992).
[31] I. Luk'yanchuk, J. de Phys. **I1**, 1155 (1991).
[32] K. Machida and M. Ozaki, Phys. Rev. Lett. **66**, 3293 (1991).
[33] D. Chen and A. Garg, Phys. Rev. Lett. **70**, 1689 (1993).
[34] C. Choi and J. Sauls, Phys. Rev. Lett. **66**, 484 (1991).
[35] J. Sauls, to be published (1993).
[36] B. Shivaram, Y. Jeong, T. Rosenbaum, and D. Hinks, Phys. Rev. Lett. **56**, 1078 (1986).
[37] J. Rammer and H. Smith, Rev. Mod. Phys. **58**, 323 (1986).
[38] G. Eliashberg, Sov.Phys.-JETP **15**, 1151 (1962).
[39] R. E. Prange and L. P. Kadanoff, Phys. Rev. **134**, A566 (1964).
[40] G. Baym and C. J. Pethick, *Landau Fermi-Liquid Theory* (Wiley, New York, 1991).
[41] G. Eilenberger, Z. Physik **214**, 195 (1968).
[42] A. I. Larkin and Y. N. Ovchinnikov, Sov. Phys. JETP **28**, 1200 (1969).
[43] A. I. Larkin and Y. N. Ovchinnikov, Sov.Phys.-JETP **41**, 960 (1976).
[44] G. Eliashberg, Sov.Phys.-JETP **34**, 668 (1972).
[45] U. Eckern, Ann. Phys. **133**, 390 (1981).
[46] J. Serene and D. Rainer, Phys. Reports **101**, 221 (1983).
[47] A. I. Larkin and Y. N. Ovchinnikov, in *Nonequilibrium Superconductivity*, edited by D. N. Langenberg and A. I. Larkin (Elsevier Science Publishers, Amsterdam, 1986).
[48] A. J. Leggett, Phys. Rev. **140**, 1869 (1965).
[49] C. Choi and P. Muzikar, Phys. Rev. **B37**, 5947 (1988).
[50] C. Choi and P. Muzikar, Phys. Rev. **B39**, 11296 (1989).
[51] T. Tokuyasu, D. Hess, and J. Sauls, Phys. Rev. **B41**, 8891 (1990).

264

[52] T. Tokuyasu and J. Sauls, Physica B **165-166**, 347 (1990).

[53] M. Palumbo, P. Muzikar, and J. Sauls, Phys. Rev. **B42**, 2681 (1990).

[54] Y. Barash and A. V. Galaktionov, Sov. Phys. JETP **73**, 939 (1991).

[55] M. Palumbo and P. Muzikar, Phys. Rev. **B45**, 12620 (1992).

[56] A. S. Mel'nikov, Sov. Phys. JETP **74**, 1059 (1992).

[57] P. Hirschfeld et al., Phys. Rev. **B40**, 6695 (1989).

[58] S. K. Yip and J. A. Sauls, J. Low Temp. Phys. **86**, 257 (1992).

[59] V. Ambegaokar, P. deGennes, and D. Rainer, Phys. Rev. **A9**, 2676 (1975).

[60] J. Kurkijärvi, D. Rainer, and J. A. Sauls, Can. Journ. Phys. **65**, 1440 (1987).

[61] V. B. Geshkenbein and A. I. Larkin, Sov. Phys. JETP Lett. **43**, 395 (1986).

[62] A. Millis, J. A. Sauls, and D. Rainer, Phys. Rev. **B38**, 4504 (1988).

[63] S. Yip, J. Low Temp. Phys. **91**, 203 (1993).

[64] S. K. Yip, Y. Sun, and J. A. Sauls, Physica B **LT20 Proceedings**, (1993).

[65] D. Van Harlingen and A. J. Leggett, LT20 Proceedings (1993).

[66] D. Vollhardt and P. Wölfle, *The Superfluid Phases of $^3$He* (Taylor & Francis, New York, 1990).

[67] R. H. McKenzie and J. A. Sauls, in *Helium Three*, edited by edited by W. P. Halperin and L. P. Pitaevskii (Elsevier Science Publishers, Amsterdam, 1990), p. 255.

[68] K. Ueda and T. Rice, Phys. Rev. **B31**, 7114 (1985).

[69] A. L. Fetter, Phys. Rev. **140**, 1921 (1965).

[70] D. Rainer and M. Vuorio, J. Phys. **C10**, 3093 (1977).

[71] C. Choi and P. Muzikar, Phys. Rev. **B39**, 9664 (1989).

[72] C. Choi and P. Muzikar, Phys. Rev. **B40**, 5144 (1989).

[73] M. Palumbo, C. H. Choi, and P. Muzikar, Physica **B165-166**, 1095 (1990).

[74] M. C. Cross, J. Low Temp. Phys. **21**, 525 (1975).

[75] N. Mermin and P. Muzikar, Phys. Rev. **B21**, 980 (1980).

[76] P. Muzikar, Phys. Rev. **B43**, 10201 (1991).

[77] E. Thuneberg, J. Kurkijarvi, and D. Rainer, J. Phys. **C14**, 5615 (1981).

[78] E. Thuneberg, J. Kurkijarvi, and D. Rainer, Phys. Rev. **B29**, 3913 (1984).

[79] C. Choi and P. Muzikar, Phys. Rev. **B41**, 1812 (1990).

# Flux motion by quantum tunneling

A. C. Mota
*Laboratorium für Festkörperphysik*
*ETH Hönggerberg*
*CH-8093 Zürich*
*Switzerland*

At low temperatures, large, temperature-independent magnetic relaxation effects are found in the high-temperature superconducting ceramics as well as in the organic superconductors, giving evidence that, at these temperatures, vortex motion involves mainly a form of macroscopic quantum tunneling. The measured tunneling rates at low magnetic induction and in the limit $T \rightarrow 0$, as well as the crossover temperatures $T_{qc}$ from the quantum to the classical regime, are in excellent agreement with the values for single vortex tunneling calculated within the framework of the quantum collective creep theory of Blatter, Geshkenbein and Vinokur. Quantum tunneling in the heavy fermion superconductor UPt$_3$, on the other hand, cannot be explained with the aforementioned theory.

## 1. Introduction

Shortly after the discovery of the high-$T_c$ superconductors, it became clear that the Shubnikov phase of these new materials presented a wealth of novel phenomena not observed before in the classical type II superconductors. One of these new phenomena, quantum creep of vortices, is the subject of this lecture.

In a material with quenched disorder, after a metastable distribution of vortices is established by a change of a magnetic field or a current, the vortices relax to their equilibrium configuration via two different mechanisms: thermal activation and quantum tunneling.

Thermally activated flux creep is well known since the work of Kim and Anderson in the early sixties. In conventional hard type II superconductors, thermally activated creep rates can be kept rather low by introducing strong pinning into the system. On the other hand, in the high-$T_c$ oxides, as a result of a combination of peculiar material parameters, vortex creep rates are high not only in the classical regime but also in the very low temperature regime, giving indication that vortex creep results from tunneling of vortices *under* the energy barriers.

Other families of exotic superconductors, the organics [1] and heavy fermions [2], also give indication of vortex motion by quantum tunneling at millikelvin temperatures.

265

*N. Bontemps et al. (eds.), The Vortex State*, 265–280.

Macroscopic vortex tunneling in bulk superconductors has been analyzed by Blatter, Geshkenbein and Vinokur [3] within the framework of the weak collective pinning theory. Quantum motion of individual vortex lines across intrinsic pinning barriers in layered superconductors has been studied by Ivlev, Ovchinnikov and Thompson [4] .

In the quantum collective creep theory [3] (QCC), the elementary tunneling process at low magnetic fields consists of a vortex segment of length $L_c$ tunneling *under* the barrier $U_c$ to its neighboring metastable state. At higher fields, the tunneling unit is a vortex bundle. The tunneling rate is given by

$$\Gamma = \Gamma_0 \exp(- S_E / \hbar) \tag{1}$$

where $S_E$ is the Euclidean action of the tunneling process. Both $\Gamma_0$ and $S_E$, depend on current, magnetic field, temperature and sample characteristics.

For weak enough pinning, the dissipative contribution to the action is always dominant. Within this limit, Blatter *et al.* calculated the quantum creep rates for the case of ohmic dissipation. At low fields, the resulting minimal action for the tunneling process is

$$\frac{S_E^{eff}}{\hbar} \approx \frac{\hbar}{e^2} \frac{\xi}{\rho_n} \left(\frac{j_0}{j_c}\right)^{1/2} \approx \frac{1}{Qu} \left(\frac{j_0}{j_c}\right)^{1/2} \tag{2}$$

with the dimensionless quantum resistance Qu defined as [5]

$$Qu = \left(\frac{e^2}{\hbar}\right) \left(\frac{\rho_n}{\xi}\right) \tag{3}$$

Here $\rho_n$ is the normal state resistivity, $\xi$ the coherence length, and $j_0$ and $j_c$ the depairing and critical current densities, respectively. The length of the tunneling unit in the QCC theory is the collective pinning length $L_c$

$$L_c = \xi \left(\frac{j_0}{j_c}\right)^{1/2} \tag{4}$$

The normalized creep rate is then

$$\frac{\partial \ln M}{\partial \ln t} \approx - \frac{\hbar}{S_E^{eff}} \tag{5}$$

From equation (2) we notice that materials with high values of the normal state resistivity $\rho_n$ and low values of the coherence length $\xi$ are good candidates for observing quantum creep. The disorder is less important since it only enters via the term $(j_c)^{1/2}$.

For anisotropic superconductors, the elasticity of the vortices is enhanced with respect to the isotropic material, leading to a reduction of the effective Euclidean action $S_E^{eff}$ and thus to an enhanced magnetic relaxation rate.

The expression for the effective action in an anisotropic superconductor with weak enough pinning is [5]

$$\frac{S_E^{eff,c}}{\hbar} \approx \frac{\hbar}{e^2} \frac{\varepsilon\xi}{\rho_n} \left(\frac{j_0}{j_c}\right)^{1/2} \approx \frac{1}{Qu} \left(\frac{j_0}{j_c}\right)^{1/2} \tag{6}$$

with the dimensionless quantum resistance now given by

$$Qu = \frac{e^2}{\hbar} \frac{\rho_n}{\varepsilon\xi} \tag{7}$$

Here $\xi$, $j_c$ and $j_0$ denote the values within the superconducting planes. $\varepsilon^2 = m/M \ll 1$ is the anisotropy parameter. We notice from (6) that $S_E^{eff,c}$ is independent of the angle between the magnetic field and the c-axis and independent of the direction of motion.

Expression (6) is valid for low fields. At higher magnetic fields, the tunneling motion of a vortex bundle is less probable so that the corresponding actions increase with respect to the values given in (2) and (6).

## 2. Experimental arrangement

Basically, two types of experimental set-ups have been used to detect long time decays of the magnetization at very low temperatures. One method uses a Hall probe technique with the specimen and the measuring system mounted on the end tip of a dilution refrigerator. The thermal contact between the specimen and the miniature Hall probe and thermometers is established by means of vacuum grease [6].

In our experimental arrangement (Fig. 1) the specimens, in the form of powder or single crystals, are mounted inside the mixing chamber of a dilution refrigerator in direct contact with the $^3$He-$^4$He mixture. Magnetic measurements are made using SQUID detectors. A cryoperm shield placed in the helium dewar reduces the earth magnetic field to less than 5 mOe in the experimental space. No superconducting shield is used around the specimens.

## 3. Experimental results

### 3.1. High-Temperature Superconductors

The first observation of temperature-independent decays of the magnetization at millikelvin temperatures was done in powdered specimens of Ba-La-Cu-O and Sr-La-Cu-O [7]. Later on, single crystals of $Y_1Ba_2Cu_3O_y$ [8] as well as $Y_1Ba_2Cu_4O_x$ in the form of non-aligned grains were investigated [9].

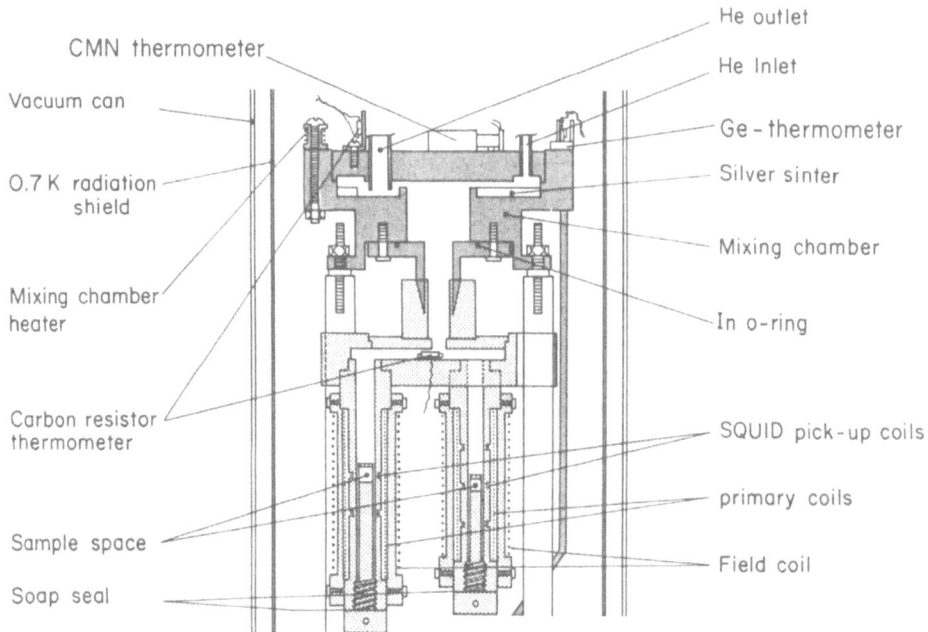

Fig. 1     Experimental cell for measurements of magnetic relaxations in the
           temperature range 0.005 K < T < 8 K.

Fig. 2 shows the measured normalized decay rates of the remanent magnetization as a function of temperature for $Y_1Ba_2Cu_4O_x$. The specimen has been cycled up to a field of 340 Oe and the relaxation of the remanent magnetization has been measured subsequently from 1 s to $10^5$ s . The field used was not big enough to establish the fully critical state in the specimen, so that the raw data were multiplied by a factor of 2 to account for this fact [10].

In Fig. 2 we clearly observe two regimes of flux motion: a low temperature regime, where tunneling of vortices occurs and the rates are independent of temperature, and a classical regime, where thermal activation also contributes to the decay from the metastable state. The crossover temperature $T_{qc}$ from the quantum to the classical regime is around 3 K.

A comparison of the measured rates with the results of the QCC theory can be made for this polycrystalline specimen since the action, as given in expression (6), is independent of the angle between the magnetic field H and the ab-plane. Due to the lack of the value of the anisotropy parameter $\varepsilon$ for $Y_1Ba_2Cu_4O_y$, we use the value $\varepsilon = 1/5$ as determined for $Y_1Ba_2Cu_3O_x$. We understand that the correct value might be somewhat smaller since the layer spacing in $Y_1Ba_2Cu_4O_y$ is about 20% larger than in $Y_1Ba_2Cu_3O_x$. The measured critical current for our specimen gave $j_c = 5 \times 10^5$ Acm$^{-2}$. Assuming the same value of the depairing current $j_0 = 10^8$ Acm$^{-2}$ as for $Y_1Ba_2Cu_3O_x$ and taking the

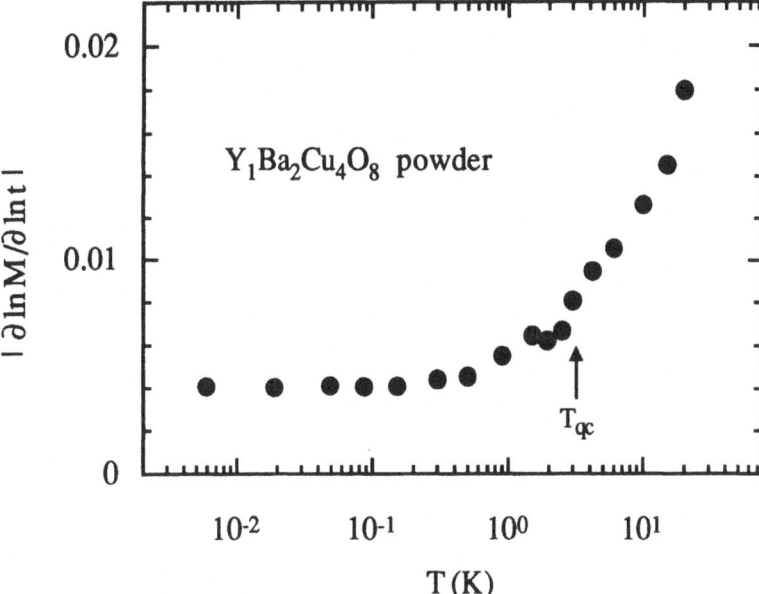

Fig. 2    Normalized decay rate | $\partial\ln M/\partial\ln t$ | of the remanent magnetization of a
$Y_1Ba_2Cu_4O_8$ powdered specimen as a function of temperature in the low-
temperature range ($T \leq 20$ K).

values for the normal state resistivity $\rho_n = 10$ $\mu\Omega$cm and for the coherence length
$\xi = 15$ Å, we obtain $S_E^{eff}/\hbar = 172$ and a relaxation rate

$$\left|\frac{\partial\ln M}{\partial\ln t}\right|^{theor} \approx 0.006$$

A zero temperature extrapolation of the measured logarithmic rate for the relaxation of
the remanent magnetization from the data shown in Fig. 2 gives

$$\left|\frac{\partial\ln M}{\partial\ln t}\right|^{exp} \approx 0.004$$

in good agreement with the theoretical value.

Recently, measurements of quantum creep have been made [11] in a single crystal of
$Bi_2Sr_2CaCu_2O_x$ for the case where the applied magnetic field is perpendicular to the ab-
plane. The crystal investigated was about 1 x 1 x 0.05 mm³ in size.

In Fig. 3 we show typical decays of the remanent magnetization after cycling the
specimen in a field $H_i = 880$ Oe. Notice that for the sake of clarity the relaxation curves
have been displaced in the vertical linear scale.

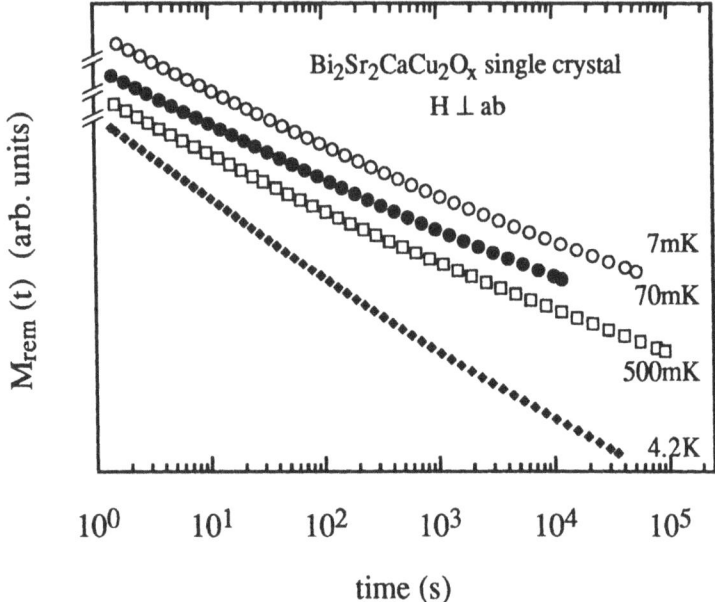

Fig. 3    Relaxation curves $M_{rem}(t)$ after cycling the crystal in a magnetic field $H_i = 880$ Oe. For the sake of clarity the relaxation curves have been displaced in the vertical linear scale.

At all temperatures, the decays are not logarithmic in our time window. Due to this fact, we have chosen to determine the rate at each temperature from the initial slope of M vs. lnt. In Fig. 4 we show the normalized rates $| \partial lnM/\partial lnt |$ as a function of temperature in the temperature range $7$ mK $\leq T \leq 15$ K. In this figure, we have plotted decay rates of the remanent magnetization after cycling the specimen in two different fields and also the field-on creep rate at $T = 4.2$ K when a field of 880 Oe is applied to the specimen. The fact that all the points fall on the same curve indicates that, within the precision of our measurements, the relaxation rates are independent of the cycling field for the fields used here. Furthermore, we do not notice any measurable difference between the relaxation rates with field-on and field-off, indicating that self-field effects, which are important in modifying the field profile at low fields, do not produce a measurable change in the relaxation rates as detected by our SQUID magnetometers.

In Fig. 4 we observe that the measured tunneling rate in the limit $T \rightarrow 0$ is

$$\left| \frac{\partial lnM}{\partial lnt} \right|^{exp} \approx 0.017$$

We can estimate the tunneling rate in $Bi_2Sr_2CaCu_2O_x$ from the QCC theory for the case of layered superconductors. In layered superconductors, flux lines are not rectilinear objects but they rather consist of so called pancake vortices joined by Josephson vortices

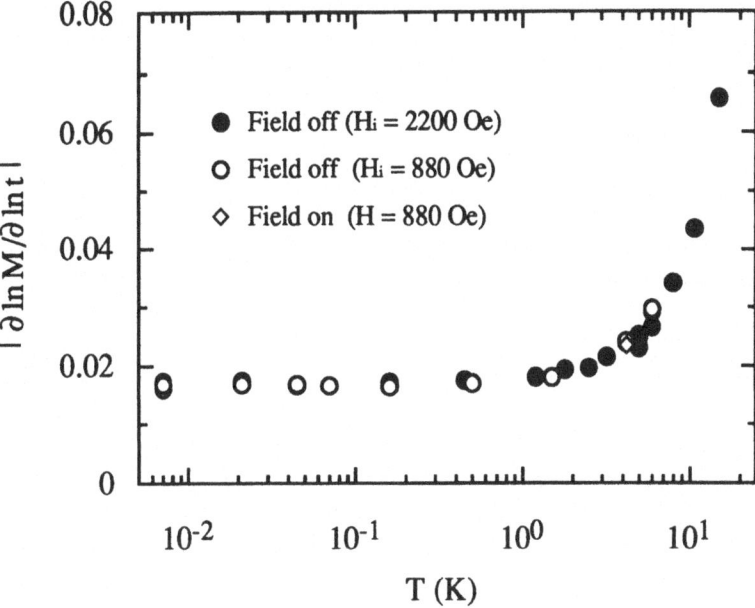

Fig. 4    Normalized decay rates as a function of the temperature of the Bi(2212) single crystal.

along the superconducting layers. In the QCC theory, the estimated quantum creep rates [5] differ depending on the values of the collective pinning length $L_c^c$ with respect to the layer separation d. For the case $L_c^c > d$, the results for single vortex quantum creep coincide with those obtained within the anisotropic continuum description (expression (6))

$$\frac{S_E^{eff,c}}{\hbar} \approx \frac{\hbar}{e^2} \frac{L_c^c}{\rho_n} \qquad (6)$$

with          $L_c^c = \varepsilon \xi (j_0/j_c)^{1/2}$          (8)

Within the QCC theory, the quantum rates depend upon the angle $\vartheta$ between the magnetic field H and the superconducting planes only when H is almost parallel to the layers, or more precisely, when $|\vartheta| < \varepsilon$.

On the other hand, for the case $L_c^c < d$ and a low magnetic field pointing along the c-axis, the pancakes are pinned individually and the effective action is

$$\frac{S_E^{eff,c}}{\hbar} \approx \frac{\hbar}{e^2} \frac{d}{\rho_n} \qquad (9)$$

Bi$_2$Sr$_2$CaCu$_2$O$_x$ is a strongly layered superconductor. Indeed, the collective pinning length is L$_c^c$ ≈ 5-8 Å and the separation between the planes is d = 15 Å. Using expression (9) to calculate the creep rate with ρ$_n$ = 30 μΩcm we obtain

$$\left|\frac{\partial \ln M}{\partial \ln t}\right|^{theor} \approx \frac{\hbar}{S_E^{eff,c}} = 0.05$$

An estimation of the crossover temperature T$_{qc}$ from the quantum to the classical regime can be made from

$$\frac{k_B T_{qc}}{U_c} = \frac{\hbar}{S_E^{eff,c} (T=0)} \qquad (10)$$

From the linear increase of the creep rate with temperature for 5 K < T < 15 K we estimate the value of the activation energy U$_c$ as U$_c$ = 240 K. The crossover temperature is then T$_{qc}$ ≈ 4 K. Around T$_{qc}$ = 4 K the quantum creep is thermally assisted. Unfortunately, the data do not allow us to obtain the correct temperature dependence of thermally assisted quantum creep.

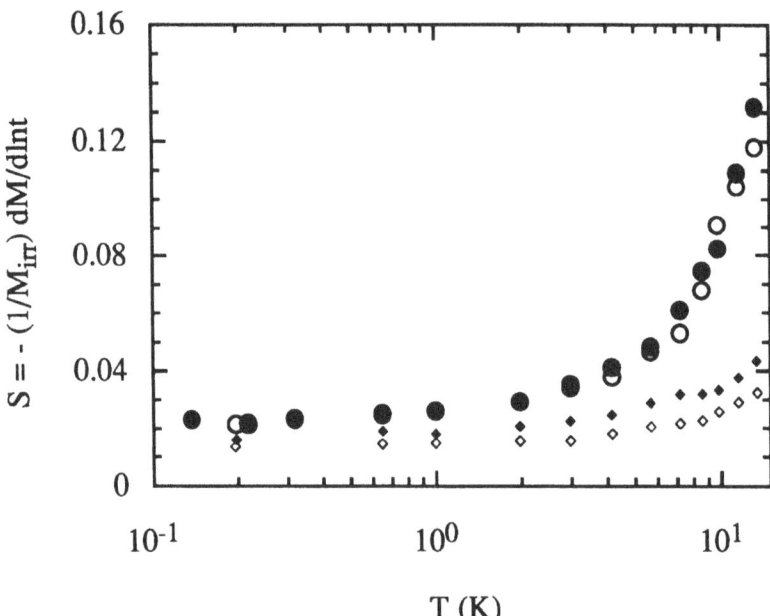

Fig. 5    Normalized relaxation rate of the irreversible magnetization in the Bi(2212) single crystal from ref. [6]. Circles: as grown crystal; squares: irradiated crystal; empty: H = 0.2 T; filled: H = 0.5 T.

Quantum tunneling of flux lines have been studied in a $Bi_2Sr_2CaCu_2O_8$ single crystal with columnar defects by Prost *et al.* [6]. At $T \to 0$, their measured creep rates are $S(T \to 0) = 0.022$ for the crystal without irradiation and $S(T \to 0) = 0.015$ for the Pb-ion irradiated crystal.

In Fig. 5 we show their experimental results. The relaxation rates of the irreversible magnetization were measured at two different fields: $H = 0.2$ T and $H = 0.5$ T. From these results it is to be concluded that quantum creep rates are not strongly reduced after the introduction of columnar defects which nevertheless substantially increase the critical current density and decrease the thermally activated creep rates.

Interesting is to compare the data for their unirradiated crystal with the data on Fig. 4. In spite of the different specimens, different experimental methods and field values, the rates agree within the experimental errors in this type of measurements.

We conclude that in the high-$T_c$ oxides the measured tunneling rates for $T \to 0$ are in good agreement with the values calculated from the QCC theory indicating that this theory correctly describes quantum creep in the cuprate superconductors.

## 3.2. Organic Superconductors

Strong, temperature independent motion of vortices has been observed in the organic superconductor $(BEDT-TTF)_2Cu(NCS)_2$. This material becomes superconducting at $T_c \approx 10.4$ K. Its crystal structure is monoclinic with room temperature lattice parameters: $a = 16.248$ Å, $b = 8.440$ Å, $c = 13.124$ Å, $\beta = 110.3°$. The BEDT-TTF molecules form dimers, which place themselves almost perpendicularly to each other. The almost two-dimensional array forming the conducting layers of this organic material along the bc-plane comes about through short intra- and interdimer contacts. The conducting layers are sandwiched along the a*-axis by insulating sheets of $Cu(NCS)_2$ anions. The resistivity is therefore highly anisotropic with $\rho_b = 0.025 - 0.100$ $\Omega$cm and $\rho_{a*} : \rho_b : \rho_c = 600 : 1 : 0.8$ [12] at room temperature. Due to the stacking pattern of the material, the crystals grow generally as thin platelets with the shape of elongated and distorted hexagons. The plane of the crystals corresponds to the bc-plane.

From literature values of upper critical field $H_{c2}$, penetration depth $\lambda$, and normal state resistivity $\rho_n$, we estimate that the anisotropy ratio $\varepsilon^2 = m/M$ is about 1/400. Values of the coherence lengths are: $\xi_b = 174$ Å, $\xi_c = 118$ Å and $\xi_{a*} = \xi_\perp = 7.74$ Å [13].

An estimation of the collective pinning length with expression (8) gives: $L_c^c \approx 300$ Å. We have used for the critical current density our measured value $j_c \approx 10^3$ Acm$^{-2}$ and estimated $j_0$ from literature values. For this material the results of the QCC theory for the limit $L_c^c > d$ can be used.

In Fig. 6 we show normalized relaxation rates of the remanent magnetization for two field orientations. For H ∥ bc-plane the cycling field was 17 Oe and for H ⊥ bc-plane the cycling field was 63 Oe. We observe that the temperature dependencies of the normalized relaxation rates are strikingly different in the two field orientations. The data taken with H ∥ bc-plane yield values that increase smoothly with temperature up to 8K. The data measured with H ⊥ bc-plane show a much stronger temperature dependence leading to a first peak around 3 K and to a second one around 7.5 K.

For this last field orientation, the relaxation curves are logarithmic only below 1 K.

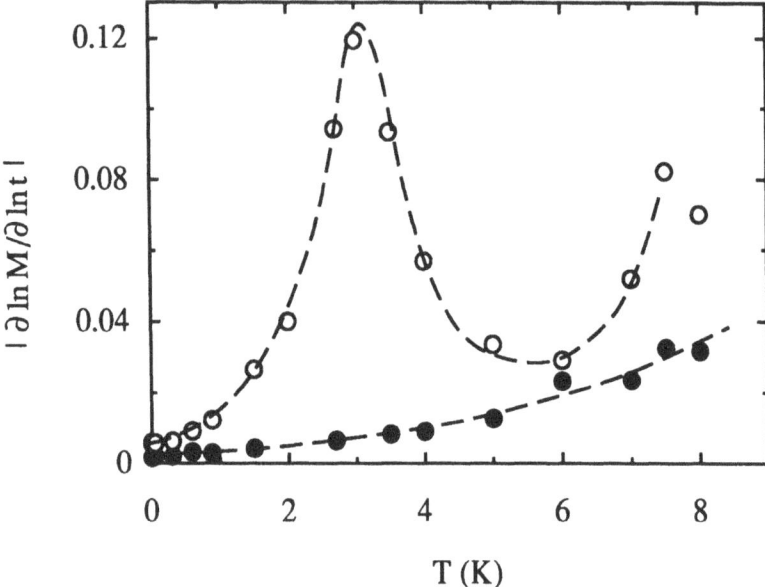

Fig. 6    Normalized decay rate | $\partial\ln M/\partial\ln t$ | of the remanent magnetization as function of temperature for $(BEDT\text{-}TTF)_2Cu(NCS)_2$ single crystals in two different field orientations: (○) $H \perp bc$ and (●) $H \parallel bc$. The dashed lines are only guides to the eye.

In Fig. 7 we show the low temperature data in a double logarithmic scale in order to emphasize the two regions of vortex motion, quantum and classical. The lines in Fig. 7 are fits given by

$$-\frac{1}{M}\frac{\partial M}{\partial\ln t} = 0.006 + 0.008 \cdot T^2 \; ; \qquad H \perp bc$$

$$-\frac{1}{M}\frac{\partial M}{\partial\ln t} = 0.002 + 0.0005 \cdot T^2 \; ; \qquad H \parallel bc$$

From the high temperature data we estimate the activation energy $U_c$ for both field directions. They are $U_c \approx 35$ meV for $H \parallel bc$-plane and $U_c \approx 2$–4 meV for $H \perp bc$-plane. The corresponding crossover temperatures are $T_{qc}^{\parallel} \approx 0.8$ K and $T_{qc}^{\perp} \approx 0.3$ K. Both crossover temperatures are indicated by the arrows in Fig. 7.

We estimate the theoretical value of the creep rates for this superconductor using expressions (6) valid for the orientation $H \perp bc$-plane.

Due to the uncertainty in some of the parameters needed to calculate the action from expression (6) we will only give an order of magnitude estimate of the $T \to 0$ relaxation rate. We use the following values: in-plane normal state resistivity $\rho_n = 100\,\mu\Omega$cm, in-

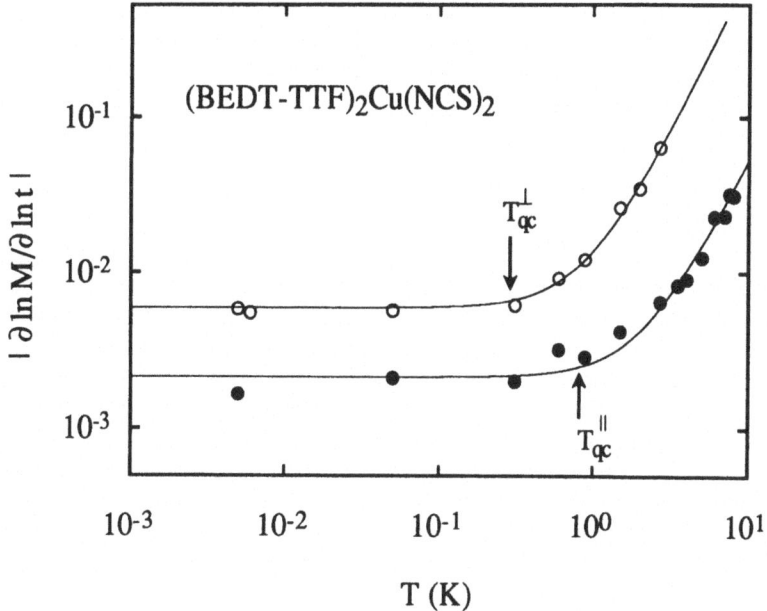

Fig. 7    Normalized decay rate | ∂lnM/∂lnt | of the remanent magnetization as
function of temperature for (BEDT-TTF)$_2$Cu(NCS)$_2$ single crystals in two
different field orientations: (○) H ⊥ bc and (●) H ∥ bc. The solid lines are fits
to the data (see text). $T_{qc}^{\perp}$ and $T_{qc}^{\parallel}$ are the calculated crossover temperatures.

plane coherence length $\xi_{bc} \approx 150$ Å and anisotropy parameter $\varepsilon = 1/20$. Our estimate of
the critical current density for this single crystal is $j_c \approx 10^3$ Acm$^{-2}$ and of the depairing
current $j_0 \approx 2 \times 10^6$ Acm$^{-2}$. With these values we obtain a quantum resistance Qu = 0.32
and a relaxation rate

$$\left. \left| \frac{\partial \ln M}{\partial \ln t} \right| \right.^{theor} \approx 0.007$$

The measured value for H ⊥ bc-plane is

$$\left. \left| \frac{\partial \ln M}{\partial \ln t} \right| \right.^{exp} \approx 0.006$$

in good agreement with the theoretical estimate.
The T → 0 relaxation rate for H ∥ bc is smaller by a factor of three than the T → 0
relaxation rate for H ⊥ bc. In order to evaluate the relaxation rate for H ∥ bc we need to
know precisely the small angle $\vartheta < 1/20$ between the applied field and the basal plane of

the specimen. According to the QCC theory a T → 0 rate for H ∥ bc smaller by a factor of three than the T → 0 rate for H ⊥ bc implies that the misalignment of the crystal with respect to the field was of the order of 0.1°. Unfortunately, we could not determine the alignment of the specimen to that accuracy.

### 3.3. Heavy Fermion Superconductors

In spite of their low transition temperatures, strong relaxation effects with creep rates which are comparable to the rates in the high-$T_c$ superconductors are observed in the vortex state of the heavy fermion superconductors $CeCu_2Si_2$ [10], $UPt_3$ [2], and $UBe_{13}$ [1]. In particular, the dynamics of the vortex system in $UPt_3$ shows unconventional behavior. In this material, the decay from a metastable vortex configuration is not described by a logarithmic or power law time dependence, but rather by a stretched exponential law. Furthermore, the strong, temperature-independent decay rates in the quantum regime cannot be explained with the quantum collective creep theory. In $UPt_3$, the quantum resistance number Qu which determines the creep rate [expression (2)] is Qu $\approx 10^{-4}$ as compared to Qu $\approx 10^{-1}$ for the high-$T_c$ superconductors.

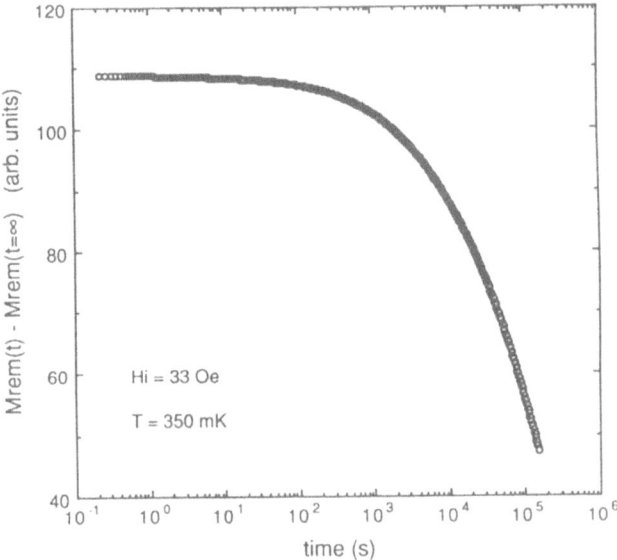

Fig. 8    Remanent magnetization reduced by the constant value $M_{rem}(\infty)$ (see text) as a function of time on a logarithmic scale.

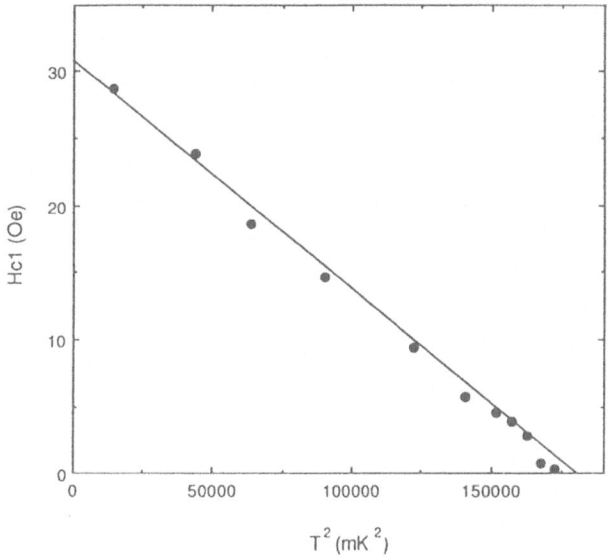

Fig. 9    Lower critical field $H_{c1}$ as a function of $T^2$ for the UPt$_3$ single crystal.

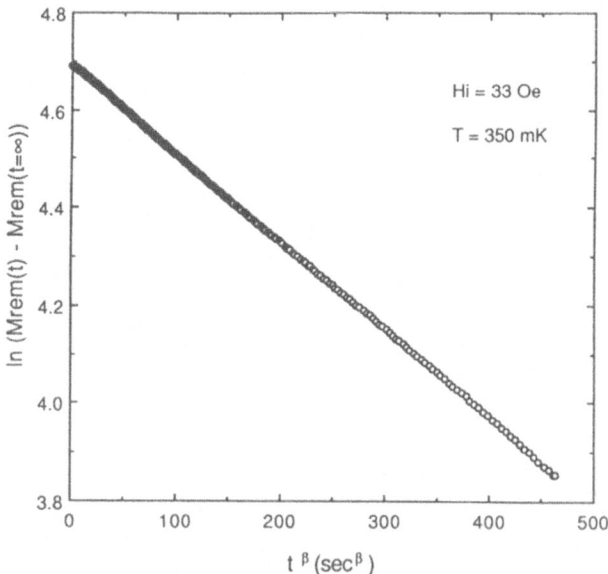

Fig. 10    Same data as in Fig. 8 plotted as $\ln(M_{rem}(t) - M_{rem}(\infty))$ vs. $t^\beta$ with $\beta = 0.515$.

In Fig. 8 we show a typical relaxation curve of the remanent magnetization $M_{rem}$ in UPt$_3$. The UPt$_3$ single crystal is in the form of a parallelepiped with dimensions 4.1 x 2.3 x 0.5 mm$^3$. The magnetic field is applied perpendicular to the c-axis. This specimen has a transition temperature $T_c$ = 430 mK defined as the half point of the diamagnetic susceptibility signal at the transition. The data shown in Fig. 8 were obtained after the zero field cooled specimen was cycled in a field of 33 Oe. This cycling field was just above $H_{c1}$. Measurements of $H_{c1}$ for this configuration, H in the basal plane, are shown in Fig. 9. The line is a fit to the data of the form $H_{c1}(T) = H_{c1}(0)[1 - (T/T_c)^2]$ with $H_{c1}(0)$ = 30.8 Oe and $T_c$ = 430 mK. Here the estimated demagnetization factor D = 0.07 has been taken into account.

The relaxation curve shown in Fig. 8 is described by a stretched exponential time dependence

$$M_{rem}(t) - M_{rem}(\infty) = [M_{rem}(0) - M_{rem}(\infty)] \exp [- (t/\tau)^\beta]$$

The characteristic time $\tau$ corresponds to the time at which the maximum logarithmic rate $| \partial M_{rem}(t) / \partial \log t |$ is observed. $\tau$ is of the order of $10^4$ - $10^5$ s. Typical values of the time-stretch exponent $\beta$ are around 0.5 - 0.6 at all temperatures. In Fig. 10 a plot of

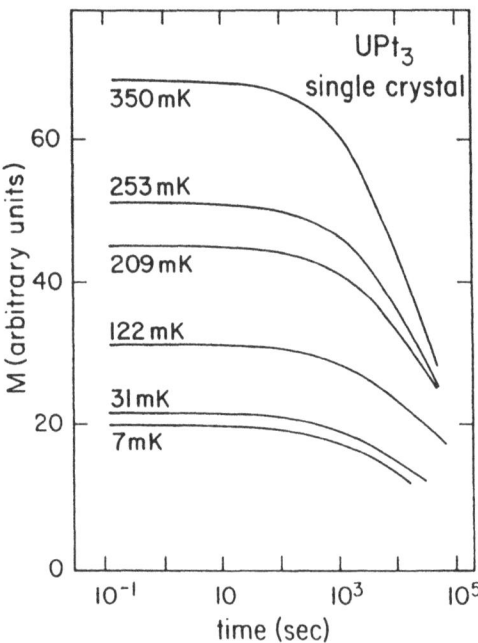

Fig. 11    Remanent magnetization as a function of time for the UPt$_3$ single crystal after cycling it isothermally up to a field of 6.6 Oe. Each curve consists typically of 800 data points which lie within the thickness of the drawn line.

$\ln[M_{rem}(t) - M_{rem}(t = \infty)]$ as a function of $t^\beta$ is given for the same data of Fig. 8. The fitting parameters here are $\beta = 0.515$ and $\tau = 1.8 \times 10^5$ s. The adjustable parameter $M_{rem}(\infty)$ accounts for the fact that at this low field the specimen is not in the completely critical state.

In Fig. 11 we show decays of the remanent magnetization at different temperatures. The cycling field here was $H_i = 6.6$ Oe, so that vortices were introduced only close to the surface of the specimen. We notice that, for each relaxation curve, the percentage decay of $M_{rem}$ in a given time interval is about the same. For example, at all temperatures $M_{rem}$ is reduced in the time interval $10^{-1}$ - $10^4$ s by 30% of its value at $t = 10^{-1}$ s. A more detailed analysis of the data is not trivial. Values of $\tau$ for the low fields investigated are of the order of our experimental observation time ($10^4$ - $10^5$ s) and they increase considerably for bigger magnetic fields. Undoubtedly, more work with higher fields and longer observation times are needed in order to understand the vortex dynamics of this unconventional superconductor.

## 4. Summary

Vortex motion by quantum tunneling has been observed in three families of superconductors: the high-$T_c$ oxides, the organics and the heavy fermions. The quantum creep rates in the first two classes of superconductors are in very good agreement with the $T \rightarrow 0$ theoretical rates calculated within the framework of the quantum collective creep theory of Blatter, Geshkenbein and Vinokur. Within their theory the creep rates are given by

$$\left| \frac{\partial \ln M}{\partial \ln t} \right| \approx Q_u \, (j_c/j_0)^{1/2}$$

with the dimensionless quantum resistance $Q_u$ given by

$$Q_u = \frac{e^2}{\hbar} \frac{\rho_n}{\epsilon \xi}$$

Typical values of $Q_u$ for the high-$T_c$ superconductors and the organics are $Q_u \approx 10^{-1}$ and typical values of $(j_c/j_0)^{1/2}$ are of the order $10^{-1}$ so that quantum creep rates are of the order of 1 %.

For the heavy fermion UPt$_3$ on the other hand, the quantum resistance number is only $Q_u \approx 10^{-4}$. Nevertheless, strong, non-thermally activated creep is observed, giving indication that the very low temperature motion of vortices is possibly related to the unconventional character of this superconductor.

## 5.  Acknowledgments

I wish to express my grateful thanks to A. Amann, K. Aupke, G. Juri, A. Pollini, T. Teruzzi and P. Visani for their valuable contribution to much of the work described here. I also wish to thank G. Blatter and V. Geshkenbein for very fruitful discussions. This work was partially supported by the Schweizerischer Nationalfonds zur Förderung der wissenschaftlichen Forschung and by the Eidgenössische Stiftung zur Förderung der schweizerischen Volkswirtschaft durch wissenschaftliche Forschung.

## References

[1]   A. C. Mota, A. Pollini, G. Juri, P. Visani, and B. Hilti, Physica A **168** (1990) 298

[2]   A. Pollini, A. C. Mota, P. Visani, G. Juri, and J. J. M. Franse, Physica B **165-166** (1990) 365

[3]   G. Blatter, V.B. Geshkenbein, and V.M. Vinokur, Phys. Rev. Lett. **66** (1991) 3297

[4]   B. I. Ivlev, Yu. N. Ovchinnikov, and R. S. Thompson, Phys. Rev. B **44** (1991) 7023

[5]   G. Blatter, M. V. Feigel'man, V. B. Geshkenbein, A. I. Larkin, and V. M. Vinokur, to be published

[6]   D. Prost, L. Fruchter, I. A. Campbell, N. Motohira, and M. Konczykowski, Phys. Rev. B **47** (1993) 3457

[7]   A. C. Mota, A. Pollini, P. Visani, K. A. Müller, and J. G. Bednorz, Phys. Rev. B **36** (1987) 4011

[8]   L. Fruchter, A. P. Malozemoff, I. A. Campbell, J. Sanchez, M. Konczykowski, R. Griessen, and F. Holtzberg, Phys. Rev. B **43** (1991) 8709

[9]   A. C. Mota, G. Juri, P. Visani, A. Pollini, T. Teruzzi, K. Aupke, and B. Hilti, Physica C **185-189** (1991) 343

[10]  A. Pollini, A. C. Mota, P. Visani, R. Pittini, G. Juri, and T. Teruzzi, J. Low Temp. Phys. **90** (1993) 15

[11]  K. Aupke, T. Teruzzi, P. Visani, A. Amann, A. C. Mota, and V. N. Zavaritsky, Physica C **209** (1993) 255

[12]  G. Saito, Physica C **162-164** (1989) 577

[13]  K. Murata, Y. Honda, H. Anzai, M. Tokumoto, K. Takahashi, N. Kinoshita, and T. Ishiguro, Synth. Metals **27** (1988) A341

# LAYERED SUPERCONDUCTORS: DO VORTICES INTERACT WITH THE CRYSTAL LATTICE ?

Denis FEINBERG
Laboratoire d'Etudes des Propriétés Electroniques des Solides
Centre National de la Recherche Scientifique
B. P. 166
38042 Grenoble Cedex 9
France

ABSTRACT. Some specific properties of vortices and vortex lines in layered superconductors are reviewed. Emphasis is put on new features originating from very short coherence lengths, especially along the normal to the layers. As far as bulk properties are concerned, vortex lines close to the layer direction can lock-in to the layer direction and experience intrinsic pinning by the layered structure itself. On the other hand, surface properties are also modified with respect to conventional superconductors. This is illustrated by entrance fields at a perfect surface, for both "Josephson" vortices and "pancake" vortices.

## 1. Introduction

In conventional type-II superconductors, all relevant lengths (coherence length $\xi$, screening length $\lambda$) are much larger than the lattice length scale. As a result, the discrete nature of the underlying crystal lattice is irrelevant, and vortices interact with the crystal only through defects like dislocations, precipitates, etc..., leading to the phenomenon (essential from the point of view of applications) of vortex pinning. On the other hand, in layered superconductors such as high-$T_c$ materials, quasi-2D organic salts, or intercalated chalcogenides, the strong electronic anisotropy leads to much shorter coherence lengths in the direction perpendicular to the layers, resulting in weakly coupled superconducting layers and possible interaction of vortices with the layer stacking [1-2]. This phenomenon is enhanced in high-$T_c$ compounds which already possess very short coherence lengths within the layers (down to 15Å). This can lead to extremely weak interlayer coupling, correctly described by Josephson tunneling, and manifesting in quasi-2D behaviour. On the other hand, due to divergence of the coherence length at $T_c$, a 3D behaviour is recovered close to $T_c$. Similar physics can be encountered in artificially prepared multilayers alternating superconducting and insulating (or metallic) layers, where tuning the individual layer thickness lead to a variety of dimensional crossovers in low- $T_c$ [3] or high-$T_c$ [4] materials.

In the first Section, the structure of the core of a vortex line parallel to the layers is described in the whole temperature range, including the 2D-3D crossover. The vortex free

*N. Bontemps et al. (eds.), The Vortex State,* 281–292.

energy is shown to be modulated along the normal to the layers. This leads to an intrinsic pinning barrier opposing vortex motion. Section 2 is devoted to a brief review of the lock-in of vortex lines onto the layer direction, which crucially depend on the same parameter. The last section treats the problem of surface (Bean-Livingston) barriers, on one hand for vortices parallel to the layers, and on the other hand for "pancake" vortices for extremely weak layer coupling.

## 2. From Josephson to Abrikosov vortex cores

The fundamentals of vortices in layered superconductors were presented in detail in J. Clem's lectures, this school. Here we shall use the Lawrence-Doniach (LD) model [5]

$$
F = s \sum_n \int d^2r \left[ a(T) \left| \Psi_n \right|^2 + \frac{1}{2} b \left| \Psi_n \right|^4 + \frac{h^2}{4m_{ab}} \left| \left( i\nabla_{//} + \frac{2e}{hc} A_{//} \right) \Psi_n \right|^2 \right.
$$

$$
\left. + f_J \left| \Psi_{n+1} - \Psi_n \exp\left( \frac{2ie}{hc} \int_{nd}^{(n+1)d} A_z \, dz \right) \right|^2 \right] + \int d^2r \, dz \, \frac{b^2}{8\pi}
$$

(1)

where the order parameter $\Psi_n(x,y)$ is defined in the layers indexed by n, with spacing s, the z axis denoting the normal to the layers. The Josephson coupling $f_J$ can be expressed in terms of an effective mass $m_c$, by $f_J = h^2/4m_c s^2$. The effective anisotropy ratio, here a phenomenological quantity, is defined by $\Gamma = (m_c/m_{ab})^{1/2}$. The coherence lengths within and across the layers are $\xi_{ab} = (h^2/4m_{ab} |a(T)|)^{1/2}$ and $\xi_c = \xi_{ab}/\Gamma$ respectively. This model exhibits a dimensionnal crossover at T* such as $\xi_c(T^*) \approx s$, the interlayer distance [6]. Below T* one finds a quasi-2D regime, between T* and $T_c$ a quasi-3D regime. T* is of the order of 80K in Y:123 ($T_c \approx 90K$) and extremely close to $T_c$ in Bi:2212 and related compounds.

At low temperatures (T < T*), the effective coherence length $\xi_c(T)$ is much smaller than s. This is meaningless in the frame of a continuous Ginzburg-Landau model, and simply points out towards a very weak coupling of the order parameter amplitudes in neighbouring layers. Moreover, as shown in J. Clem's lectures, a vortex line parallel to the layers has no Abrikosov core. This can be understood by the following argument: the core radius is usually obtained when the screening currents become of the order of the depairing current $J_d$. Here the screening lengths are $\lambda_{ab}$ and $\lambda_c = \Gamma\lambda_{ab} = (m_c c^2/8\pi\Psi_0^2 e^2)^{1/2}$ for currents flowing in or across the layers respectively. However the maximum current between layers is the critical Josephson current, $J_0 = \Phi_0 c/8\pi\lambda_c^2 s$, which in the quasi-2D regime is much smaller than $J_d \approx \Phi_0 c/12\sqrt{3}\pi\lambda_c^2\xi_c$. Therefore, within the layers, the depairing current is not attained and the order parameter amplitude is essentially unaffected. Eliminating the uniform amplitude $|\Psi_n| = \Psi_0 = (|a(T)|/b)^{1/2}$ from Equation (1) leads to a more tractable formulation involving only the phase $\phi_n$ of the order parameter, defined by $\Psi_n = |\Psi_n|\exp(i\phi_n)$. It is more convenient to use the gauge-invariant phase $\gamma_n$ and the phase differences $\Delta\gamma_n$ defined as

$$\Delta\gamma_n = \gamma_n - \gamma_{n+1} - \frac{2\pi}{\Phi_0} \int_{ns}^{(n+1)s} A_z dz \qquad (2)$$

with the Josephson current $J_z = J_0 \sin(\Delta\gamma_n)$. The solution for a single vortex was studied by Bulaevskii [6], and Clem and Coffey [7]. Far from the vortex core, anisotropic currents flow, like in an Abrikosov vortex. On the contrary, close to the vortex centre, in a region of section s times $\Gamma$s, phase differences between two layers are not small, and the variation of $J_z$ looks like that of a Josephson fluxon in a single junction. The lateral size of this region which replaces the usual core is fixed by the "Josephson length" $\Gamma$s, rather than by the true screening length $\lambda_c$ which is much larger [8]. The Josephson length basically governs all the dimensional crossovers occurring in layered superconductors: the physics at scales smaller than $\Gamma$s is essentially 2D, while at larger scales it can be mapped onto an effective 3D system with anisotropy parameter $\Gamma$ [2,17-18].

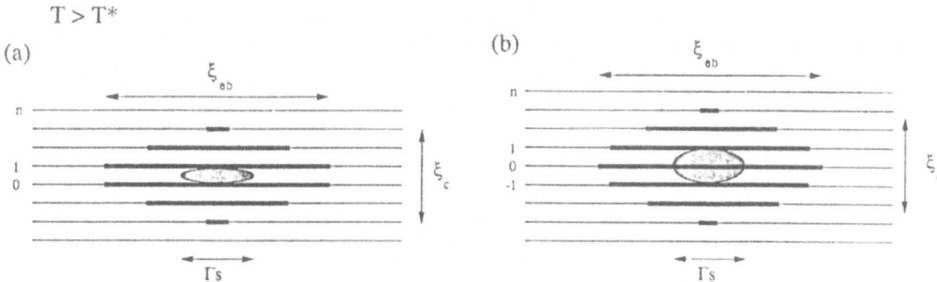

Figure 1. Core of a vortex line parallel to the layers, quasi-3D regime. a) minimum energy; b) maximum energy. Shaded areas denote the Josephson cores and thick lines the regions in the layers wher the order parameter amplitude is depressed.

At temperatures $T \geq T^*$, $\xi_c(T) \geq s$ and the order parameter amplitude must be weakened within the vortex core. This situation has been recently investigated by a variational method, starting from the LD functional [9]. It combines the phase ansatz used by Clem et al. for the quasi-2D case [10] with the ansatz introduced earlier by Clem, who proposed a simple but very accurate form for the amplitude variation $|\Psi|(\rho)$ in a 3D superconductor [11]. Assuming the vortex core center to sit midway between adjacent layers n = 0 and 1 (Figure 1a), one takes

$$h(y,z) = \frac{\Phi_0}{2\pi\Gamma\lambda_{ab}^2} K_0(\tilde{R}) \qquad \tilde{R} = \sqrt{\tilde{u}_0^2 + \Gamma^{-2}\tilde{y}^2 + \tilde{z}^2} \qquad (3)$$

and

$$\frac{\Psi_n(y)}{\Psi_0} = \frac{\tilde{\rho}_n}{\sqrt{\tilde{\rho}_n^2 + \tilde{\xi}^2}} \qquad\qquad \tilde{\rho}_n = \sqrt{\Gamma^{-2}\tilde{y}^2 + (n-1/2)^2\tilde{s}^2} \qquad (4)$$

where $K_0$ is a zeroth-order Bessel function and all lengths (with a tilde) are normalized to $\lambda_{ab}$. Each of these ansatz contains a different length scale: the first one concerns the phase difference variation $\Delta\gamma_0$ inside the Josephson core, as well as the field h, and fixes the Josephson core size, with $u_0 \approx s/2$. The second one concerns the amplitude variation, and fixes the Abrikosov core size, i.e. $\xi \approx \xi_c$. The solution proceeds through an exact transformation of the free energy (1), similar to that of Ref.10, but keeping explicitly the amplitudes $|\Psi_n|$ (for more details see Ref.9). An accurate result can be obtained in the limit $\xi_c \gg s$, where the leading corrections to the 3D Abrikosov vortex are obtained. These corrections originate from nonlinear Josephson terms and reflect the presence of a Josephson core which persists even in this quasi-3D regime. In this case, contrarily to the quasi-2D situation, the Josephson core is smaller than the Abrikosov core.

Due to the layered structure, translational invariance is lost and the vortex line energy is periodically modulated along the z direction, in a way similar to Peierls-Nabarro barriers for dislocations in crystals. The above configuration indeed corresponds to a free energy minimum. One can repeate the calculation with the vortex center sitting now on the layer n = 0 (Figure 1b). Modifying the ansatz (3-4), one obtains the maximum vortex line energy as a function of its coordinate z. The result can be summarized in a free energy per unit length, depending on z [9,12]

$$\varepsilon(z) = \left(\frac{\Phi_0}{4\pi\lambda_{ab}}\right)^2 \left(\mathrm{Ln}\,\frac{\lambda_{ab}}{s} + \alpha(z)\right) \qquad (5)$$

The core energy modulation is entirely contained in the constant $\alpha(z)$, which varies between $\alpha(ns + s/2)$ and $\alpha(ns)$. One finds the core-lattice interaction parameter

$$\alpha_1 = \alpha(0) - \alpha(1/2) \approx 0.285\left(\frac{s}{\xi_c(T)}\right)^2 + 11.2\left(\frac{\xi_c(T)}{s}\right)^{1/2}\exp\left(-2\pi\sqrt{2}\,\frac{\xi_c(T)}{s}\right) \qquad (6)$$

The exponential contribution, due to the amplitude variation, is dominated by the first one which comes from the Josephson core. The algebraic dependence in the small parameter $s/\xi_c$ can be easily understood by noticing that the area of the Josephson core is roughly a fraction $(s/\xi_c)^2$ of the total core, and that in the unstable configuration (Figure 1b), large phase differences are generated between n = −1 and 0 as well as between n = 0 and 1. Thus an extra area of Josephson core is created, which also scales as $(s/\xi_c)^2$. Assuming that $\xi_c(T) = \xi_c(0)(1-T/T_c)^{-1/2}$ close to $T_c$, one finds that the dominant contribution to $\alpha_1$ is $\alpha_1 \approx 0.57\,(T/T^*)$. This linear temperature variation shows that the effects of the layered structure persist even close to $T_c$. This striking conclusion contrasts with an exponential result, based on a perturbative treatment of the layering effect [14]. We believe that such an approach cannot deal with the intrinsic nonlinearities of the Josephson terms, and fails to account for the presence of the Josephson core.

## 3. Kinks and the lock-in transition

The strong modification of vortex cores in layered superconductors leads to qualitatively new features. Within the Lawrence-Doniach model, the core of a vortex line piercing the layers is discontinuous and made of "pancake" or 2D vortex cores localized in the superconducting layers (see J. Clem's lectures) [15-16]. If the field is tilted with respect to the normal to the layers, the coupling between those 2D vortices is due, first, to the dipolar coupling via the magnetic field generated by the vortices, and secondly, to the Josephson coupling. Only the latter coupling is able to tilt vortex lines [16-17]. However, as stressed above, at short length scales it is negligible and the dipolar coupling dominates. The problem of a tilted vortex lattice in the LD model had been considered recently [17-18]. For $\tan\theta < \Gamma$, where $\theta$ is the angle between the field and the normal to the layers, the free energy is affected only in the logarithmic cutoffs. On the other hand, for fields close to the layer direction ($\tan\theta > \Gamma$), true pieces of Josephson cores start to develop between the 2D cores, leading to a "staircase" effective core (Figure 2). Such a vortex lattice can still be approximately treated within the London approximation with anisotropy $\Gamma$, but assuming an effective core made of pieces parallel to the layers and kinks. In the quasi-2D regime ($T < T^*$), the length of a kink is the Josephson length $\Gamma s$. This picture is valid provided $H << \Phi_0/\pi\Gamma s^2$, which means that the vortex spacing in the z-direction is much larger than s. In the opposite case, the possibility of a combined lattice with coexistence of vortices parallel and normal to the layers has been evoked [17-18].

A similar picture emerges in the quasi-3D regime $T^* < T < T_c$. However in this case the behaviour is essentially that of a 3D anisotropic system, modified as in Section 2 by the nonlinear Josephson coupling at scale $\Gamma s < \xi_{ab}$. Treating the core-lattice interaction $\alpha_1$ as a perturbation, one sees that a tilted vortex line will form kinks [9,13]. The length of the kinks results from the balance between the vortex line tension and the core energy modulation, like in classical problems as dislocations or domain walls in presence of a periodic potential. As a result one finds for the kink length $L_K$ and energy $E_K$ [9]

$$L_K = \Gamma s \left( \frac{2\left(\text{Ln}\,\Gamma + \alpha\right)}{\pi\,\alpha_1} \right)^{1/2} \qquad E_K \approx 4s \left( \frac{\Phi_0}{4\pi\lambda_{ab}} \right)^2 \left( \frac{\alpha_1\left(\text{Ln}\,\Gamma + \alpha\right)}{\pi} \right)^{1/2} \qquad (7)$$

where $\alpha$ is the average core constant. Both $L_K$ and $1/E_K$ diverge at $T_c$. Indeed, due to the linear temperature variation of $\alpha_1$, $L_K$ is of the order of the coherence length $\xi_{ab}(T)$, up to a numerical factor. Extrapolation at $T^*$ shows that $L_K$ is of the order of the Josephson length, while the kink energy is of the order of the energy of a pancake vortex. Therefore the quasi-2D and the quasi-3D results can be matched together in a consistent way. This shows the power of the London approach in extreme type-II superconductors, provided the elastic parameters as well as the vortex-lattice interactions are correctly taken into account to determine the shape of the effective core.

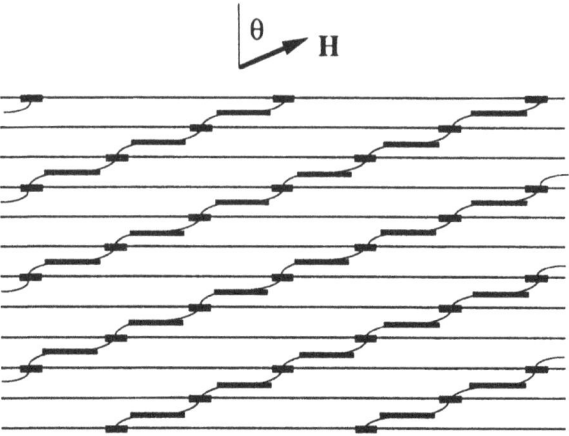

Figure 2. Staircase vortex lines close to the layer direction. In the quasi-2D regime the cores alternate Josephson and pancake cores, while in the quasi-3D regime they alternate quasi-3D cores (see Figure 1a) and kinks.

For a field slightly tilted with respect to the layers, the competition between the magnetic and the core-lattice interaction energies manifests itself in a lock-in state for $\theta_l < \theta < \pi/2$ [12-13]. In this angular range, vortices prefer to align parallel to the layers, with cores as described in Section 2. The lock-in angle $\theta_l$ is such as the normal field component $H \cos\theta_l$ is of the order of $H_{c1c}$, the first critical field along the c-axis, for $T < T^*$ [19], and lower for $T > T^*$. In the former case, this is easily understood since $\theta_l$ marks the onset of perfect screening of the field component normal to the layers, i. e. $B_c$ becomes zero. Moreover, the disappearance of kinks or 2D vortices results in an anomaly in the magnetization, compared to the behaviour in a 3D anisotropic superconductor. The transverse component of the magnetization is enhanced at the lock-in angle, due to the larger deviation between the field and the vortex line directions [9,20]. This can be directly measured by torque magnetometry [21-22], and the lock-in anomaly has been indeed detected by this very sensitive technique in Y:123 [23-24] as well as in more anisotropic materials as Bi:2212, Tl: 2223 or organic salts [25] (Figure 3). Let us mention also dc [26] and ac[27] transport as well as NMR [28] evidences for lock-in phenomena.

This simple picture must be completed along some important directions. First, like in first critical field measurements, demagnetizing effects are very crucial [13,29]. This is essential in the slab or film geometry where the field component $H_c$ is strongly enhanced at $H_{c1c}$, making the *applied* $H_{c1c}$ nearly zero, and leading to an undetectable lock-in anomaly.

Secondly, pinning of kinks is at the origin of an *irreversible* magnetization component which may completely mask the reversible one. Up to now, only the "lock-out" transition has been observed ($\theta$ decreasing from $\pi/2$ to 0), pointing towards a strong pinning of kinks, by defects as twin boundaries, etc... One may simply consider that, as the field is tilted from the layer direction, reversible penetration of kinks (lock-out) would appear when $H_c \approx H_{c1c}$, but that this process is impeded by pinning. Then, in a way similar to the Bean picture for irreversible magnetization, a "critical state" of kinks may be formed. An nonhomogeneous density of kinks on vortex lines would lead to curved vortices, as represented on Figure 4. The torque data close to the layer direction can in principle be directly reinterpreted in terms of a magnetization cycle, concerning the c-component only, in presence of a nearly constant ab component. It is also possible that surface barriers play a dominant role. In this context, some elements will be given in the following Section.

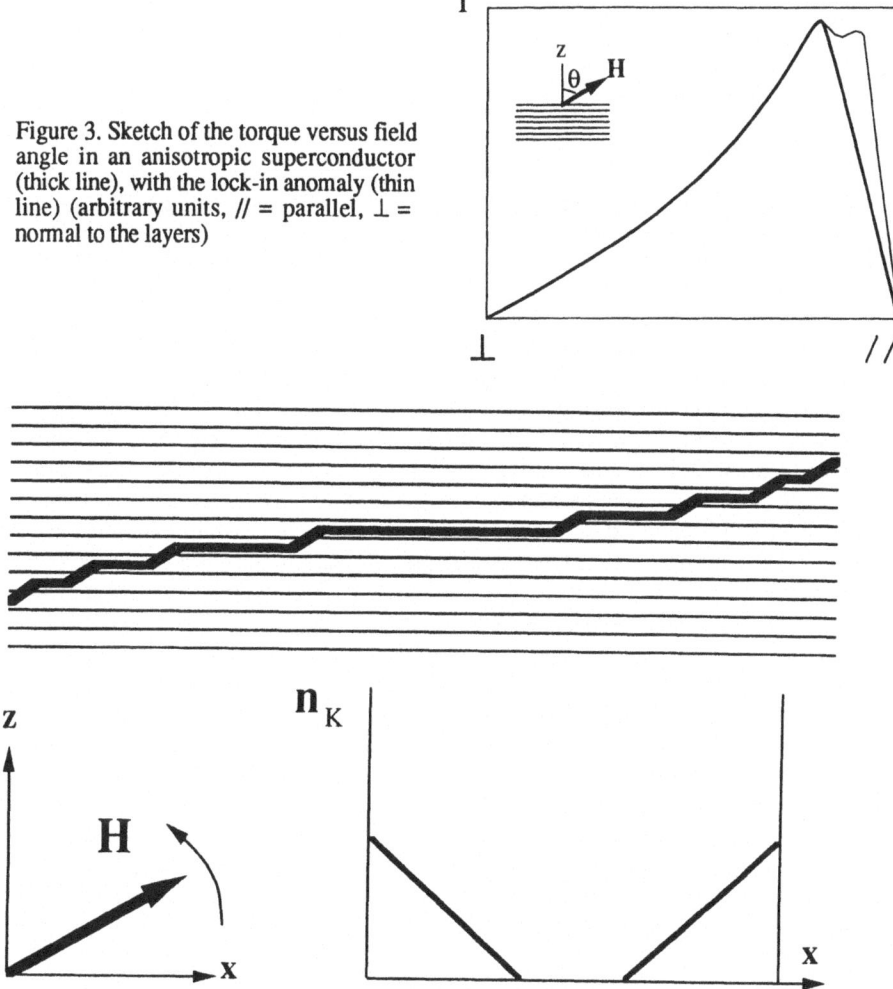

Figure 3. Sketch of the torque versus field angle in an anisotropic superconductor (thick line), with the lock-in anomaly (thin line) (arbitrary units, // = parallel, $\perp$ = normal to the layers)

Figure 4. Curved vortex line due to pinning of kinks, as the field is tilted from the layer direction: a Bean critical state of kinks results with a nonuniform kink density $n_K$. In this example the center of the line remains locked in the layer direction.

Third, thermal fluctuations usually play an important role in the interesting field and temperature range. They involve wandering of Josephson cores, essentially confined by the layers, and thermal excitations of kinks, a possible process owing to the relatively small kink energy $E_K$, which is less than some hundred Kelvins (Figure 5). Even if some of these points have been already addressed, no fully consistent picture emerges at the moment.

288

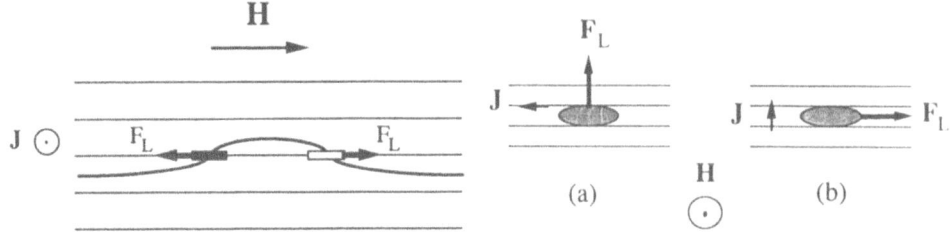

Figure 5. Activated pair of kinks under a parallel field, with the Lorentz force due to a current in the layers.

Figure 6. Forces exerted on a vortex core parallel to the layers. a) in-layer current, intrinsic pinning. b) current normal to the layers: weak pinning.

Further work is needed in this direction, in particular to explain the puzzling transport anomalies observed in this angular range [30]. Those anomalies probably come from the different dissipation channels induced by the motion of Josephson vortices and 2D vortices respectively, leading to deviations from the behaviour expected for a 3D anisotropic superconductor with isotropic pinning. In fact, for vortex lines parallel to the layers, the core energy barrier opposing the motion across the layers result in critical currents (Figure 6), which may be qualified as "intrinsic", contrarily to those due to pinning by crystal defects [13]. Intrinsic pinning of vortices by the layered structure is a generic and essential feature of such systems. Here pinning operates along the whole vortex length, and strong pinning is possible, contrarily to weak collective pinning by the disorder. Some analogy exists with pinning by other extended structures as twin boundaries, or irradiation tracks (see lectures by D. Nelson and E. H. Brandt, this school). Assuming strong pinning, at low temperatures the pinning energy is a large (up to 10%) fraction of the total free energy, and intrinsic critical currents can be in the same ratio with respect to depairing currents, i. e. $J_c \leq 10^8$ A. cm$^{-2}$. However, as with other kinds of pinning, the critical currents fall down with temperature, due here to thermal activation of kinks. If the field is slightly tilted from the layer direction, the dissipation channel due to equilibrium kinks (Figure 2) may dominate, so the dissipation scales with the component $H_c$, but is still very low. Concerning practical applications, this range of field orientation may provide good performances at high temperature and fields.

## 4. Surface barriers in layered superconductors

In a type-II superconductor, at fields larger than $H_{c1}$, vortex lines are thermodynamically stable. However, vortices involve topological singularities of the superconducting wavefunction and can only appear in the bulk as lines or closed loops. Usually, their free energy is prohibitively high and thermal activation of vortices is unlikely (this may be questioned in quasi-2D superconductors where small loops can be formed, leading to 2D vortex-antivortex pairs in case of extremely weak layer coupling). This implies that vortex lines must enter from the surface. For a perfect surface, the free energy of vortices at the surface is higher than in the bulk, owing to the vortex-surface interaction (see lectures by A. A. Abrikosov, this school), so that the entrance field $H_s$ for surface penetration is higher than $H_{c1}$ [31,33]. An argument similar to Landau's argument for superfluids can be given for an estimate of $H_s$: in the Meissner state, the current flowing at the surface normal to the y-direction (Figure 7a), is given by curl h = $(4\pi/c)$ J, thus J = $Hc/4\pi\lambda$ where $\lambda$ is the London screening length and H the external field. When J becomes of the order of the depairing current $J_d$, the Meissner state becomes unstable against the formation of a vortex core. This shows that nucleation occurs at $H_s = H_s^0 \approx (4\pi\lambda/c)J_d \approx H_c$, where $H_c = \Phi_0/2\sqrt{2}\pi\lambda\xi$ is the thermodynamical critical field. By the same token, one obtains that nucleation occurs at the distance $\xi$ from the surface (Figure 7a).

In anisotropic 3D superconductors, the above result is not changed, ad shown in Ref. 32. This is due to the fact that $H_c$ does not feel the anisotropy. On the other hand, nucleation directly probes the scale of the coherence length, and one may expects anomalies to occur when the vortex core is deeply modified by the layered structure. This is indeed the case for a Josephson vortex line parallel to the layers, and a surface normal to them (Figure 7b). In the quasi-2D regime (T < T*), as mentioned in Section 1, the maximum current tolerated is the Josephson critical current $J_0 = \Phi_0 c/8\pi\lambda_c^2 s$ instead of the depairing current. Using $\lambda_c$ as the screening length in the y-direction, one finds

$$H_s = \frac{\Phi_0}{4\pi\lambda_c s} \qquad (8)$$

This result is confirmed by a more precise calculation [34] following the lines of Ref.33. On the other hand, in the quasi-3D regime, nucleation is dominated by the depairing current and the 3D result holds, i.e. $H_s = H_s^0 = \Phi_0/4\pi\lambda_c\xi_c$. Equation (8) shows that below T*, since $\xi_c < s$, the entrance field is lowered compared to the 3D result. Moreover, it has a different temperature dependence (square root instead of linear, if the Ginzburg-Landau variation of $\lambda$ and $\xi$ is obeyed). Unambiguous observation of this effect would be a direct probe of the change of nature of the vortex core in a layered superconductor.

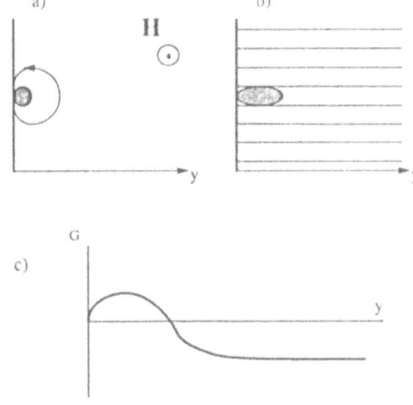

Figure 7. Surface barrier, field parallel to the surface (x0z). a) Isotropic, 3D. b) layered, quasi-2D. c) Vortex Gibbs energy as a function of the vortex-surface distance.

As mentioned in the previous Section, an important problem is the entrance field for kinks or pancake vortices. While the bulk properties of those objects have been studied by a number of authors (see J. Clem's lectures), the surface problem has been recently addressed in the case of zero Josephson coupling [35]. Let us consider a single pancake at a distance $y_0$ from the surface (Figure 8). The standard method for a vortex line introduces an image vortex line which simulates the boundary condition by cancelling exactly the field at the surface. This cancellation is due to translational symmetry. Then, by linear superposition, the effect of the external field must be added to obtain the vortex free energy near the surface. However, even in zero field, for a pancake vortex and its image, one can show that the boundary conditions are not fulfilled, owing to the lack of symmetry. An additional field must be added to restore the boundary conditions, or in other words to ensure that the field goes properly to zero at infinity. The total field created by the pancake near the surface, in presence of an external field normal to the layers is then

$$\mathbf{h} = \mathbf{h}_v + \mathbf{h}_{\bar{v}} + \mathbf{h}_M + \mathbf{h}'$$

(9)

Here $\mathbf{h}_v$ and $\mathbf{h}_{\bar{v}}$ are respectively the fields created by the pancake and its image, $\mathbf{h}_M = \mathbf{H} \exp(-y/\lambda_{ab})$ is the field induced by $\mathbf{H}$ in the Meissner state, and $\mathbf{h}'$ is the additional field. The effect of the latter is actually of order $s/\lambda_{ab}$, and can be neglected. Equation (9) allows to calculate the Gibbs energy [35]

$$G = \frac{\Phi_0^2 s}{16\pi^2 \lambda_{ab}^2} \, Ln \, \frac{2y_0}{\xi_{ab}} + \frac{\Phi_0 s}{4\pi} H\left(e^{-y_0/\lambda_c} - 1\right)$$

(10)

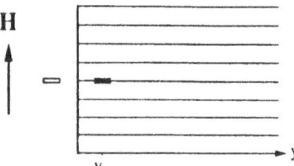

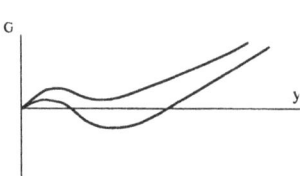

Figure 8. Surface barrier for a single pancake, field normal to the layers. The image pancake is represented. The curve G(y) has a minimum at distance $\approx \lambda_{ab}$.

The first term of G makes the difference with the case of a vortex line: as shown in Refs.15-16, the interaction of the pancake with its image is logarithmic, instead of exponential. The resulting function $G(y_0)$ is represented on Figure 8. It increases monotonously at distances larger than $\lambda_{ab}$, and presents for $H > H_s$ a stable minimum at a distance $y_m$ of order $\lambda_{ab}$, with

$$H_s \approx H_{c1c} + \frac{\Phi_0}{4\pi\lambda_{ab}^2} Ln(Ln \frac{\lambda_{ab}}{\xi_{ab}})$$

(11)

which is only slightly higher than $H_{c1c}$. As in the case of the vortex line penetration, the energy barrier against penetration vanishes at the field $H_s^0 \approx H_c$. However, and this is the essential difference, the energy barrier is finite (of the order of the single pancake free energy), while it is infinite for a vortex line. At high enough temperatures, this renders possible thermally activated penetration at fields nearly equal to $H_{c1}$, while at low temperatures the effective entrance field is again of order $H_c$. One thus expects an upturn in the temperature variation of $H_s$. Such an effect seems to have been observed in Tl:2212 [36], but a more detailed analysis is necessary. One should remark that pancake nucleation at the distance $y_m$ has the same probability in every layer, and that when a sufficient number of pancakes have been nucleated, their magnetic interactions will lower the barrier for entrance of other pancakes. An avalanche penetration of pancakes should take place until a full vortex line is nucleated. This shows the strongly irreversible nature of this process, and might explain the experimental evidence for surface barriers at fields of order $H_{c1}$. To be more realistic, in case of very weak Josephson coupling between layers, the object nucleated would be a pancake with a pair of Josephson "strings" attached to the surface. For large anisotropies, such an extremely curved vortex half-loop would be thermally activated.

To end this short discussion, one can remark that the knowledge of the entrance fields for both Josephson vortices and pancake vortices is the starting point for the study of entrance fields for staircase vortex lines, as suggested in Section 3.

## 5. Conclusion

Recent studies have shown that, in natural or artificial layered superconductor, vortex lines strongly interact with the periodic layered structure. These new features show up when taking as a reference the behaviour of a 3D anisotropic superconductor. Most deviations occur in the field angular range close to the layer direction. They indirectly probe the interaction of vortex cores with the lattice, which result in the appearance of Josephson vortices and 2D vortices. This interaction is especially important in the quasi-2D regime $T < T^*$, but still persists, though weaker, in the quasi-3D regime $T^* < T < T_c$. A great deal of work is still necessary to clarify the competing mechanisms involved in this angular regime which offers the largest possible complexity in the physics of high-temperature superconductors. The phase diagram should include possible Kosterlitz-Thouless transition of 2D vortices, melting of the Josephson vortex lattice, and glassy behaviour.

## References

1. J. Friedel, J. Phys. France 49 (1988) 1561; J. Phys. Condens. Matter I, 1 (1989) 7757.
2. D. Feinberg, J. Phys. Condens. Matter I, in press (a review).
3. C. S. L. Chun, Guo Guang Zheng, J. L. Vicent and I. K. Schuller, Phys. Rev. B 29 (1984) 4915..
4. Qi Li, C. Kwon, X. X. Xi, S. Bhattacharya, A. Walkenhorst, T. Venkatesan, S. J. Hagen, W. Jiang and R. L. Greene, Phys. Rev. Lett. 69 (1992) 2713.

5. W.E. Lawrence and S. Doniach, in *Proc. 12$^{th}$ Int. Conf. on Low Temperature Physics, Kyoto (1971)*, ed. E. Kanda (Keigaku, Tokyo, 1971) p 361.
6. L. N. Bulaevskii, Zh. Eksp. Teor. Fiz. 64 (1973) 2241 [Sov. Phys. JETP 37 (1973) 1133]. Ibid., Int. J. Mod. Phys. B 4 (1990) 1849 (a review).
7. J. R. Clem and M. W. Coffey, Phys. Rev. B 42 (1990) 6209.
8. S. Doniach in *"High Temperature Superconductivity", Los Alamos Symposium 1989*, edited by K. S. Bedell, D. Coffey, D. E. Meltzer, D. Pines and J. R. Schrieffer (Addison Wesley, Redwood City, 1990), p. 406.
9 D. Feinberg and A. M. Ettouhami, Int. J. Mod. Phys. B 7 (1993) 2085.
10. J. R. Clem, M. W. Coffey and Zhidong Hao, Phys. Rev. B 44 (1991) 2732.
11. J. R. Clem, J. Low Temp. Phys. 18 (1975) 427.
12. D. Feinberg and C. Villard, Mod. Phys. Lett. B 4 (1990) 9.
13. D. Feinberg and C. Villard, Phys. Rev. Lett. 65 (1990) 919.
14. A. Barone, A. I. Larkin and Yu. N. Ovchinnikov, J. of Supercond., 3 (1990) 155.
15. A. Buzdin and D. Feinberg, J. Phys. (Paris) 51 (1990) 1971.
16. J. R. Clem, Phys. Rev. B 43 (1991) 7837.
17. D. Feinberg, Physica C 194 (1992) 126.
18. L. N. Bulaevskii, M. Ledvij and V. G. Kogan, Phys. Rev. B 46 (1992) 366.
19. L. N. Bulaevskii, Phys. Rev. B 44 (1991) 910.
20. D. Feinberg, Solid St. Comm. 76 (1990) 789.
21. V. G. Kogan, Phys. Rev. B 38 (1988) 7049.
22. D. E. Farrell, C. M. Williams, S. A. Wolf, N. P. Bansal and V. G. Kogan, Phys. Rev. Lett. 61 (1988) 2805.
23. 20. D. E. Farrell, J. P. Rice, D. M. Ginsberg and J. Z. Liu, Phys. Rev. Lett. 64 (1990) 1573.
24. P. Pugnat, G. Fillion, S. Khene, H. Noël, P. Schuster and B. Barbara, *Proc. 20$^{th}$ Int. Conf. on Low Temperature Physics, Eugene (1993)*.
25. F. Steinmeyer, R. Kleiner, P. Müller and K. Winzer, *Proc. 20$^{th}$ Int. Conf. on Low Temperature Physics, Eugene (1993)*.; F. Steinmeyer, R. Kleiner, P. Müller, H. Müller and K. Winzer, Phys. Rev. Lett., in press.
26. W. K. Kwok, U. Welp, V. M. Vinokur, S. Fleshler, J. Downey and G. W. Crabtree, Phys. Rev. Lett. 67 (1991) 390.
27. P. A. Mansky, P. M. Chaikin and R. C. Haddon, Phys. Rev. Lett. 70 (1993) 1323.
28. S. M. De Soto, C. P. Slichter, H. H. Wang, U. Geiser and J. M. Williams, Phys. Rev. Lett. 70 (1993) 2956.
29. S. S. Maslov and V. L. Pokrovsky, Europhys. Lett. 14 (1991) 591.
30. Y. Iye, T. Tamegai and S. Nakamura, Physica C 174 (1991) 227.
31. C. P. Bean, and J. B. Livingston, Phys. Rev. Lett. 12 (1964) 14.
32. V. P. Damjanovic and A. Yu. Simonov , J. Physique. Condens. Matter I, 1 (1991) 1639.
33 P. G. De Gennes, *Superconductivity of Metals and Alloys* (Benjamin, N. Y., 1966).
34. A. Buzdin and D. Feinberg, Phys. Lett. A 165 (1992) 281.
35. A. Buzdin and D. Feinberg, Phys. Lett. A 167 (1992) 89.
36. V. N. Kopylov, A. Koshelev, I. F. Schegolev and T. G. Togonidze, Physica C 170 (1990) 291.

# VORTEX STATE AND DIMENSIONALITY

V.V.MOSHCHALKOV, and Y.BRUYNSERAEDE
*Laboratorium voor Vaste Stof-Fysika en Magnetisme, K.U. Leuven, Celestijnenlaan 200D, B-3001 Leuven, Belgium*

V.V.METLUSHKO, and G.GÜNTHERODT
*2.Physikalisches Institut, RWTH Aachen, Templergraben 55, W-5100 Aachen, Germany*

I.N.GONCHAROV and A.Yu.DIDYK
*Joint Institute for Nuclear Research, Dubna, 141980, Russia*

P.WAGNER, and H.ADRIAN
*Institut für Festkörperphysik, TH Darmstadt, Hochschulstr.8, 6100 Darmstadt, Germany*

H.THOMAS and K.WINZER
*1.Physikalisches Institut, Universität Göttingen, Bunsenstr. 9, W-3400 Göttingen, Germany*

ABSTRACT. We report on detailed studies of the superconducting critical parameters of high $T_c$ cuprates with different anisotropy $\gamma$. For this purpose we have taken single crystalline samples of $YBa_2Cu_3O_x$, $Pb_2Sr_2RE_{1-x}Ca_xCu_3O_{8+y}$ (RE=Y, Dy), $Tl_2Ba_2Ca_2Cu_3O_{10}$ and $Bi_2Sr_2CaCu_2O_{8+z}$ cuprates with the systematically increasing parameter $\gamma$, which makes it possible to realize different coupling strengths between the superconducting $CuO_2$ planes. The superconducting critical parameters have been derived from the analysis of the magnetization data M(T,H). The reversible contribution $M^{REV}$(H,T) measured for H∥c and H⊥c is used to determine the lower critical fields $H_{c1}$(T). The temperature dependences $H_{c1}$(T) can be successfully interpreted in the framework of the models taking into account the layered structure of high $T_c$ cuprates with the proximity induced superconductivity. The irreversible contribution $M^{IRR}$ is used to obtain the critical current $j_c$, the flux creep rate d(lnM)/d(lnt) for H∥c. The influence of the decoupling of the $CuO_2$ superconducting planes on the superconducting critical parameters is discussed. Different pinning regimes have been identified. The vortex confinement by the irradiation induced amorphous columnar tracks or by defects formed during the film growth and its influence on $j_c$ and d(lnM)/d(lnt) have been analyzed.

## 1. Introduction

High $T_c$ cuprates are layered compounds which can be considered as a stack of weakly coupled superconducting $CuO_2$ planes. The strength of the coupling determines the anisotropy $\gamma=m_c/m_{ab}=(\lambda_c^2/\lambda_{ab}^2)$ and the effective dimensionality of these layered materials. The reported values of the parameter $\gamma$ vary in a wide range: $\gamma\approx25$ in $YBa_2Cu_3O_7$ (Y123) to $\gamma\approx3000$ [1] and even $\gamma\approx8.1\cdot10^5$ [2] in $Bi_2Sr_2CaCu_2O_8$ (Bi2212). In the present paper we report on the detailed studies of the superconducting critical parameters in single crystals of high $T_c$ cuprates with the different anisotropy $\gamma$. These parameters were derived from the analysis of the magnetization data M(T,H), measured in the range T≤130K, H≤50kG in a Quantum Design SQUID. First, from the two branches $M^+$ and $M^-$ of the hysteresis loops we determined the reversible contribution

*N. Bontemps et al. (eds.), The Vortex State, 293–302.*

$M^{REV}(T,H)=0.5(M^++M^-)$ which was used to determine the temperature dependences of the lower critical fields $H_{c1}(T)$. Secondly, we focused on the behavior of the critical currents $j_c(T,H)$ and the flux creep rate $S\equiv d(lnM)/d(lnt)$, which were obtained from the irreversible contribution $M^{IRR}=0.5(M^+-M^-)$. We described the main features of the $j_c(T,H)$ and $S(T)$ dependences introducing different pinning regimes predicted by the collective pinning theory [3, 4]. Thirdly, the proposed identification of different pinning regimes has been confirmed by the successful interpretation of the modifications of the $j_c(T,H)$ and $S(T)$ curves which are caused by introducing columnar defects in single crystals or growth defects in thin films.

## 2. Lower Critical Fields

To determine the $H_{c1}$ values from the reversible magnetization we used the extrapolation method [5,6], which predicts the simple relation between the deviation $\Delta M$ from the linear M vs H Meissner response with the slope $dM/dH=-1/4\pi$ and the applied field: $\Delta M \propto (H_a-H_{c1})^2$. As a result,the extrapolation to zero of the dependence $\Delta M^{1/2}$ vs $H_a$ gives the $H_{c1}$ value. This method is based on the analysis of the expected bulk sample behavior and it neglects the flux line penetration at sample corners, sharp edges, along defects, etc., where the demagnetizing fields are larger and therefore field starts to penetrate well below the $H_{c1}$ value of a bulk sample.

The insert in Fig.1a demonstrates a good linearity of the $\Delta M^{1/2}$ vs $H_a$ plot which confirms the expected $\Delta M \propto (H_a-H_{c1})^2$ relation between $\Delta M$ and $H_a$. Fig.1a shows the temperature dependences of the low critical fields obtained by the extrapolation method. These data clearly deviate from a conventional parabolic behavior both for $Pb_2Sr_2Y_{1-x}Ca_xCu_3O_{8+y}$ (Pb2213) and $Tl_2Ba_2Ca_2Cu_3O_{10}$ (Tl2223) single crystals.

Fig.1b shows temperature dependences of lower critical fields for Pb2213 also obtained by the extrapolation method. We should note here that our $H_{c1}^{\perp c}(T)$ data give the highest possible estimate of $H_{c1}^{\perp c}$, which means that the real $H_{c1}^{\perp c}$ values might be even lower. The anisotropy parameter

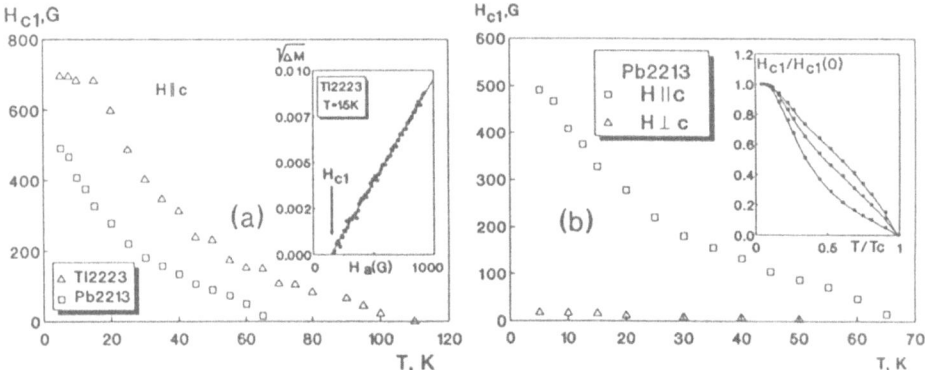

*Fig.1a. Temperature dependences of the low critical field $H_{c1}$ of a Pb2213 and Tl2223 single crystals. Insert: $M^{1/2}$ vs the field $H_a$, H∥c, T=15K for Tl2223*

*Fig.1b. Temperature dependences of the low critical field $H_{c1}$ of a Pb2213 single crystal for H∥ c and H⊥c. Insert: $H_{c1}(T)$ curves calculated for S-N-S multilayers with different weight coefficients σ of N-layers (curve 1-σ =5; 2- σ=10; 3-σ=100) (data taken from Golubov et al. in Ref.7)*

$$\gamma^{1/2} \equiv (m_c/m_{ab})^{1/2} = H_{c1}{}^{\|c}/H_{c1}{}^{\perp c} = \lambda_c/\lambda_{ab} \geq 31$$

gives the effective mass ratio $m_c/m_{ab} \geq 960$ indicating a well defined regime of a weak coupling between superconducting $CuO_2$ bilayers along the c-axis. The quasi two-dimensional (2D) character of high-Tc oxides has been emphasized before for $Bi_2Sr_2CaCu_2O_y$ ($\gamma \approx 3 \cdot 10^3 \div 8.1 \cdot 10^5$ [1, 2]), $Tl_2Ba_2CaCu_2O_x$ ($\gamma \approx 5 \cdot 10^3$ [8]) and for $YBa_2Cu_3O_z$ ($\gamma \approx 25$ [9]).

The anomalous temperature dependence of $H_{c1}{}^{\|c}(T)$ is also a signature of the layered structure of high-Tc oxides, which may be effectively treated as S-N-S or S-I-S multilayers with the superconducting $CuO_2$ layer separated by nonsuperconducting layers [7]. The contribution arising from these layers leads to an appearance of a positive $H_{c1}(T)$ curvature at low temperatures. We have made a qualitative comparison of our measured $H_{c1}{}^{\|c}(T)$ curve for Pb2213 single crystal with the model calculation [7] for the S-N-S multilayers (see inset in Fig.1b). The theoretical fit may describe an anomalous positive curvature of $H_{c1}{}^{\|c}(T)$, though the constraints used in Ref.7 might not be suitable for the case of high-$T_c$ materials, where the spacer layers may be insulating.

Due to ambiguities in the determination of $H_{c1}{}^{\perp c}(T)$, we are not discussing the $H_{c1}{}^{\perp c}(T)$ dependence given in Fig.1b. Nevertheless, we are sure that our data give only an upper limit for $H_{c1}{}^{\perp c}(T)$, i.e. the real anisotropy may be, in fact, even essentially higher than $\gamma \approx 960$ mentioned above.

## 3. Critical Currents and Flux Creep

For H||c the $j_c$ values have been calculated from $M^{IRR}$ using the isotropic Bean model. Fig.2a demonstrates the evolution of the $j_c(T)$ dependences in high $T_c$ cuprates with a different anisotropy $\gamma$. A remarkable transformation of the $j_c(T)$ curves is clearly seen as one goes from isotropic cubic $Ba_{1-x}K_xBiO_3$ (BKBiO) ($\gamma \approx 1$) to nearly 2D Bi2212 ($\gamma \approx 3 \cdot 10^3 \div 8.1 \cdot 10^5$ [1, 2]). In particular, a kink on the $j_c(T)$ curve is developed as anisotropy increases. This transformation of $j_c(T)$ dependences is also accompanied by dramatic changes in the normalized flux creep rate $S(T)$ (Fig.2b). For Bi2212 the $j_c(T)$ kink occurs at the same temperature $T_{CR} \approx 22K$ in the vicinity of which the sharp $S(T)$ peak is observed (Fig.2b). This peak is strongly suppressed and pushed towards higher temperatures as anisotropy $\gamma$ decreases.

In what follows, we shall use the specific $j_c(T,H)$ (Fig.4-6) and $S(T)$ (Fig.4) features to identify different pinning regimes in Bi2212. However, prior to the identification it is worth making a remark about the relation between $j_c(T)$ and $S(T)$. In spite of a high flux creep rate in high $T_c$ cuprates, especially in Bi2212, we still think that the $j_c(T)$ value (Fig.2,3) determined from the hysteresis loop with a typical delay of a few minutes is a good first approximation of the intrinsic $j_c(T)$ behavior. This argument is based on the experimental observation of a persistence of the $j_c(T)$ shape, if to define the $j_c(T)$ values after different delay times (Fig.3). Though the $j_c(T)$ curve is shifted as the delay time grows, the main $j_c(T)$ features, such as the nearly exponential $j_c$ decrease at $T < T^{CR}$, the presence of the kink at $T = T^{CR}$, are clearly conserved.

The four temperature intervals (I-IV) shown in Fig.4a define the boundaries of different pinning regimes derived in the collective pinning theory [3, 4]. Before comparing our experimental results with this theory, we have to define dimensionality (3D or 2D) which should be used. The pure 2D case is valid when the flux line lattice constant $a_v$ becomes smaller than the Josephson bending length $R_J = \gamma^{1/2} \cdot d$, where $\gamma$ is the anisotropy factor $\gamma = m_c/m_{ab}$ and d is the separation between the $CuO_2$ planes. With d=1.5nm and $\gamma \approx 3000$ [1] for BSCCO, the lowest limit of the crossover field is $H_{2D} = \Phi_0/R_J{}^2 \approx 3kG$ ($\Phi_0$ is flux quantum) and therefore the average trapped field $H_{av}$ in the sample (with the field profile H=0 at the boundary and $H=H_t$ in the middle) is in our case $H_{av} \approx H_t/2 < H_{2D} \approx 3kG$. As a result, we may analyze our data in the framework of the 3D collective pinning model [3]. One should however keep in mind that the restoring force $F_p$ for the pinned flux lines is very weak in anisotropic superconductors. The latter markedly highlights the effect of the thermal fluctuations on the flux line dynamics. The four different temperature intervals I-IV shown in Fig.4a may be identified as follows:

(i) *In the temperature region $T<T_V$ (interval I) the pinning of individual flux lines takes place* with the characteristic energy $U=U_c\approx H_c^2\xi^2(\xi/L_c)\approx 80(\xi/L_c)[K]$ [4], where $H_c$ is the thermodynamical critical field, $\xi$ is the coherence length in the a,b-plane and $L_c$ is the length of a single vortex line pinned collectively by many point defects. The necessary bending of the flux line is easily provided by the extreme weakness of the pinning force $F_p$. The flux creep is described by the interpolation formula [10]:

$$M(t) = \frac{M_{co}}{\left(1+\dfrac{\mu k_B T \ln(t/t_{eff})}{U}\right)^{1/\mu}} \qquad (1)$$

where $M_{co}$ and $t_{eff}$ are the initial magnetization and characteristic time, respectively, with $\mu=1/7$ and $U=U_c$ [3, 4]. From the measured rate $S(T)\approx k_B T/U$ we find that due to the very high anisotropy the length of the single hopping vortex line, which gives the observed low temperature value $U(T<T_V)\approx 120K$, is very short $(L_c\approx\xi\approx 2d)$,, i.e. very close to the 3D-2D crossover. The temperature dependence of $j_c$ should interpolate between $j_c=const$ ($T\ll T_V$) and the exponential dependence [3]:

$$j_c = \frac{3\sqrt{3}j_0 B}{32\pi H_{c2}}\left[1+\frac{T}{T_T^*}\right]^{1/2}\cdot\exp\left[-\pi\sqrt{2}\left[\frac{B}{H_{c2}}\right]^{3/2}\frac{G_m}{G}\left[1+\frac{T}{T_T^*}\right]^3\right] \qquad (2)$$

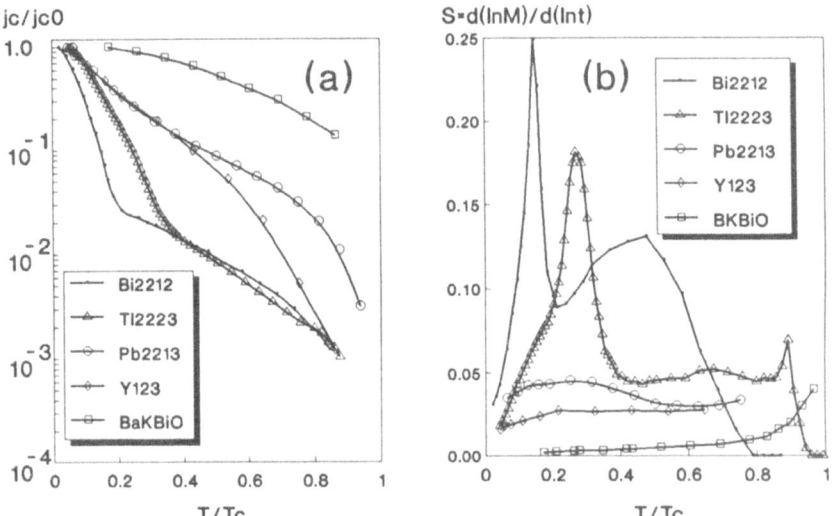

**Fig.2.** *Temperature dependences of the critical current $j_c$ (a) and normalized relaxation rate S (b) for $Bi_2Sr_2CaCu_2O_x$, $Tl_2Ba_2Ca_2Cu_3O_{10}$, $Pb_2Sr_2Y_{1-x}Ca_xCu_3O_{8+y}$, $YBa_2Cu_3O_y$, $Ba_{1-x}K_xBiO_3$*

where $j_0$ is the depairing current, G is the characteristic strength of the pinning potential, $G_m=(H_c^2\gamma^2)^2\gamma$ is the value of the parameter G corresponding to an extremely strong short-scale disorder, B is the magnetic induction, $T_T^*$ is the temperature at which the thermal fluctuations are larger than the coherence length: $\langle u^2\rangle \geq \xi^2$ and $H_{c2}$ is the upper critical field [3]. Such an interpolation is indeed in agreement with our experimental observations (see the solid lines (Eq.2) in Fig.5 for $j_c>j_c^{CR}$).

(ii) *The temperature range $T_V<T<T_{CR}$ (interval II) corresponds to the collective pinning of the small flux bundles with a strong dispersion.* The term "collective" here emphasizes, in fact, that (1) *many pinning centers are involved* and (2) *several flux lines* are forming a flux bundle. In this case the current $j_c$ (Fig.4a) still follows an exponential dependence (Eq.2), and the effective pinning potential grows due to the collective nature of the pinning, i.e. we should take $U\approx U_c(a_v/L_c)^{1/5}$ [4]. Inserting $L_c\approx2d\approx3nm$ and $a_v\approx200nm$ we would expect an enhancement of U by a factor of 2.5, which is not far from the experimental value $U(T>T_{CR})/U(T<T_{CR})\approx3$-4. Since $L_c\approx4a_v/\gamma$, then this ratio, characterizing the jump of the two S(T) slopes at $T<T_{CR}$ and $T>T_{CR}$, is reduced to $(\gamma/4)^{1/5}$. The very low power 1/5 seems to explain a fast suppression of the S(T) peak with a growth of the anisotropy $\gamma$ (see Fig.2b). It should be noted here that the derivation of U just as the inverse of the slope of lnM vs lnt is only an estimate, since the real relation between M and t is much more complicated (see the interpolation formula, Eq.1) and the transition from the temperature interval I to the interval II is also accompanied by the change of $\mu$ from 1/7 to 3/2 [4].

(iii) *When the current $j_c$ crosses a certain threshold level $j_c^{CR}\approx(4$-5$)\cdot10^4 A/cm^2$, we enter the temperature region $T_{CR}<T<T_M$ (interval III).* At these temperatures the strong exponential decrease of $j_c$ with T is converted into a much weaker temperature dependence. The behavior of both $j_c(T)$ and S(T) vs T *can be related in this case to the large-flux-bundle weakly dispersive collective pinning*. The theory [3] predicts a power law dependence for $j_c<j_c^{CR}$:

$$j_c \approx \frac{10j_0}{\kappa^2}\left[\frac{G}{G_m}\right]^2\left[\frac{H_{c2}}{B}\right]^3\left[\frac{T_T^*}{T+T_T^*}\right]^{11/2}$$

(3)

where $\kappa$ is the Ginzburg-Landau parameter. The kink in $j_c(T)$ at $T=T_{CR}$ results from the transition from the exponential (Eq.2) to the power law (Eq.3) dependence (see Fig.5). From this point of view $j_c^{CR}$ should correspond to $j_0/\kappa^2$ [3], where $j_0$ is the depairing current in Eq.2. The

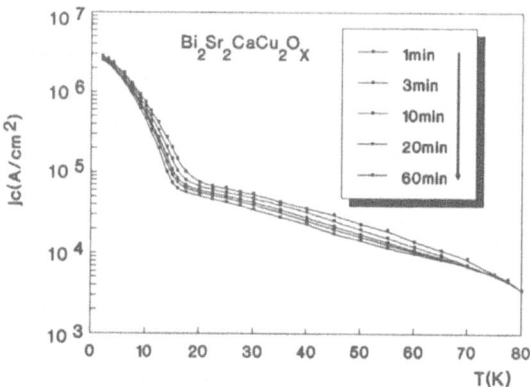

*Fig.3. Temperature dependence of the critical current $j_c$ in $H_a=0$ for $Bi_2Sr_2CaCu_2O_x$*

experimental value of the crossover current (Fig.4a-c,5) $j_c^{CR} \approx (4-5) \cdot 10^4 A/cm^2$ is indeed nearly independent of the irradiation dose and fits quantitatively the theoretical estimate [3]: $j_c^{CR} \approx j_0/\kappa^2$ $\approx 10^4$-$10^5 A/cm^2$ with $\kappa$=80-200 [3]. The power law (Eq.3) gives a reasonably good fit of our experimental data for currents $j_c(T) < j_0/\kappa^2$ (see the solid line (Eq.3) in Fig.5).

For $j_c < j_0/\kappa^2$ the flux creep potential U and the exponent $\mu$ are also renormalized due to a much weaker dispersion of the elastic moduli for the motion of a large flux bundle [4]: $\mu$=7/9 and $U \approx [U_c(a_v/L_c)^{1/5}](\lambda/a_v)^3$. This potential barrier may be further increased at $T \rightarrow T_M$ due to the formation of plastic deformations of the vortex lattice with $U >> [U_c(a_v/L_c)^{1/5}](\lambda/a_v)^3$ [11]. Such an increase of U may cause a substantial suppression of S(T) at temperatures $T \rightarrow T_M$ (see Fig.4).

(iv) *The temperature range $T > T_M$* seems to correspond to the *regime of the molten flux lattice* [12] with a reversible behavior of the vortex fluid.

## 4. Critical Currents, Flux Creep and Current-Voltage Characteristics in Irradiated Bi2212 Single Crystals and Thin Films

The evolution of the S(T) and $j_c(T)$ curves as a function of the irradiation dose (Fig.4a-c) supports the interpretation of the different pinning regimes I-IV. First of all the irradiation leads to a strong expansion of the low temperature individual flux-line pinning regime due to a confinement of the flux lines by available columnar defects (Fig.4b,c). Theoretically this may be easily explained as a result of an increase of the characteristic temperature $T_V^* = (G \cdot C_1/6)^{1/3}$ [3], where $C_1$ is the vortex line tension and G is the strength of the pinning potential. The characteristic temperature $T_V^* \approx T_V$ [3] is apparently enhanced by creating artificial columnar tracks suitable for matching by the individual flux lines, since the distance ($\approx$30nm) between columnar defects in our case is much smaller than the flux lattice period $a_v$ ($\approx$200nm). The crossover temperature T between regimes I and II is found from the relation $<u^2>^{1/2} \approx \xi$. Therefore, the reduction of $<u^2>$ with the irradiation should also lead to an essential growth of $T_{CR}$, since for larger G values a higher temperature $T_T^* \approx T_{CR}$ is needed. This also explains the increase of $T_{CR}$ with the irradiation dose.

The influence of a magnetic field on the critical current density is shown in Fig.6. These data clearly indicate that $j_c$ is much less influenced by H in irradiated single crystals and in thin films compared to the unirradiated single crystal which may be related to a matching of the flux lines to the introduced pinning centers.

To study also the onset of the resistive state in the picovolt range we used current-voltage characteristics determined from the magnetization measurements, assuming that j is proportional to the width of the hysteresis loop, and the electric field E is calculated using Faraday's inductance law [13]:

$$E = \frac{3\mu_0 A}{2\pi^2 R^2} \cdot \frac{P_m}{dt} \qquad (4)$$

Here Pm is the magnetic moment, R is the radius of the sample, $\mu_0 = 4\pi \cdot 10^{-7}$ H/m and the constant A=3.328 [14].

Fig.7 shows the current-voltage characteristics at different temperatures deduced from magnetization measurements on the Bi2212 film. The linearity of the log-log plots indicates that $E \propto j^n$, with the exponent n being strongly dependent on temperature and magnetic field.

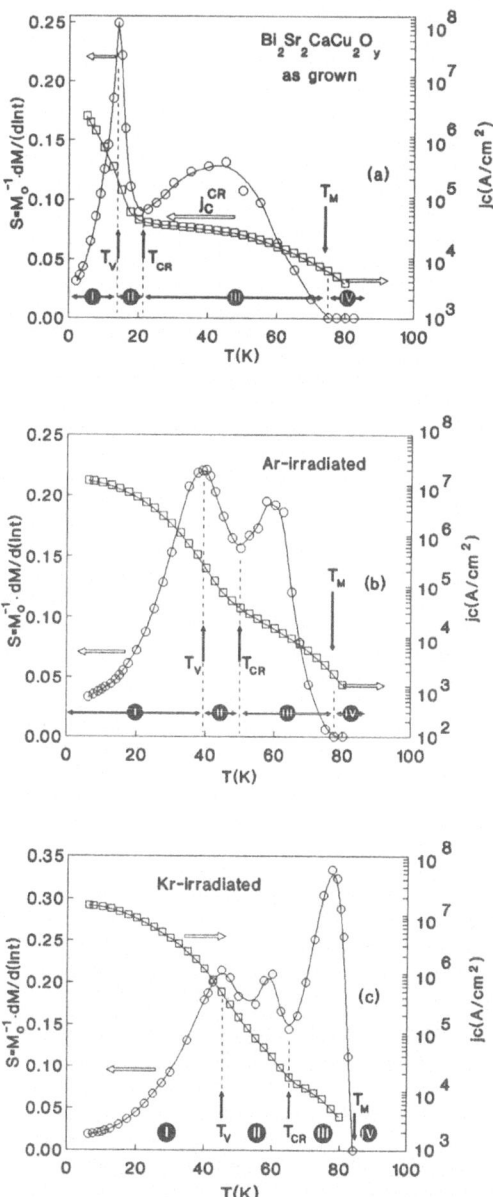

**Fig. 4.** *Temperature dependences of the normalized flux creep rate S(T) and the critical current density $j_c(T)$ for an as-grown (a), an Ar - (b) and a Kr-(c) irradiated $Bi_2Sr_2CaCu_2O_8$ single crystal with the doses $\Phi(Ar) \approx 5 \cdot 10^{10} ions/cm^2$ and $\Phi(Kr) \approx 10^{11} ions/cm^2$*

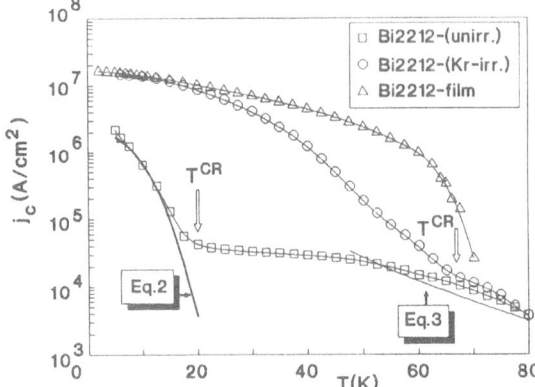

**Fig. 5.** *Temperature dependences of the critical current density $j_c$ of the Bi2212 as-grown and Kr-irradiated single crystal, and of the dc-sputtered Bi2212 thin film. Thick solid and solid lines are the $j_c$ values calculated for as-grown Bi2212 single crystal, using Eq.2 ($j_c > j_o/\kappa^2$) and Eq.3 ($j_c < j_o/\kappa^2$), respectively*

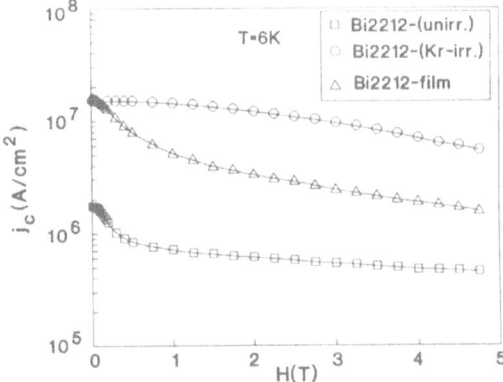

**Fig. 6.** *Magnetic field dependences of the critical current density $j_c$ measured at $T=6K$ for an as-grown, an Ar and a Kr-irradiated $Bi_2Sr_2CaCu_2O_8$ single crystal*

These data have been used to derive the temperature and field dependence of the exponent n as shown in Fig.8a and Fig.8b, respectively. In the as-grown single crystal an abrupt change of the slope of n vs T is observed in the vicinity of the crossover temperature $T=T^{CR}$. In irradiated single crystals the vortices are strongly confined to amorphous tracks and the crossover temperature $T^{CR}$ increases to about 65K (see Fig.8a). Finally, in the thin film the n(T) is monotonously decreasing with temperature and the crossover temperature $T^{CR}$ is completely smeared out due to the presence of many different types of pinning centers formed during the growth process.

For the as-grown Bi2212 single crystal the n(H) curve (Fig.8b) is characterized by a substantial enhancement of n in the field range $H \leq 1.5T$. This means that in applied magnetic field the I-V characteristics become steeper. The maximum at $H=H^{CR}$ may be related to the

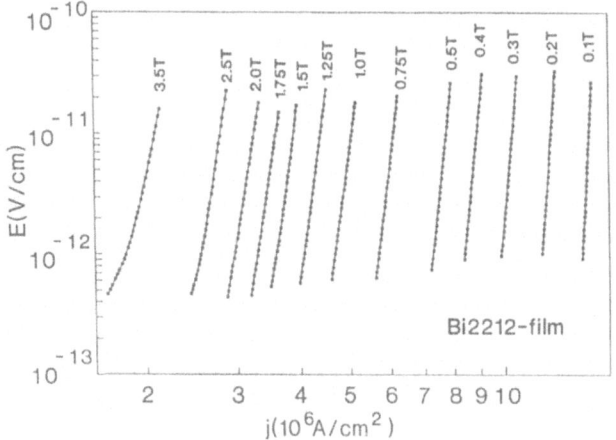

**Fig.7.** *Voltage-current characteristics measured magnetically for the Bi2212 thin film as a function of field, T=6K.*

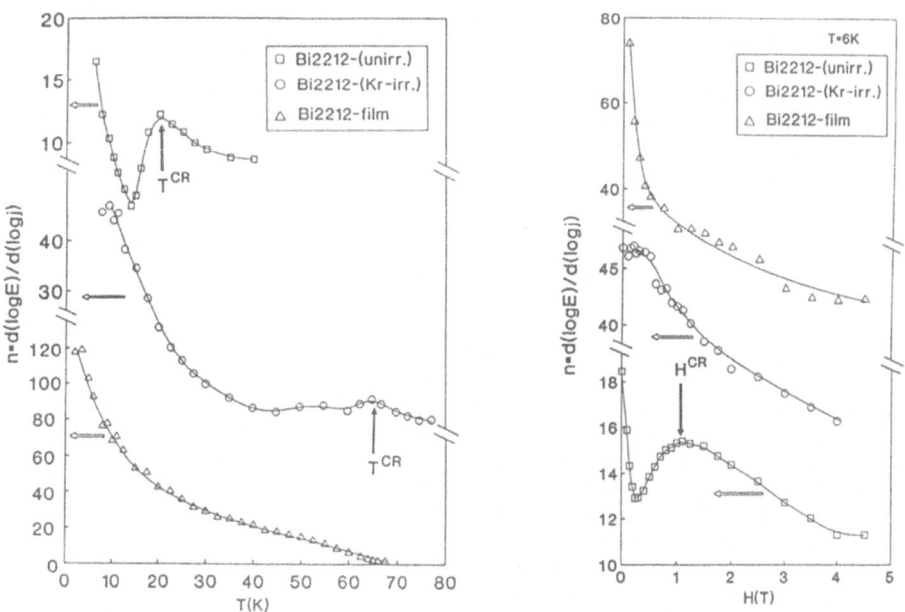

**Fig.8a.** *The exponent n, relating voltage with current ($V \propto I^n$), vs temperature for the as-grown (□), Kr-irradiated (O) single crystal and for the Bi2212 thin film (Δ).*

**Fig.8b.** *The exponent n, relating voltage with current ($V \propto I^n$), vs and magnetic field for the for the as-grown (□), Kr-irradiated (O) single crystal and for the Bi2212 thin film (Δ).*

crossover field $H^{CR}$ given by [15]: $H^{CR}=\Phi_0/(\gamma d^2)$, where $d\approx 15\text{Å}$ the separation distance between the $CuO_2$ planes [2]. The estimate for $H^{CR}$ lies in the range 0.3 - 1.2T.

For the thin Bi2212 film n is monotonously decreasing with temperature and field from $n\approx 120$ (T=6K and H=0) to $n\approx 1.8$ (T=67K and H=0), in sharp contrast to the as-grown single crystal. The shape of the n(H) curve of the Bi2212 film is in good agreement with the behavior observed for Bi2212 single crystals near $T_c$ [16-17]. It is also clear from Fig.8b that in the presence of a magnetic field, the absolute values of n are systematically higher for the thin film compared to the as-grown single crystal. The difference may be due to more efficient pinning by the defects produced during the film growth process. Comparing the $j_c$(T,H) curves for the irradiated single crystal (Fig.5,6) with a single type of well-defined pinning centers and for the film with a distribution of the pinning centers we can speculate that by creating *a single type* of the pinning centers in a single crystal it is possible to obtain the same $j_c$ values as in the thin film only at low temperatures. At temperatures above 30K, however, as it is evident from $j_c$(T) of the thin film (Fig.5), the $j_c$ enhancement may be realized only by introducing a broad distribution of pinning center parameters.

In conclusion, we have demonstrated that in Bi2212 and Tl2223 single crystals temperature and field dependences of the critical current and the flux creep rate can be successfully interpreted in terms of the collective pinning theory. These high $T_c$ materials are characterized by the presence of oxygen vacancies which act as point defects and at the same time their anisotropy $\gamma$ strongly reduces the tilt modulus thus making flux lines sufficiently flexible to be pinned by these point defects.

*This work is supported by the Belgian High Temperature Superconductivity and Concerted Action Programs (at K.U.Leuven) and by BMFT/FZ13N5487A (at RWTH Aachen) and by BMFT/13N5748A and by DFG/SFB252 (at TH Darmstadt).

**References**

[1] D.E.Farrel et al., Phys.Rev.Lett. **63**, (1989) 782.
[2] F.Steinmeyer et al., will be published in Physica B
[3] M.V.Feigel'man and V.M.Vinokur, Phys.Rev.B. **41**, (1990) 8986.
[4] M.V.Feigel'man et al., Phys.Rev.Lett. **63**, (1989) 2303.
[5] V.V.Moshchalkov et al., Physica C **175**, (1991) 407.
[6] M.Naito et al., Phys.Rev.B **41**, (1990) 4823.
[7] A.A.Golubov and V.M.Krasnov, Physica C **196**, (1992) 177.
[8] L.A.Gurevich et al., Physica C **195**, 323 (1992).
[9] D.E.Farrell et al., Phys.Rev.Lett. **61**, 2805 (1988).
[10] A.P.Malozemoff, Physica C **185-189**, (1991) 264.
[11] E.H.Brandt, Physica C **195**, (1992) 1
[12] D.R.Nelson, Phys.Rev.Lett. **60**, (1988) 1973
[13] G.Ries et al., Supercond.Sci.Technol. **5**, (1992) S81.
[14] A.A.Zhukov et al., Cryogenics **33**, (1993). 142
[15] D.S.Fisher et al., Phys.Rev.B **43**, (1991). 130
[16] A.Ando et al., Phys.Rev.Lett. **67**, (1991). 2737
[17] A.K.Pradhan et al., Physica C **204**, (1993). 325

# ENHANCEMENT OF SUPERCONDUCTIVITY BY PHOTOEXCITATION

Y. BRUYNSERAEDE, E. OSQUIGUIL, M. MAENHOUDT and B. WUYTS
*Laboratorium voor Vaste Stof-Fysika en Magnetisme*
*Katholieke Universiteit Leuven*
*Celestijnenlaan 200 D, B-3001 Leuven, Belgium*

G. NIEVA and J. GUIMPEL
*Centro Atómico Bariloche*
*8400 S.C. de Bariloche, RN-Argentina*

D. LEDERMAN and IVAN K. SCHULLER
*Physics Department - 0319, University of California - San Diego*
*La Jolla, CA 92093-0319, USA*

ABSTRACT. We report on extensive investigations of photoinduced phenomena in oxygen deficient YBCO films. It is clearly shown that persistent photoinduced phenomena are present for all oxygen contents x < 7, but are absent in films with an optimum doping level (i.e. maximum $T_c$). The illumination of fully oxidized Pr doped YBCO films does not produce persistent photoinduced effects at any doping level. However, by deoxidizing these films, persistent photoconductivity and photoinduced enhanced superconductivity are clearly observed. On the other hand, we could not observe persistent photoinduced phenomena after the illumination of underdoped, overdoped and/or optimum doped $Bi_2Sr_2CaCu_2O_{8+y}$ and $La_{2-x}Sr_xCuO_{4\pm\delta}$ films. Detailed x-ray diffraction studies show that light causes a large contraction of the c-axis lattice parameter in oxygen depleted YBCO films, which is similar to the change observed when oxygen deficient YBCO is annealed at room temperature after quenching from high temperatures. These observations not only indicate that oxygen vacancies are essential for the existence of persistent photoinduced phenomena, but they also strongly suggest that the chain planes play a crucial role in the microscopic mechanism responsible for the persistent photoinduced effects.

## 1. Introduction

It is well known that the photoexcitation of insulating and semiconducting materials gives rise to changes in their transport properties which are caused by the photoproduction of excess carriers. The existence of an insulating phase in high $T_c$ superconductors triggered the search, a few years ago, for such effects in these copper-oxide materials [1].

It has been shown that the pulsed laser illumination of $YBa_2Cu_3O_x$ (YBCO) insulating single crystals induces an increase in the non-equilibrium carrier density for short periods of time [2]. These experiments opened the possibility to dope this material through the insulator-metal (I-M) transition without changing its oxygen concentration [1, 2]. On the

303

*N. Bontemps et al. (eds.), The Vortex State, 303–312.*
*© 1994 Kluwer Academic Publishers.*

other hand, the work initiated by Kudinov et al. [3] with continuous wave laser illumination experiments performed in insulating YBCO thin films, led to the discovery of a persistent photoinduced enhancement of the conductivity [3, 4]. Both type of experiments indicated that photodoping is indeed an alternate way to induce superconductivity in YBCO insulating samples.

Motivated by these results, we demonstrated recently the existence of an interesting effect which only occurs in high temperature superconductors: the persistent photoinduced enhancement of superconductivity in metallic- oxygen-deficient YBCO thin films [5]. We also showed that the enhancement of superconductivity is accompanied by a corresponding decrease in the resistivity and Hall coefficient during photoexcitation, indicating a persistent increase in the carrier density [6].

In this paper we give a brief review and address important questions concerning the photoexcitation mechanism in YBCO films. In particular we discuss its dependence on the I-M transition, oxygen deficiency and the resulting structural changes. We show that although the proximity of the I-M transition enhances the photoinduced effects, the photoinduced enhancement of superconductivity can still be observed up to oxygen contents just below $x = 7$. In addition we show that the persistent photoinduced changes in the normal state and superconducting state, as well as in the structure of YBCO films, are strikingly similar to those changes induced by the ordering of oxygen ions in the basal planes.

## 2. Experimental

Fully oxidized c-axis oriented YBCO films were deposited on MgO(100) substrates using a planar magnetron sputtering system in the 90° off- axis configuration. A detailed description of the film production and their properties has been published elsewhere [7]. Oxygen deficient YBCO films, with a nominal oxygen content $6.4 \leq x \leq 7$, are prepared using a method based on the oxygen partial pressure-temperature phase diagram for bulk material [8]. We adopted for this work the phase diagram reported in [9]. One 1000 Å thick YBCO film was patterned using conventional photolithography and wet etching, into a four point pattern with dimensions 35 $\mu$m x 1 mm. The illumination of the oxygen deficient YBCO film was performed during 8 hours using an Ar-ion laser ($\lambda = 514$ nm). The measured power density at the surface of the film was 1.83 W/cm$^2$ corresponding to a total photon density $Q \cong 1.4 \times 10^{23}$ ph/cm$^2$. The resistance versus temperature curves, R(T), were measured before and after illumination during the warming up cycle and up to temperatures not exceeding 250 K, in a He flow cryostat with optical access.

## 3. Results

The normalized resistive transitions before illumination measured in one YBCO film with different oxygen contents is shown in Fig. 1a. As x is reduced, the transitions move uniformly to lower temperatures but show a pronounced rounding near the normal state. Since the anisotropy of YBCO increases as oxygen is removed from the basal planes [10], this rounding might be due to an increased contribution of two-dimensional fluctuations at low x. The critical temperature $T_c$, defined as the temperature at which dR/dT is maximal, is plotted in Fig. 2b as a function of the oxygen content.

Fig. 2 shows the transitions to the superconducting state measured before and after the laser illumination, for several oxygen contents: near the metal-insulator transition (Fig. 2a and 2b); in the 60 K plateau region (Fig. 2c), and in the 90 K plateau region (Fig. 2d and 2e). In contrast to the results of ref. [11], persistent photoinduced changes in $T_c$ and $\rho_{ab}$

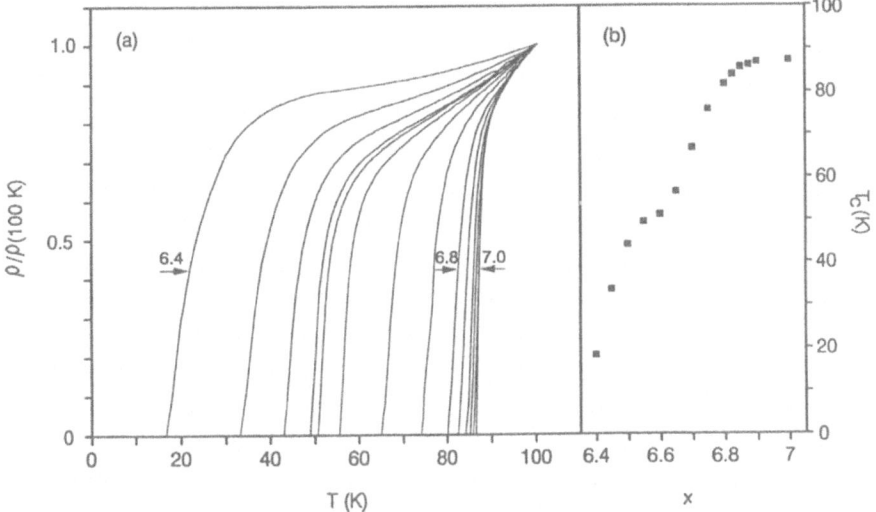

*Figure 1: a) Normalized resistive transitions for a 1000 Å thick YBCO film with different oxygen contents x. Between x = 6.4 and x = 6.8 the oxygen content has been varied in steps of 0.05, while between x = 6.8 and x = 6.9 the step is 0.025.*
*b) Critical temperature as a function of the oxygen content.*

are evident, for all oxygen concentrations. Clearly, the illumination produces a decrease of the resistivity throughout the entire temperature range for all x < 7. Note also that due to the parallel shift of the transition towards higher temperatures the same absolute shift of $T_c$ ( i.e. $\delta T_c = T_c$(after illumination) - $T_c$(before illumination)) is obtained irrespective of the definition of $T_c$. After 3 days of annealing at room temperature, the $T_c$ and $\rho(T)$ values measured before the illumination are recovered.

In Fig. 3 we plotted the *total* photoinduced shifts of respectively the critical temperature, $\delta T_c$, and the in-plane conductivity, $\delta\sigma_{ab}$ (see inset), as a function of x obtained after the illumination of the same YBCO film. The results ($\delta T_c$) are compared to the enhancement of $T_c$, $\Delta T_c$, produced by room temperature annealing of quenched oxygen deficient YBCO single crystals as reported by Veal et al. [12]. *Clearly both quantities, $\delta T_c$ and $\Delta T_c$, display a similar functional dependence on x.* Three different regions can be recognized: i) for x < 6.6, there is a fast increase in $\Delta T_c$ and $\delta T_c$ as the I-M transition is approached, ii) for $6.6 \leq x \leq 6.75$ the total increase in $T_c$ is almost independent of the oxygen content giving rise to a plateau in both the $\delta T_c(x)$ and $\Delta T_c(x)$ curves, and iii) for x > 6.75, $\delta T_c$ and $\Delta T_c$ decrease continuously as x is increased and extrapolate to zero at x = 7, where neither persistent photoinduced effects nor oxygen ordering effects are observed. Since an increase in the orthorhombicity [12], which is also reflected in a contraction of the c-axis lattice parameter [13], has also been observed during the room temperature annealing, we monitored eventual changes in the c-axis lattice parameter from $\Theta - 2\Theta$ x-ray scans of the [00 10] peak. These measurements were performed before, during, and after illumination

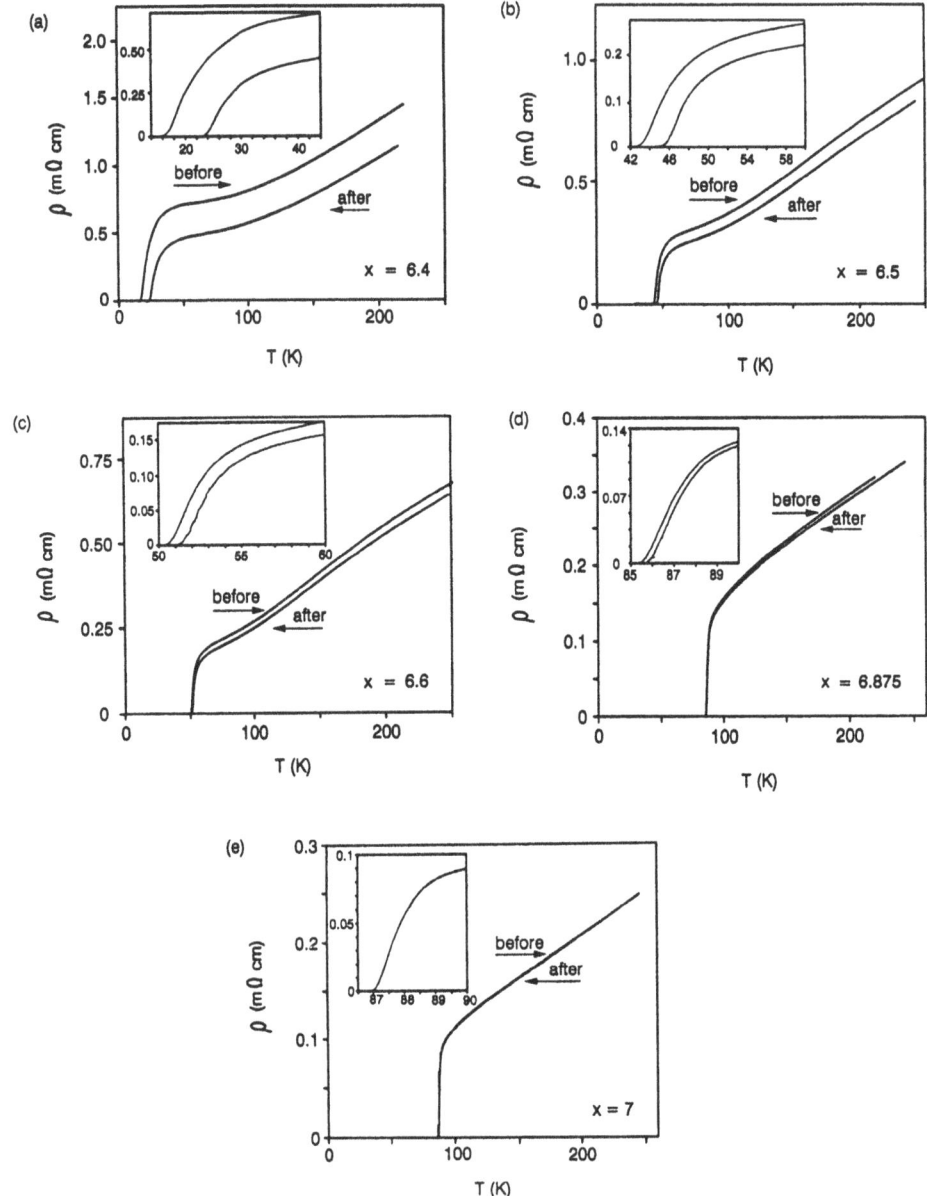

*Figure 2: Electrical resistivity ρ vs temperature before and immediately after laser illumination for one YBCO film with different oxygen stoichiometries, a) x = 6.4, b) x = 6.5, c) x = 6.6, d) x = 6.875 and e) x = 7. The insets show the region near $T_c$ in an expanded scale.*

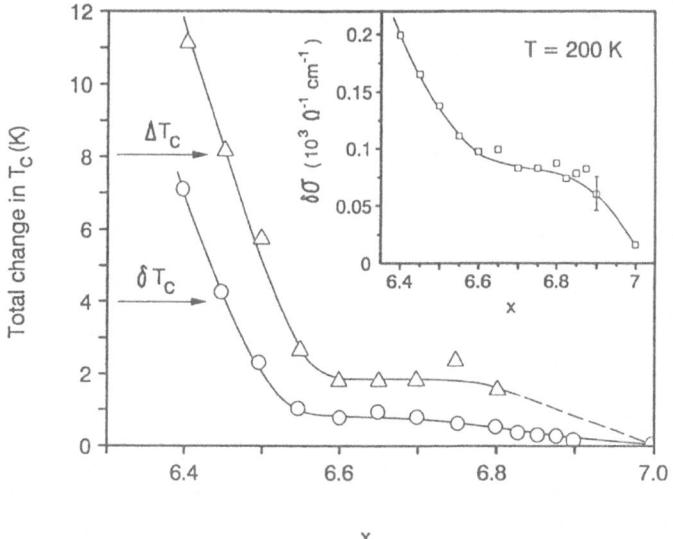

*Figure 3: Total enhancement of the critical temperature after the illumination, $\delta T_c$, as a function of the oxygen content. The data are compared to the total increase of $T_c$ due to oxygen ordering effects, $\Delta T_c$, in YBCO single crystals (ref. 12). The inset shows the total increase of the conductivity at $T = 200$ K after the illumination. All the solid lines are guides to the eye.*

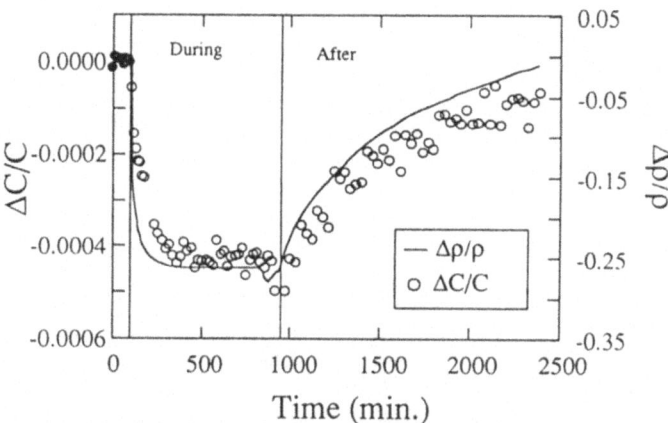

*Figure 4: Fractional changes $\Delta c/c$ of the c-axis lattice parameter ($\circ$) and resistivity $\Delta\rho/\rho$ (solid line) versus time in a semiconducting ($x = 6.5$) 1000 Å thick YBCO film, before, during and after illumination.*

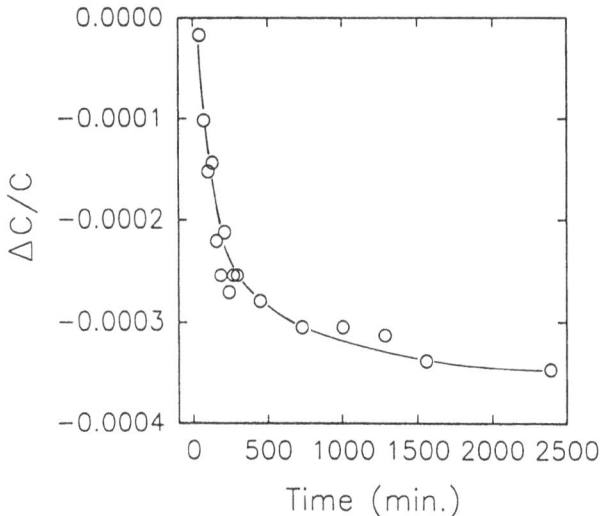

*Figure 5: Fractional change $\Delta c/c$ of the c-axis lattice parameter as a function of annealing time at room temperature after quenching an $x = 6.55$ YBCO film (after ref. 13). The solid line is a guide to the eye.*

with a halogen lamp at room temperature of an YBCO film with x = 6.5. The position of the peak was determined by fitting it to two Gaussians, corresponding to the $K_{\alpha 1}$ and $K_{\alpha 2}$ Cu x-ray wavelengths. Dry nitrogen gas was sprayed on the sample during the experiment to prevent excessive heating. The sample resistivity was simultaneously measured with a 4-point method. The resulting structural changes are displayed in Fig. 4, which shows the fractional changes in the c-axis lattice parameter ($\Delta c/c$) and resistivity ($\Delta \rho/\rho$) as a function of time, during and after the illumination. Notice that $\Delta c/c < 0$, which is the opposite of what would be expected from simple thermal expansion effects. This clearly indicates a contraction of the c-axis lattice parameter during the illumination, which can be compared with the contraction observed during annealing at room temperature of a quenched x = 6.55 film [13] (see Fig. 5).

## 4. Discussion

We now discuss several possible explanations of the observed persistent photoinduced phenomena. The first is related to the photoexcitation phenomena observed in low $T_c$ granular In-CdS [14] films and in Sn-Cds-Sn junctions [15], i.e. changes in the conductivity of the photosensitive (CdS) intergranular material. Assuming that the superconductivity in the YBCO films is granular in character, an analogous effect might be present during the laser irradiation. It should be noted, however, that granular In-CdS films exhibit a decrease in the width of the normal to superconducting transition, but no change in the onset critical temperature after illumination. In contrast, our results show a parallel shift of the R(T) curves and a clear increase in $T_c$ after illumination. Moreover, a rather peculiar behaviour of the intergranular material must be assumed in order to explain the oxygen content dependence of the photoinduced phenomena using this granular model.

Another possibility comes from the analogy with the well known photoinduced phenomena in semiconductors. According to this model an electron- hole pair is photoinduced by absorption of one photon in the $CuO_2$ planes. While the hole increases the carrier density in these planes, the electron is transferred via some mechanism to the CuO chain planes, *where it is trapped by an oxygen vacancy* [11, 16]. Experimentally we find that at each oxygen content, $T_c$ and $\sigma_{ab}$ saturate at different values, $T_c^{sat}$ and $\sigma_{ab}^{sat}$ respectively. This seems to contradict the assumption that oxygen vacancies may act as effective traps for the photogenerated electrons during the long time photoexcitation. Indeed, the number of oxygen vacancies per unit cell in oxygen deficient $YBa_2Cu_3O_x$ is $\delta = 7\text{-x}$ ($0 \leq \delta \leq 1$) and the number of holes per unit cell necessary to increase $T_c$ from zero ($\delta \geq 0.6$) to 90 K ($\delta = 0$) is approximately 0.15 [17]. It follows that the number of oxygen vacancies per unit cell for any oxygen content is large enough to trap all the electrons needed to raise the $T_c$ of the material to 90 K, an effect that has never been observed in the continuous wave laser experiments. Nevertheless, the possibility might exist that there is an oxygen dependent distribution of trapping energies, which together with the recombination dynamics could give rise to the saturation of the photoenhancement of $T_c$ and $\sigma_{ab}$.

The third possibility is based on the similarities shown in Figs. 3, 4 and 5, which suggest that during the long time laser illuminations a rearrangement of oxygen ions in the basal plane of YBCO may occur [18]. In general, the photoproduction of electron-hole pairs can give rise to local redistributions of charge inside the material with a consequent modification of the local electric fields, which in turn can induce the movement or diffusion of atoms. The occurence of *photoassisted oxygen ordering* may be described as follows. It has been shown [19] that the ordering of oxygen atoms decreases the fraction of three fold coordinated Cu1 and increases the fraction of two and four fold coordinated Cu1. It has also been shown [20] that the most probable oxidation state for Cu1 ions is 2+ when it is three fold (chain end) or four fold (Cu in a chain) coordinated, while it is 1+ when the Cu1 is two fold coordinated (empty chain). Therefore, we assume that in YBCO the produced photoelectrons change the valence state of three fold coordinated Cu1 from $Cu^{2+}$ to $Cu^+$ through the reaction $e^- + Cu^{2+} \longrightarrow Cu^+$, while the hole contributes to increase the carrier density. After the capture of the photoelectron the three fold coordination for a $Cu^+$ ion is energetically unfavorable [20] (see Fig. 6.a), and therefore the system will locally relax to a new configuration in which longer chains are produced at the expense of shorter chains or wrongly positioned oxygen atoms (Fig. 6.b). The new ordered configuration does not correspond to an equilibrium state since it involves an increase in the lattice elastic strain energy. However, in the absence of enough thermal energy this new situation persists in time preventing the recombination of the electron-hole pair. The system disorders again when the temperature is raised (Fig. 6.c), with characteristic times equal to the ones observed in phenomena related to the oxygen ordering process and following the same time dependence [11].

This simple physical picture accounts, in a natural way, for i) the existence of persistent photoinduced phenomena for all x < 7; ii) the saturation of the persistent photoinduced phenomena; iii) the peculiar behaviour of $\delta T_c$ as a function of the oxygen content (Fig. 3); iv) the contraction of the c-axis lattice parameter observed during the illumination of an insulating YBCO film (Fig. 4); and v) the non-exponential decay of the photoinduced phenomena.

310

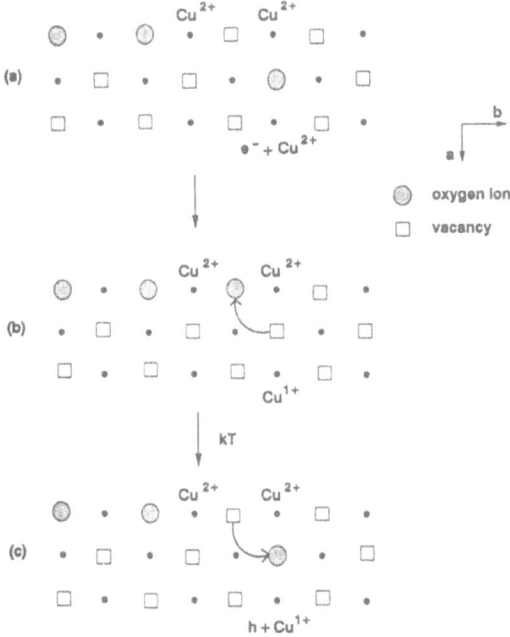

*Figure 6: Schematic representation of the photoassisted growth of chains: (a) a photoelectron is captured at the end of a short chain; (b) the oxygen ion in the short chain finds a more stable environment in a longer chain; (c) when the temperature is raised the oxygen ion comes back to its original position and a hole is removed from the Cu2O₂ layers.*

We note here that in ref. [6] we raised some arguments against the occurence of photoassisted oxygen ordering. Our arguments were based on the observation at room temperature of an *increase* in the Hall mobility $\mu$ during the illumination in an x = 6.5 film, while a *decrease* was seen during the annealing at room temperature of a different quenched (non-illuminated) film with a different x = 6.6. In Fig. 7 we show that the function $\mu(x)$ has a maximum at $x_{max} \simeq 6.55$ measured at T= 300 K. Note that an increase in $n$ for x < $x_{max}$ gives rise to an increase in $\mu$ (since to increase $n$ is equivalent to increase x), while the opposite is true for x > $x_{max}$. Aging and illumination experiments carried out in this film with x = 6.7, show indeed that $\mu$ *decreases in both cases*, while during the illumination of the same film with x = 6.5 the mobility *increased*. In view of this new experimental facts, the argument used in ref. [6] is not longer valid.

Finally, if this photoassisted oxygen ordering mechanism is correct, persistent photoinduced phenomena should only be present in high $T_c$ materials containing chain planes in their crystalline structure. So far, we could not observe persistent photoinduced phenomena after the illumination of underdoped, overdoped and optimally doped $Bi_2Sr_2Ca_1Cu_2O_{8\pm y}$ [21] and $La_{2-x}Sr_xCuO_{4\pm\delta}$ [22] films, nor in fully oxidized Pr doped YBCO phase spread alloy films [23] for any Pr doping level [24].

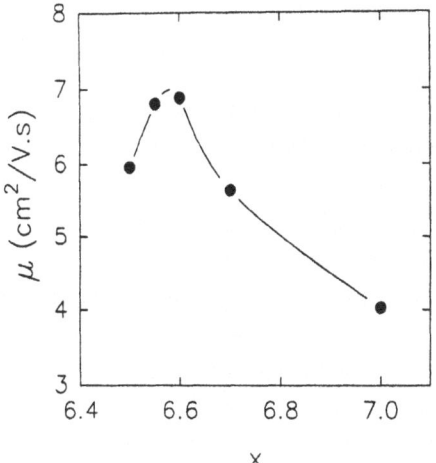

*Figure 7: The Hall mobility $\mu = \frac{R_H}{\rho_{ab}}$ as a function of the oxygen content in an YBCO film. The solid line is a guide to the eye.*

## 5. Summary and conclusions

In summary, we have shown the existence of measurable persistent photoinduced effects in YBCO films for all oxygen contents below x = 7. We have also shown that the total enhancement of $T_c$ after illumination depends on x in a similar way as that observed for the $T_c$ enhancement due to ordering effects in oxygen deficient YBCO single crystals. This, together with the similarity between the photoinduced structural changes and those caused by room temperature annealing of quenched samples, indicate that during the long time photoexcitation a rearrangement of oxygen ions in the basal plane of YBCO is likely to occur.

## Acknowledgements

We are indebted to P. Wagner and H. Adrian for providing us with the Bi-2212 films and to J.-P. Locquet for providing the LaSrCuO films and for active collaboration. Work supported by the Belgian High Temperature Superconductivity Incentive (E.O. and M.M.) and Concerted Action Programs at KUL and ONR grant N00014-91J-1438 at UCSD.

## References

[1] G. Yu, A. J. Heeger, G. Stucky, N. Herron, and E. M. McCarron, Solid State Commun. **72**, 345 (1989).

[2] G. Yu, C. H. Lee, A. J. Heeger, N. Herron, and E. M. McCarron, Phys. Rev. Lett. **67**, 2581 (1991); G. Yu, C. H. Lee, A. J. Heeger, N. Herron, E. M. McCarron, Lin Cong, J. C. Spalding, C. A. Nordman, and A. M. Goldman, Phys. Rev. B **45**, 4964 (1991).

[3] A. I. Kirilyuk, N. M. Kreines and V. I. Kudinov, Pis'ma Zh. Eksp. Teor. Fiz. **52**,

696 (1990); (JETP Lett. **52**, 49 (1990)).

[4]  V. I. Kudinov, A. I. Kirilyuk, N. M. Kreines, R. Laiho, and E. Lähderanta, Phys. Lett. A **151**, 358 (1990).

[5]  G. Nieva, E. Osquiguil, J. Guimpel, M. Maenhoudt, B. Wuyts, Y. Bruynseraede, M. B. Maple, and Ivan K. Schuller, Appl. Phys. Lett. **60**, 2159 (1992).

[6]  G. Nieva, E. Osquiguil, J. Guimpel, M. Maenhoudt, B. Wuyts, Y. Bruynseraede, M. B. Maple, and Ivan K. Schuller, Phys. Rev. B. **46**, 14249 (1992).

[7]  B. Wuyts, Z. X. Gao, S. Libbrecht, M. Maenhoudt, E. Osquiguil, and Y. Bruynseraede, Physica C **203**, 235 (1992).

[8]  E. Osquiguil, M. Maenhoudt, B. Wuyts, and Y. Bruynseraede, Appl. Phys. Lett. **60**, 1627 (1992).

[9]  M. Tetenbaum, L. A. Curtiss, B. Tani, B. Czech, and M. Blander, Physica C **158**, 371 (1989).

[10]  E. Osquiguil, Z. X. Gao, M. Maenhoudt, B. Wuyts, S. Libbrecht, and Y. Bruynseraede, J. Phys.: Cond. Matter **5**, A385 (1993) and references therein.

[11]  V. I. Kudinov, A. I. Kirilyuk and N. M. Kreines, Pis'ma Zh. Eksp. Teor. Fiz. **56**, 101 (1992); (JETP Lett. **56**, 102 (1992)); V. I. Kudinov, I. L. Chaplygin, A. I. Kirilyuk, N. M. Kreines, R. Laiho and E. Lähderanta and C. Ayache, Phys. Rev. B **47**, 9017 (1993).

[12]  B. W. Veal, A. P. Paulikas, Hoydoo You, Hao Shi, Y. Fang and J. W. Downey, Phys. Rev. B **42**, 6305 (1990).

[13]  S. Libbrecht, E. Osquiguil, B. Wuyts, M. Maenhoudt, Z. X. Gao, and Y. Bruynseraede, Physica C **206**, 51 (1993).

[14]  G. Deutscher and M. L. Rappaport, Phys. Lett. **71A**, 471 (1979).

[15]  I. Giaver, Phys. Lett. **20**, 1286 (1968); R. C. Dynes and T. Fulton, Phys. Rev. B **3** 3015 (1971).

[16]  N. M. Kreines, and V. I. Kudinov, Mod. Phys. Lett. B, **6**, 289 (1992).

[17]  H. Zhang, and H. Sato, Phys. Rev. Lett. **70**, 1697 (1993).

[18]  E. Osquiguil, M. Maenhoudt, B. Wuyts, Y. Bruynseraede, D. Lederman, and Ivan K. Schuller (submitted for publication).

[19]  G. Ceder, R. McCormack, and D. de Fontaine, Phys. Rev. B **44**, 2377 (1991).

[20]  J. M. Tranquada, S. M. Heald, A. R. Moodenbaugh, and Youwen Xu, Phys. Rev. E **38**, 8893 (1988); H. Rushan, G. Zizhao, Y. Daole, and L. Qing, Phys. Rev. B **41**, 6683 (1990); A. Latgé, E. V. Anda, and J. L. Morán-López, Phys. Rev. B **42**, 4288 (1990).

[21]  P. Wagner, H. Adrian, and C. Tomé-Rosa, Physica C **195**, 258 (1992).

[22]  J.-P. Locquet, C. Gerber, A. Cretton, Yvan Jaccard, E. Williams, and E. Mächler, Appl. Phys. A **57**, 211 (1993).

[23]  D. Lederman, T. J. Moran, J. Hasen, and Ivan K. Schuller, Appl. Phys. Lett. **63**, 1276 (1993).

[24]  D. Lederman, J. Hasen, Ivan K. Schuller, E. Osquiguil, and Y. Bruynseraede (submitted for publication).

# SURFACE ESR PROBING OF SUPERCONDUCTORS

N. BONTEMPS, P.Y. BERTIN, P. MONOD
*Laboratoire de Physique de la Matière Condensée (URA 1437 du CNRS)*
*Ecole Normale Supérieure, 24 rue Lhomond, 75005 Paris*
*France*

ABSTRACT. A paramagnetic probe located at a controlled distance from the surface of a superconductor has been used in order to investigate the field distribution arising from short and long range self organization of the vortices in the quasi-equilibrium regime. The characteristic lengthscale over which the field fluctuates may be derived from the dependence of the electron spin resonance linewidth upon the distance of the probe to the surface. Indeed, a periodic array of vortices produces outside a superconductor a field distribution whose second moment decreases exponentially over a distance which is essentially the period of the vortex lattice divided by $2\pi$. Field fluctuations over larger spatial lengthscales decay over larger distances. Disorder changes the exponential decay into an algebraic one. Results are presented for $YBa_2Cu_3O_7$ ceramics, $Nb_3Sn$ thin films, and $Bi_2Sr_2CaCu_2O_8$ single crystals.

Imaging of vortices first realised by Essman and Träuble [1] stands as one of the most striking and direct evidence for the existence of the Abrikosov triangular lattice. Unfortunately, this technique is restricted to fields low enough such that the inter vortex distance is larger than the London penetration depth, in order to ensure sufficiently large field gradients, hence enough contrast to make the emergence of vortices at the surface visible. Very recent electron holography results, using Lorentz microscopy, have allowed direct observation of the vortex lattice in $Bi_2Sr_2CaCu_2O_8$, but also in a low field (15 G) [2]. Magneto-optical imaging at fields larger than $H_{c1}$, which does not achieve an intervortex scale spatial resolution, investigates field inhomogeneities at the surface on macroscopic scales (a few μm) [3], related to flux penetration in the critical state.

These methods of investigation of the field inhomogeneity at the surface of a superconductor should be compared to the more classical techniques of Muon Spin Resonance (μSR) [4] and Nuclear Magnetic Resonance (NMR) [5]. In both cases, what is being probed is the inhomogeneity in the bulk material: indeed, by monitoring the precession rate of the spins of either the implanted muon or the present nuclei, one finds after proper Fourier transform the probability distribution of the local field averaged over the whole sample. However, in general, no simple deconvolution can provide the possible different sources of inhomogeneity when they are of comparable magnitude.

313

*N. Bontemps et al. (eds.), The Vortex State, 313–328.*

Following the initial suggestion of Shoenberg (1950) for the intermediate state [6], it was recalled in 1971 by Goren and Tinkham [7] that the field distribution generated by a *periodic array of vortices at some distance z from the surface* of a superconducting sample, is controlled by the first harmonic of the Fourier series which describes the *surface* distribution, as soon as this distance z becomes large compared to the inverse wavevectors related to the higher harmonics. .

This paper reports on a method which takes advantage of this remark: we have developed the ability of evaporating thin (typically 200 up to 5000Å) layers of organic free radicals at a controlled distance from the surface. Such free radicals exhibit a narrow Electron Spin Resonance (ESR) line in an homogeneous field, and any quasi static field distribution provides some extrawidth which reflects this distribution. As the layer is located further and further from the surface, the contributions of the first harmonics associated with the distribution are exponentially damped [8]. The period associated to a vortex lattice is of the order of 1000 Å in a field of a few kG. Flux modulation may occur on a much larger scale if related to e.g. the microstructure, as we will show for ceramic samples. Demagnetizing field effects in non ellipsoïdal samples may also generate field inhomogeneities, however on the sample size scale. These various lengthscales are reflected in the decay of the field distribution as one gets further and further from the surface.

We shall first give an estimate of the width, e.g. the second order moment (2nd Mt), of the field distribution at the surface, with respect to the bulk field distribution in the case of a periodic lattice of vortices. If the array is no longer periodic (e.g. in case of static disorder), the Fourier spectrum of the distribution is no longer discrete: we will derive the possible consequences on the decrease of the 2nd Mt with the distance z from the surface in some specific cases.

We will then present our experimental results on samples where the array of vortices may be arranged very differently: $YBa_2Cu_3O_7$ ceramic samples whose granular structure may be expected to play a role on the spatial field distribution , $Nb_3Sn$ thin films as an example of an isotropic superconductor, and $Bi_2Sr_2CaCu_2O_8$ single crystals, an extremely anisotropic superconductor.

## 1.   Surface distribution

### 1.1 PERIODIC ARRAY OF VORTICES

Let us assume a periodic 2D array of flux lines, with a spatial period **a** along the x axis and **c** along the y axis. Outside the superconductor, the associated field is controlled by Laplace equation.

$$\nabla^2 \hat{B} = 0 \qquad (1)$$

If at the surface,

$$\hat{B}_z(x,y,z=0) = \sum_{n,m=0}^{\infty} b_{nm} \cos[npx]\cos[mqy] \qquad (2)$$

where $p = \dfrac{2\pi}{a}$, $q = \dfrac{2\pi}{c}$, then at a distance z from the surface, eq.1 yields:

$$\hat{B}_z(x,y,z) = \sum_{n,m=0}^{\infty} e^{-z\sqrt{(n^2p^2+m^2q^2)}} b_{nm} \cos[nqx]\cos[mpy] \qquad (3)$$

We are interested here in the 2nd Mt $\overline{\Delta B^2}$ of the distribution described by eq.2 and how it reflects the 2nd Mt of the field distribution in the bulk.

At the distance z, from eq.(3) one finds:

$$\overline{\Delta B^2} = \sum_{n,m=1}^{\infty} e^{-2z\sqrt{(n^2p^2+m^2q^2)}} \frac{b_{nm}^2}{2} \qquad (4)$$

As expected, the higher the harmonic in the Fourier series, the closer one should get to the surface in order for this component not to be exponentially damped. As an example which will be particularly relevant to our experiment, in a field of 3.3 kG, the intervortex distance hence the spatial period of the flux line lattice (if any) is approximately 800 Å. Because of the $2\pi$ factor, the distance from the surface beyond which only the first harmonic of the Fourier expansion will play a significant role is ~130 Å. As we shall discuss further, the experimental sensitivity precludes the use of layers thinner than 200Å. It is therefore a reasonable approximation to indeed restrict our calculation to the first harmonic of the field distribution, namely p=1, q=0; p=0, q=1; p=1, q=1:
Eq. 4 reduces thus to:

$$\overline{\Delta B^2} = \frac{1}{2}[e^{-2zp} b_{10}^2 + e^{-2zq} b_{01}^2 + e^{-2z\sqrt{p^2+q^2}} b_{11}^2] \qquad (5)$$

To simplify further the calculation, we neglect as in ref.7 the p=1, q=1 contribution which represents ~ 25% of the p=1, q=0 or p=0, q=1 terms and we shall consider a square array of vortices, hence:

$$\hat{B}_z(x,y,z) = \hat{B}_0 + b_{10}(x,y,z)(\cos kx + \cos ky) \qquad (6)$$

where now $p = q = k = \dfrac{2\pi}{a}$.

We now recall the starting point of the calculation of ref.7 in order to compute $b_{10}$.

Inside the superconductor, $\hat{B}$ obeys the London equation:

$$\hat{B} - \lambda^2 \nabla^2 \hat{B} = n\Phi_0 \sum_i \delta(r - r_i)\hat{z} \qquad (7)$$

$\lambda$ is the London penetration depth, n the density of flux quanta per surface unit, $\Phi_0$ the flux quantum ($B = n\Phi_0$), and the second term of eq.7 represents the array of flux lines, approximating the core to a $\delta$ function. Keeping the first order terms of the Fourier series expansion of the $\delta$ function and of $\hat{B}$, and using the conditions : $\hat{\nabla}.\hat{B} = 0$ and the continuity of the field at the surface, Goren and Tinkham derive for the z component of $b_{10}$:

$$b_{10} = \frac{2n\Phi_0}{K^2\lambda^2}[1 + \frac{k}{K}\coth(\frac{Kd}{2})]^{-1} \qquad (8)$$

where $K^2 = k^2 + \frac{1}{\lambda^2}$ and d is the thickness of the superconducting sample

As already mentioned, we work in an applied field of 3.3 kG, hence $1/k \sim 130$Å. $\lambda$ is much larger than 130Å (typically 1500 Å in YBa$_2$Cu$_3$O$_7$ [9]) and therefore K~k. Also, the samples that we have examined are 20μm thick or more, and $\coth(\frac{Kd}{2}) \sim 1$. Eventually, we get:

$$b_{10} \sim \frac{\Phi_0}{4\pi^2\lambda^2} \qquad (9)$$

The second moment of a field distribution associated with a spatial variation given by eq.6 is:

$$\overline{\Delta B_s^2} = b_{10}^2 \qquad (10)$$

The index s refers to the fact that eq.10 yields the second moment of the field distribution at the surface (e.g. at a distance z close enough so that the first harmonics have not yet died out, but far enough so that the higher ones are negligible). This should be the case within the probe layer thickness **e** that we use, namely **e**=200Å (with respect to the inverse wavevector a/2π=130Å).

The value of the 2nd Mt that one derives from Eq.9 and 10 must be further corrected from the finite thickness effect of the probe layer. We compute finally:

$$\overline{\Delta B_s^2}(e \sim 200\text{Å}, \frac{a}{2\pi} \sim 130\text{Å}) \sim \frac{1}{\pi}\frac{1}{16}\frac{\Phi_0^2}{\pi^4\lambda^4} = 0.0002\frac{\Phi_0^2}{\lambda^4} \qquad (11)$$

where the $\sim \frac{1}{\pi}$ prefactor appears after proper integration of the damping of the first harmonic within the 200Å layer. We recall that this is an underestimate of the actual value since we have completely neglected higher harmonics.

Eq. 11 must now be compared to the value $\overline{\Delta B_b^2}$ of the 2nd Mt in the bulk [10]:

$$\overline{\Delta B_b^2} = 0.00371 \frac{\Phi_0^2}{\lambda^4} \qquad (12)$$

for a triangular lattice. For a square lattice, 0.00371 must be replaced by 0.00387 [11]. Therefore the 2nd Mt of the surface distribution is reduced with respect to the bulk by a factor of ~18. As an example, for a square lattice with $\lambda(T=0K)=1500$Å:

$$\overline{(\Delta B_s^2)}^{\frac{1}{2}} = 12 \text{ G}, \qquad \overline{(\Delta B_b^2)}^{\frac{1}{2}} = 55 \text{ G} .$$

1.2 EFFECT OF DISORDER

1.2.1 *Line vortices*

According to E.H. Brandt, the field fluctuation of randomly positioned stiff flux lines is always larger than that of a perfect flux line lattice (FLL). An order of magnitude of the ratio of the 2nd Mt, in the case of disorder, to the 2nd Mt for a FLL is ~20 [11].

How this result, supposed to be valid in the bulk, is reflected in the surface distribution can be discussed as follows. Because of disorder, the Fourier spectrum of the field distribution is no longer made out of discrete frequencies, but calls for some continuous function b(k); b(k) must exhibit a lower cut-off, $k_{min}\sim2\pi/d$ corresponding to the sample size and an upper limit corresponding approximately to the average intervortex distance, $k_{max}\sim2\pi/a$, which is the most likely smallest scale in absence of pinning (this could be however wrong in the case of an entangled vortex state). The 2nd Mt writes :

$$\overline{\Delta B_s^2} \sim \int_{k_{min}}^{k_{max}} b^2(k)e^{-2kz}dk \quad (13)$$

One may expect two consequences : the weight of the harmonic associated with the period of the former FLL may decrease. On the other hand, the random fluctuations of the lattice structure give rise to lower harmonics associated to the disorder over longer wavelenths. These extra terms will decay more slowly as a function of distance from the surface and the calculation done in the previous paragraph must be implemented with these terms.

Correlatively, the decay will be no longer exponential but slower, since it will involve a continuous set of exponentials decaying upon larger distances.

This can be illustrated in a simple way by noticing that the decay along z is only contributed by wave vectors whose inverse compares to z. Let us assume for simplicity that the b(k) function behaves as $k^{-\alpha}$. From eq.13, one gets an algebraic dependence of $\overline{\Delta B_s^2}$ versus z:

$$\overline{\Delta B_s^2} \sim z^{1-2\alpha} \qquad (14)$$

Note that the magnitude of $\alpha$ is critical since it may produce a slow decrease with z, in contrast with the exponential decay. This result, established in an oversimplified framework, is actually general. It was shown by E. Bouchaud and J.P. Bouchaud that while the field produced by a periodic array of edge dislocations or dipoles decreases exponentially with the distance, disorder yields fluctutations of this field which decay with a power law dependence [12].

### 1.2.2 *Point vortices*

Point vortices or "pancakes" are the proper representation in 2D geometry. According to E.H. Brandt, the field fluctuation of randomly positioned point vortices is always smaller than that of a FLL (~0.3) [11].

One can develop a simple symmetry argument in order to relate the field distribution at the surface to the distribution within the bulk: in the latter case, the field distribution is built from the superposition of the contributions of each individual pancake. One may consider the full space as two complementary half spaces. Suppose one of this half space is being removed. Because pancakes are uncorrelated between one superconducting plane and the other, doing so does not alter the pancake screening currents and therefore the field produced by the remaining half space is half of its previous value. The 2nd Mt is hence 4 times smaller. This remark is at odds with a 3D FLL, where the presence of a surface changes the current distribution close to the surface [13].

## 2. Experimental procedure for ESR surface probing of the field distribution

The paramagnetic probe is the bisphtalocyanin-lutetium molecule (Pc$_2$Lu)which contains one S = 1/2 delocalised electron per formula unit. It gives rise to a single, temperature independent ESR line whose derivative is 1 to 3 gauss peak to peak wide and is only slightly sample dependent [14]. The Pc$_2$Lu molecules are conveniently evaporated using a conventional evaporator in a vacuum of $10^{-6}$ Torr at 450°C, to form an amorphous (green) film containing 5 $10^{13}$ spins per nanometer thickness and per $cm^2$. In order to avoid as much as possible any edge effect, only the central part of the sample is covered with $Pc_2Lu$ using an appropriate mask (2 or 2.5 mm diameter), which represents ~$10^{12}$ spins per nanometer: this is hardly 10 times the ultimate sensitivity of our spectrometer for a typical 1 G broad ESR line. In order to maintain a reasonable signal-to-noise ratio, we have therefore worked on 200Å $Pc_2Lu$ layers.

Controlling the distance of the probe to the surface is achieved by deposition of a buffer layer of ajustable thickness before $Pc_2Lu$ evaporation. For thin buffer layers (between 500 and 2000Å), we have used evaporated neutral, diamagnetic hydrogen-phtalocyanin ($PcH_2$). For thicker buffer layers (1 to 10 µm), we have used layers of neutral resin (polyphenylquinoxaline) prepared by a spinning technique.

The experimental geometry is shown on fig.1.

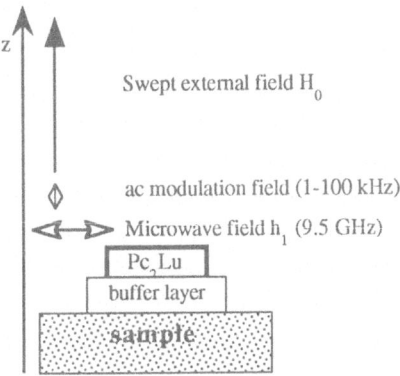

Fig.1 : Sketch of the $Pc_2Lu$ and buffer layers on top of the superconducting sample, and of the various applied magnetic fields: a quasi-static field $H_0$ (3300±250 Oe), slowly swept, a small ac field (up to 20 Oe) and the microwave magnetic field (≤200 mOe).

The measurements are done at 9.5 GHz hence at 3.3 kG for a g=2 spin, in a standard ESP 300E Brüker spectrometer, equipped with a helium flow cryostat. The temperature of the sample is measured with a suitable probe (Platinum or carbon resistor) placed close to the sample [8].The resonance line is measured while sweeping slowly (20s or more) a small field range (<250G). This constrains the experiment to the so-called reversible regime, where such a sweep is not balanced by the diamagnetic response of the sample in the critical state and where the width of the line does not depend on the sweep time and on the sweep direction. These criteria are fulfilled in the experiments that we shall describe but turn out to be a rather severe limitation: the available temperature range is extremely narrow in $Nb_3Sn$ (T> $0.97T_c$) [15] and spans only a few degrees in $YBa_2Cu_3O_7$ single crystals [16]. In $YBa_2Cu_3O_7$ ceramics, it is not so stringent and one can get down to ~75K. In the case of $Bi_2Sr_2CaCu_2O_8$ crystals, the reversible regime extends down to ~30K at 3.3 kG [17].

Note that a small ac modulation field (at 1 or 100 kHz) is superimposed parallel to the quasi-static field for sake of lock-in detection. The signal is then the first derivative with respect to the field of the microwave absorption. One can relate, for a given lineshape, the so-called peak-to-peak width $\Delta H_{pp}$ of this signal to the square root of the 2nd Mt. Throughout this study, we have made the approximation (which is not very critical) of a gaussian lineshape, hence

$$\mu_0 \Delta H_{pp} = 2\sqrt{\overline{\Delta B_s^2}} \quad (15)$$

At 3.3 kG, in the reversible regime, where the sample exhibits a finite resistance, eddy currents are induced by the modulation field. These eddy currents in turn generate a field opposite to the driving one. The resulting total ac modulation field which is actually seen by the spins within the probe layer becomes smaller and smaller as the resistance decreases with temperature. Such "skin depth" effects have beeen pointed out by V. Geshkenbein et al [18] and observed experimentally by D.G. Steel et al [19]. We have confirmed ourselves this effect at a 3.3 kG static field on $Bi_2Sr_2CaCu_2O_8$ samples: the amplitude of the ac field at the surface experiences the same relative attenuation at 100 kHz as at 1 kHz, provided that the temperature has decreased by the amount required for the resistance at 3.3 kG to drop by a factor of 10, which is expected in the case of conventional skin depth effect [20,21].

Thus, this phenomenon further restricts our measurements to a temperature range where the ac modulation field is not cancelled, namely T>50 K for $Bi_2Sr_2CaCu_2O_8$.

## 3. $YBa_2Cu_3O_7$ ceramics

In this section, we focus on the search for a field variation on a scale much larger than the intervortex distance. The probe layer was in that case 5000Å thick and a resin buffer layer ranging from 2.5 to 10 μm was used [8].
Experiments were performed on $YBa_2Cu_3O_7$ (4mm diameter, 1 mm thick) disc shaped pellets (sets S1 and S2) synthetised in different ways (solid state reaction of standard precursors or freeze dried precursors [22]), and exhibiting totally different apparent grain sizes D ( S1 : D=20±10 μm, and S2 : D≤ 3 μm , as observed by optical microscopy).

We report on fig.2 the peak-to-peak linewidth as a function of temperature for various resin spacers.

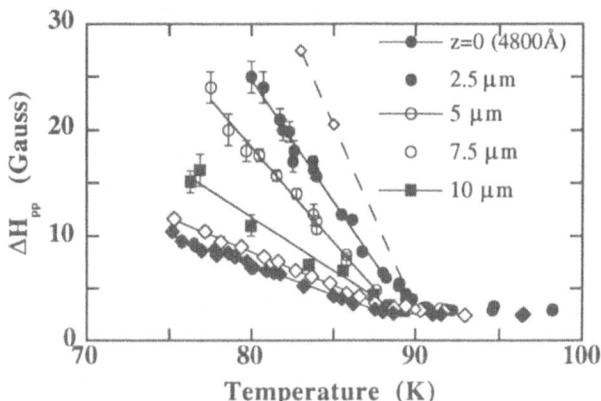

Fig.2-a: peak-to peak linewidth of the surface ESR on $YBa_2Cu_3O_7$ ceramics prepared from standard oxide precursors for different thicknesses of (non magnetic) buffer layers. The open diamonds refer to μSR results taken from ref.4 and represent the equivalent width seen by the muons in the bulk of ceramics.

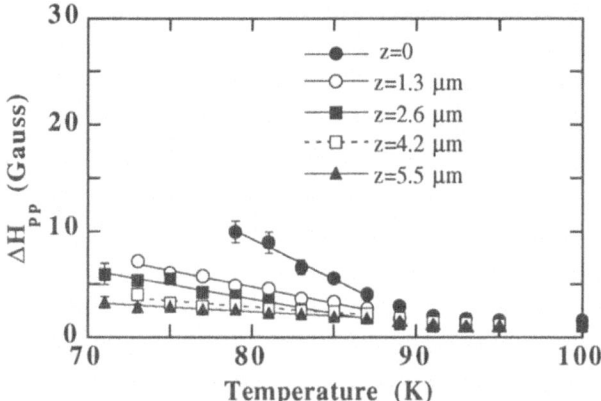

Fig.2-b: peak-to peak linewidth of the surface ESR on $YBa_2Cu_3O_7$ ceramics prepared from freeze dried precursors for different thicknesses of (non magnetic) buffer layers.

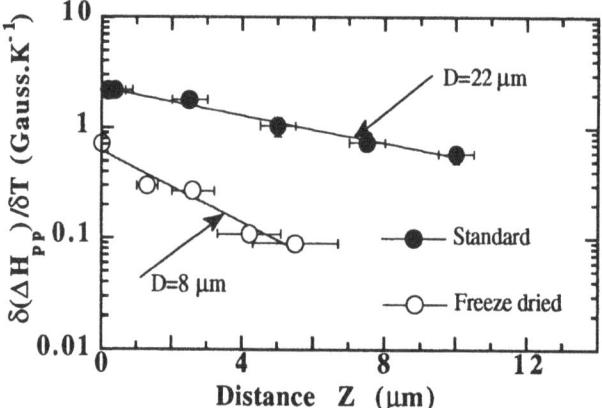

Fig.3: Variation of the broadening rate with distance in the 2 sets of YBa$_2$Cu$_3$O$_7$ ceramics. The size D of the grains is determined by a fit to a numerical simulation of the field generated by a square array of random dipoles

The slope $\delta(\Delta H_{pp})/\delta T$ deduced from fig.2 is plotted as a function of the spacer thickness in fig.3.

The origin of this field distribution is clearly not the Abrikosov vortex lattice since the vortex spacing is about 800Å at 3.3 kG. The contribution of the vortex lattice is damped beyond any observation within a 5000Å layer. It cannot be either the macroscopic demagnetizing field inhomogeneity; we have indeed checked that the EPR width is only 10% smaller when the static field is parallel to the plane of the disc, in which case the demagnetizing field correction is negligible.

A likely explanation of our findings calls for the magnetic dipole contribution from individual ceramic grains. We have performed a simulation of the field distribution created above an array of randomly oriented dipoles on a square lattice (restricted to nearest and next nearest neighbours). The simulation relates the temperature variation of the EPR linewidth to the temperature variation of the magnetic moment of each grain, hence of the magnetization. The computed 2nd Mt exhibits a decrease with z, which can be approximately fitted to an exponential (solid line in fig.2). Adjusting the experimental and the computed data yields a period of 22±3 μm and of 8±2 μm, compatible in the former case with the apparent grain size, but suggesting a different source of disorder for the latter (scanning electron microscopy pictures show in particular 15 μm inclusions of green phase).

We note that the pure exponential decay which fits best the data would yield a period of 44 and 17 μm respectively. The decay associated with the actual period (22 and 8 μm) should be faster if it were merely related with the first harmonic of the field distribution, associated with a strictly periodic array of dipoles. Since our random picture involves disorder, it is no surprise that the actual decay is slower.

### 4. Nb₃Sn granular films

This conventional superconductor has been selected because its London penetration depth is comparable to the one in $YBa_2Cu_3O_7$, namely $\lambda \sim 1100 \text{Å}$ [23]. The films are 2500Å thick and have been evaporated on a zirconia subtrate.

As previously mentioned, the difficulty with such a material is that the reversible regime at 3.3 kG is restricted to a very narrow range, 0.5 K, in temperature below $T_c = 18K$. This was shown for Nb₃Sn wires [15] and we have confirmed it by magnetization measurements on our samples.

We have measured the peak-to-peak linewidth for layers of 200Å $Pc_2Lu$ deposited either directly on the surface of the film or on top of a $PcH_2$ 1000Å thick buffer layer. The results are shown on fig.4 for the 2 distances investigated.

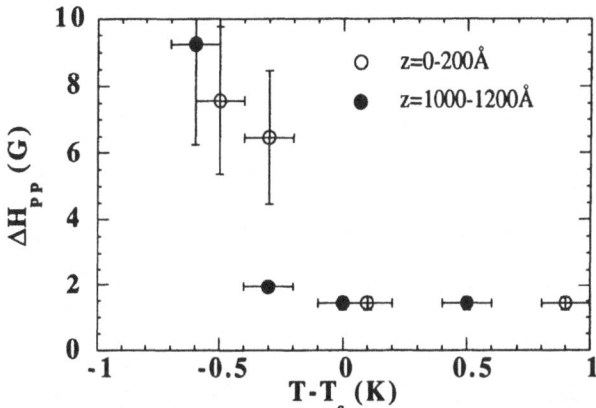

Fig.4 : Peak-to-peak linewidth of the $Pc_2Lu$ probe on a Nb₃Sn 2500Å thin film

We observe a significant broadening in this narrow temperature range, but the most striking result is that there is hardly any difference ($2 \pm 1$ G) for $0 \leq z \leq 200 \text{Å}$ and $1000 \leq z \leq 1200 \text{Å}$, although the latter value is significantly larger than 130Å associated to a possible FLL. Let us also compare the absolute expected value $\Delta H_{pp}$ expected for a perfect FLL to the measured one. Assuming a simple $[1-(T/T_c)^4]^{-1/2}$ Gorter-Casimir temperature variation for $\lambda(T)$ and using eq.11, we compute $\Delta H_{pp} \sim 1.5G$, at $T-T_c=0.5K$, to be compared to $7.5 \pm 2$ G as measured. We have checked experimentally the demagnetizing effect on the broadening of the line and found it negligible [21]. Therefore the experimental width is too large with respect to the one possibly generated by a perfect FLL.

The reversible regime in $Nb_3Sn$ was associated with flux melting [15]. Our experimental observations are entirely consistent with a disordered vortex state, which would give rise to a significant enhancement of the 2nd Mt, and which would explain its anomalous slow decay, but cannot unambiguously confirm a melted flux regime. An alternative explananation could be the microstructure of the $Nb_3Sn$ film, which should then be associated with lengthscales at least of the order of 1 $\mu m$.

## 5. $Bi_2Sr_2CaCu_2O_8$ single crystals

Similar measurements have been run on several series of $Bi_2Sr_2CaCu_2O_8$ single crystals of approximate size 3x3x0.02 mm with $T_c=(87\pm1K)$[24]. These samples exhibit an extremely smooth surface( average rugosity ~10Å) , as checked by Atomic Force Microscopy [21]. We have measured the linewidth of the ESR lines for 200Å $Pc_2Lu$ layers located right on the surface and on top of $PcH_2$ buffer layer of 500Å. We have also measured, using a DPPH crystal of approximate size 50μm placed at various points of the $Bi_2Sr_2CaCu_2O_8$ crystal, the field inhomogeneity due to the sample shape [25,26].

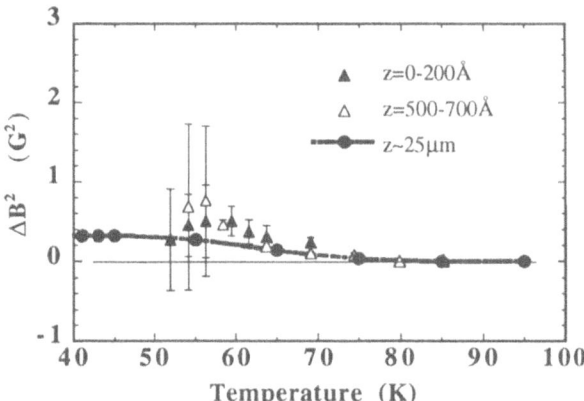

Fig.5: 2nd Mt of the field distribution at the surface of $Bi_2Sr_2CaCu_2O_8$ single crystals. The line displays the part of the 2nd Mt which is due to shape effects, obtained from the ESR linewidth of the ~ 50μm DPPH crystal.

The broadening (if any) is extremely small. We have plotted on fig.5 the computed 2nd Mt (Gauss$^2$), assuming a gaussian field distribution convoluting a lorentzian natural line shape (Voigt profile) [25].

It is clear from fig.5 that down to 60K, most, if not all the broadening is due to the inhomogeneous demagnetizing field. In this temperature range, an upper limit for $\overline{\Delta B_s^2}$ is set by our experimental accuracy, e.g. ~0.2 G$^2$. This implies, if we assume a regular vortex lattice, $\lambda_{ab}(T=60K) \geq 8000$Å, relying on eq.11. We would expect an even larger value for a disordered one.

Below 60K, although the error bars become large, there may exist for this present set of data an extra broadening, ~$0.5^{+0.8}_{-0.5}$G$^2$ at 55K..This value is to be compared to the hypothetical value of 50 G$^2$ which is calculated from eq.11, using $\lambda_{ab}(T=0)=1900$ Å and the Gorter-Casimir temperature variation for $\lambda_{ab}(T)$ [27].

*Overall, fig.5 displays neither the magnitude nor the temperature variation of $\overline{\Delta B_s^2}$ that one would expect from a regular FLL [4,27].*

Muon spin resonance experiments also show a highly unusual behavior at the same field, but seem to recover a more conventional one at 15kG, from which a value of $\lambda_{ab}= 4200$Å was inferred [28]. Such experiments are run on a mosaic of single crystals and include the contribution of the inhomogeneity of the demagnetizing field. Above 60K, where we do not observe any broadening, the muon result at 3 kG may be assigned to the demagnetizing field inhomogeneity [23].

Finally, from a set of identical measurements on well centered layers, where demagnetizing field effects are entirely cancelled [21], we have definitely found an upper limit of 0.2 G$^2$ down to 50 K for $\overline{\Delta B_s^2}$ .

The question which has to be answered is the reason for such an homogeneous field at the surface above 50-60K (and the possible building up of a small field inhomogeneity below these temperatures).

It has been suggested in the case of the μSR experiment, that the small 2nd Mt results from vortex motion on "a time scale comparable with the muon lifetime" (2.2 μs) [28]. In the ESR experiment, the condition for motionnal narrowing is set by the expected inhomogeneous width $\Delta\Omega$ (~7 G) which has to be narrowed. The condition $\Delta\Omega\tau\sim 1$ yields $\tau\sim 3.10^{-9}$ s.This implies that vortex motion has to occur on a $2.10^{-10}$ s time scale in order to get the amount of narrowing which is observed, a value inconsistent with ref.28. If one assumes that two vortices interchange within this time interval, this leads to a velocity of

$5.10^2$ m.s$^{-1}$. We note that the closest compatible time scale is the one associated with the microwave field period ($\sim 10^{-10}$ s).)

We think more appropriate to incorporate for the $Bi_2Sr_2CaCu_2O_8$ compound, the physics of Josephson coupled layers [29]. Vortex lines are then built up by stacks of 2D pancakes. We suggest that above 50-60K, the thermal energy overcomes the magnetic coupling between the layers, leading to a disordered state of pancake vortices. Such a situation has been decribed by J. Clem [30]. This, as previously discussed, reduces dramatically the field distribution within the bulk [11] as well as outside the superconductor.

## Conclusion

Probing the mixed state by surface ESR turns out to be essential whenever one suspects that the vortex state is controlled not only by the classical Abrikosov flux line lattice, but by other mechanisms which may alter the lenghscale over which the magnetic field fluctuates. Surface probing is then unique inasmuch as it allows to monitor one parameter, namely the distance to the surface. From the magnitude of the ESR broadening and from its possible variation, it is possible to pinpoint the possible origin of the field distribution. This was shown for the case of sintered samples. However, disorder tends to smear the spectacular consequence of Laplace equation, e.g. the exponential decay of the width of the distribution over its typical lengthscale.

Although the physics close to $T_c$ in $Nb_3Sn$ thin films could not be clarified, we have ruled out the existence of a regular flux line lattice. In $Bi_2Sr_2CaCu_2O_8$ single crystals, we have shown that at 3.3 kG, a consistent explanation of the highly homogeneous field that we observe above 50K calls for thermally decoupled pancake vortices.

## Acknowledgements

This work has been supported in part by the Direction de la Recherche et des Etudes Techniques, France (DRET), by the Ministère de la Recherche et de la Technologie, France (MRT) and by the New Energy and industrial technology Development Organization, Japan (NEDO).
We are indebted to Dr J.P. Bouchaud for drawing our attention on the role played by disorder. We thank Pr J. Simon helpful suggestions.
We are especially grateful to Pr. D. Davidov who initiated this work with us.
We wish to thank Dr. J. Arabski, P. Bassoul, M. Bouvet, R. Even, C. Fretigny, C. Lacour, L. Lombardo, and P. Martin for their help in providing us with samples, in preparing the probe and buffer layers and in performing the experiments.

# References

1. V. Essmann and H. Träuble, Phys. Lett. $\underline{24}$A (1967) 526; see e.g. D.J. Bishop, P.L. Gammel and C.A. Murray, this issue

2. K. Harada, T. Matsuda, H. Kasai, J.E. Bonevich, T. Yoshida, U. Kawabe and A. Tonomura, Phys. Rev. Lett. $\underline{71}$ (1993) 3371

3. see e.g. N. Moser, M.R. Koblischka, H. Kronmüller, B. Gegenheimer and H. Theuss, Physica C $\underline{159}$ (1989) 117; C. A. Duran, P.L Gammel, R. Wolfe, V.J. Fratello, D.J. Bishop, J.P. Rice, and D.M. Ginsberg, Nature $\underline{357}$ (1992) 474

4. B. Pümpin, H. Keller, W. Kündig, W. Odermatt, I.M. Savic, J.W. Schneider, H. Simmler, P. Zimmermann, E. Kaldis, S. Rusiecki, Y. Maeno and C. Rossel, Phys. Rev. B $\underline{42}$ (1990) 8019; for a review, see H. Keller, in Proceedings of the International School of Cristallography: Materials and Crystallographic aspects of high $T_C$ Superconductivity, Erice (Italy) Ed. E. Kaldis (Kluwer Academic Publishers, 1994)

5. M. Mehring, F. Hentsch, H. Mattausch and A. Simon, Sol. State Commun., $\underline{75}$ (1990) 753; M. Mehring, in "Earlier and recent aspects of Superconductivity", Ed. J.G. Bednorz, K.A. Müller, Springer-Verlag (1991)

6. D. Shoenberg, "Superconductivity", p.105, Ed. N. Feather, D. Shoenberg, Cambridge monographs on physics (1960)

7. R. Goren and M. Tinkham, J. of Low Temp. Physics $\underline{5}$ (1971) 465 (we have carefully corrected in section 1.1 a few typographical errors present in the appendix A of this reference)

8. N. Bontemps, D. Davidov and P. Monod, Phys. Rev. B $\underline{43}$ (1991) 11512 and references therein.

9. J.E. Sonier, R.F. Kiefl, J.H. Brewer, D.A. Bonn, J.F. Carolan, K.H. Show, P. Dosanjh, W.N. Hardy, Ruixing Liang, W.A. MacFarlane, P. Mendels, G.D. Morris, T.M. Riseman and J.W. Schneider, Phys. Rev. Lett. $\underline{72}$ (1994) 744

10. E.H. Brandt, Phys. Rev. B $\underline{37}$ (1988) 2349

11. E.H. Brandt, Phys. Rev. Lett. $\underline{66}$ (1991) 3213

12. E. Bouchaud, J.P. Bouchaud Phil. Mag. Lett. $\underline{65}$ (1992) 339

13. J. Pearl, J. Appl. Phys. $\underline{37}$ (1966) 4139

14. J.J. André, K. Holczer, M.T. Riou, C. Clarisse, R. Even, M. Fourmigue, J. Simon, Chem. Phys. Lett. $\underline{115}$ (1985) 463

15. M. Suenaga, A.K. Ghosh, Y. Xu and D.O. Welch, Phys. Rev. Lett. $\underline{66}$ (1991) 1777

16. Y. Xu and M. Suenaga, Phys. Rev. B $\underline{43}$ (1991) 5516 and references therein

17. Y. Xu, M. Suenaga, Y. Gao, J.E. Crow and N.D. Spencer, Phys. Rev. B $\underline{42}$ (1990) 8756; L. W. Lombardo, D.B. Mitzi and A. Kapitulnik, Phys. Rev. B $\underline{46}$ (1992) 5615

18. V.B. Geshkenbein, V.M. Vinokur and R. Fehrenbacher, Phys. Rev. B $\underline{43}$ (1991) 3748

19. D.G. Steel and J.M. Graybeal, Phys. Rev. B $\underline{45}$ (1992) 12643

20. A. Kapitulnik in "Phenomenology and applications of high Temperature superconductors", The Los Alamos symposium-1991, p.34, Ed. K.S. Bedell, M. Inui, D.E. Meltzer, J.R. Schrieffer, S. Doniach, Addison-Wesley Publishing Company

21. P.Y. Bertin, thesis, Paris 1994 (unpublished)

22. C. Lacour et al, Phys. Status Solidi A 123 (1991) 241

23. M. Suenaga, D.O. Welch, R.L. Sabatini, O.F. Kammerer and S. Okuda, J. Appl. Phys. 59 (1986) 840

24. D.B. Mitzi, L.W. Lombardo, A. Kapitulnik, S.S. Laderman and R.D. Jacowitz, Phys. Rev. B 41 (1990) 6564

25. P.Y. Bertin, N. Bontemps, L. Lombardo, A. Kapitulnik and R. Even, J. of Alloys and Compounds, 195 (1993) 491

26. D. Davidov, P. Monod and N. Bontemps, Phys. Rev. B 45 (1992) 8036

27. Zhengxiang Ma, R.C. Taber, L.W. Lombardo, A. Kapitulnik, M.R. Beasley, P. Merchant, C.B. Eom and J.M. Phillips, Phys. Rev. Lett. 71 (1993) 781

28. D.R. Harshman, R.N. Kleiman, M. Inui, G.P. Espinosa, D.B. Mitzi, A. Kapitulnik, T.Pfiz and D.L. Williams, Phys. Rev. Lett. 67 (1991) 3152

29. R. Kleiner and P. Müller, Phys. Rev. B 49 (1994) 1327; F.X. Regi, J. Schneck, H. Savary, R. Mellet, P. Müller and R. Kleiner (to be published, J. Physique, France)

30. J. Clem, Phys. Rev.B 43 (1991) 7837

The manufacturer's authorised representative in the EU is Springer
Nature Customer Service Centre GmbH, Europaplatz 3, 69115 Heidelberg,
Germany. If you have any concerns regarding our products, please
contact ProductSafety@springernature.com

Printed and bound by CPI Group (UK) Ltd, Croydon, CR0 4YY

29/04/2026

02099460-0010